FLOW THROUGH OPEN CHANNELS

RAJESH SRIVASTAVA

Professor
Department of Civil Engineering
Indian Institute of Technology (IIT) Kanpur

OXFORD
UNIVERSITY PRESS

OXFORD
UNIVERSITY PRESS

Oxford University Press is a department of the University of Oxford.
It furthers the University's objective of excellence in research, scholarship,
and education by publishing worldwide. Oxford is a registered trade mark of
Oxford University Press in the UK and in certain other countries.

Published in India by
Oxford University Press
YMCA Library Building, 1 Jai Singh Road, New Delhi 110001, India

ISBN-13: 978-0-19-569038-5
ISBN-10: 0-19-569038-9

Typeset in Times Roman
by The Composers, New Delhi 110063
Printed in India by Magic International (P) Ltd., Greater Noida

Preface

This book is primarily intended for use as a textbook for an undergraduate level course dealing with the phenomenon of flow through open channels. It may also be useful, in combination with other books and reference material, for a graduate level course. Practising engineers, though not specifically considered a target audience, may find some sections useful to refresh their knowledge or to learn about new developments.

The subject of open channel flow is quite important for students of civil (and, to a lesser extent, mechanical and chemical) engineering. This can be judged by the fact that all graduate level programmes concerned with hydraulics and/or water resources engineering have a one-semester course on open channels in their syllabi and the undergraduate syllabi have it as either a one-semester course or a major portion of a one-semester course. With the recent emphasis on interlinking of rivers in India, it becomes more important for the future and practicing engineers to have a clear understanding of the concepts of open channel flow.

About the Book and Its Approach

There are a number of good books available on the subject. However, most of the classic books have not been updated recently. It may not be a serious limitation when one considers the basic concepts since these do not change frequently.[1] The way we apply these concepts to practical problems, however, changes quite rapidly with advances being made in the field of experimental and computational hydraulics. One of the objectives of this book is to discard the material that may have been useful earlier (and has been traditionally included in even the recent books) but has now become obsolete.[2] The topics covered in this book are, of course, available in numerous other texts since the practical problems related to open channels and the basic concepts used to solve them remain the same. What differs is the approach adopted in describing these topics.

---

[1]The basic concepts do change, however. For example, Liggett (1993) proposes a change in the way we look at the critical depth and Froude number, two important basic concepts in open channel flows.

[2]An example of such traditional methods is the use of varied flow function in computation of flow profiles. With easy availability of fast computing power, using a table of function values has become inconvenient.

The usual approach of first discussing the basic equations and then their applications has been slightly modified in this book, wherever possible. First a problem is described and then the concepts needed to be applied for its solution are elaborated. It would motivate the readers to study the concept since its utility is recognized beforehand. Similarly, the approach of discussing a general equation first and then simplifying it for a specific purpose is not followed. Starting with simple equations applicable to a specific problem, these equations are generalized as more complex problems are encountered. Introducing the intricate problems in step-by-step manner will help the readers in better understanding of the topics along with the appreciation of the assumptions involved in simplifying a problem and the effect of those assumptions on the solution.

Content and Coverage

The book is divided into nine chapters.

Chapter 1 introduces the subject and classifies the flow into various categories so that these can be described in subsequent chapters.

Chapter 2 discusses the simplest case of uniform flow with an introduction to various geometric elements needed to describe the flow, derivation of the resistance equations, computation of normal depth, and design of channels.

Chapters 3 and *4* deal with nonuniform flows, with the former dealing with gradually varied flow and the latter with rapidly varied flow. In Chapter 3, we derive the differential equation of gradually varied flow, introduce the concepts of critical flow and specific energy, classify the surface profiles, and look at various methods of computation of profile length. In Chapter 4, the rapidly varied flow in hydraulic jump is described with an introduction to the concept of specific force. Other rapidly varied flow situations, e.g., weir, spillway, sluice gate, and free overfall are also described in this chapter.

Chapter 5 describes various channel transitions, including change in bed elevation, width, and alignment with emphasis on the use of the specific energy diagram. A detailed description of channel transitions in supercritical flows is also given in this chapter.

Chapter 6 deals with spatially varied flow with derivation of the governing equations for increasing and decreasing discharge conditions, and analyses of flow over a side weir and flow through a side channel spillway.

Chapter 7 discusses unsteady flow, starting with the simplest case when an unsteady flow problem can be converted into an equivalent steady-state problem. The governing equations (Saint-Venant equations) are then derived and the methods of solution are discussed.

Chapter 8 describes flow in mobile bed channels and *Chapter 9* discusses transport of pollutants in open channels. Both these topics merit a separate book but we provide a very brief introduction here considering their importance. Chapter 8 introduces the forces on sediment particles and initiation of motion and moves onto discuss various regimes of flow, resistance to flow due to grain and form roughness, and design of channels using the tractive force and regime approach.

Chapter 9 describes the convection, diffusion, and dispersion processes of transport of pollutants in open channels. Some simple cases of transport with their analytical solutions and some complex cases with numerical solutions are also discussed.

Solved examples are used throughout the book to explain the practical application of the knowledge gained through reading. Futhermore, practice problems are provided to enable the readers to apply their learning to a variety of situations and to encourage them to think beyond the matter covered in the text.

References are provided at the end of each chapter. To harness the power of the Internet, our effort has been to provide online references in most cases.

On a personal note, open channel flow was the first subject I taught when I started my teaching career at the University of Roorkee almost 25 years ago. Over the years, I have taught and studied this course a number of times and kept on learning new things. However, during the process of writing this book, I realized how little I knew (and how little I still know!) about this vast field. I have thoroughly enjoyed every minute of writing this book and even if the readers enjoy it only half as much as I did, I would consider my efforts to be well-rewarded. Suggestions for improvement of the book are welcome and may be sent to me through e-mail (rajeshs@iitk.ac.in). Updated information about the book is available on the webpage http://home.iitk.ac.in/~rajeshs/ftoc.htm.

Rajesh Srivastava

Contents

1 | Introduction

1.1 Definition of Open Channel Flow

As the name implies, the term *open channel flow* represents flows through channels that are open to the atmosphere. One should note, however, that flow in a closed conduit (e.g., a circular pipe) may also be classified as open channel flow, if the fluid level falls below the crown of the pipe and atmospheric pressure exists on the surface. Existence of the free surface, thus, is what distinguishes the open channel flows from the closed conduit flows (or pressure flows). Therefore, free surface flows would probably be a more appropriate term for open channel flows. We will, however, use the more common nomenclature and simply use the term open channel flows. For flow through a culvert[1] draining water across and under a road (or for a storm-water drain), the flow may be an open channel flow for small discharges and may become a closed conduit flow for larger discharges. Sometimes, part of the channel may have a closed conduit flow and part may behave as an open channel (e.g., when the headwater level is above the top of the pipe and the tailwater level is below the crown). Figure 1.1 shows the flow through a culvert for different headwater and tailwater elevations, which give rise to different flow conditions.

Examples of open channel flow include flow in rivers, canals, laboratory flumes, storm-water drains, flow over weirs[2] and spillways,[3] and overland runoff (Fig. 1.2). The presence of a free surface makes the open channel flow more complicated to analyse than closed conduit flow since the cross-sectional area depends on the flow depth. Further complications arise due to the fact

---

[1] A culvert is a conduit or channel crossing under a road or embankment.
[2] A weir is a small dam-like obstruction in a stream or river to raise the water level or divert its flow. It is generally used for discharge estimation by measuring the height of water over the weir.
[3] A spillway is a passage for surplus water to run over or around an obstruction (generally a dam).

(a)
(b)
(c)
(c)
(b)
(a)
(c)
(b)

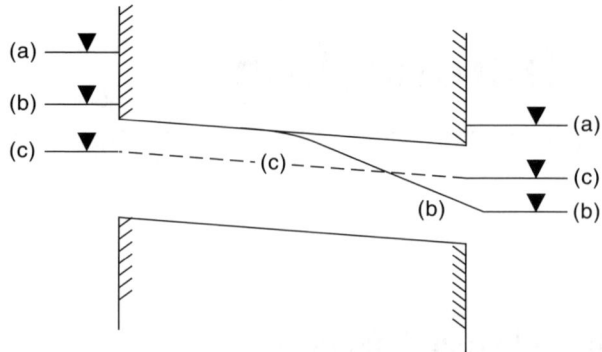

Fig. 1.1 Flow through a culvert: (a) closed conduit or pressure flow, (b) partly closed and partly open flow, and (c) open channel or free surface flow

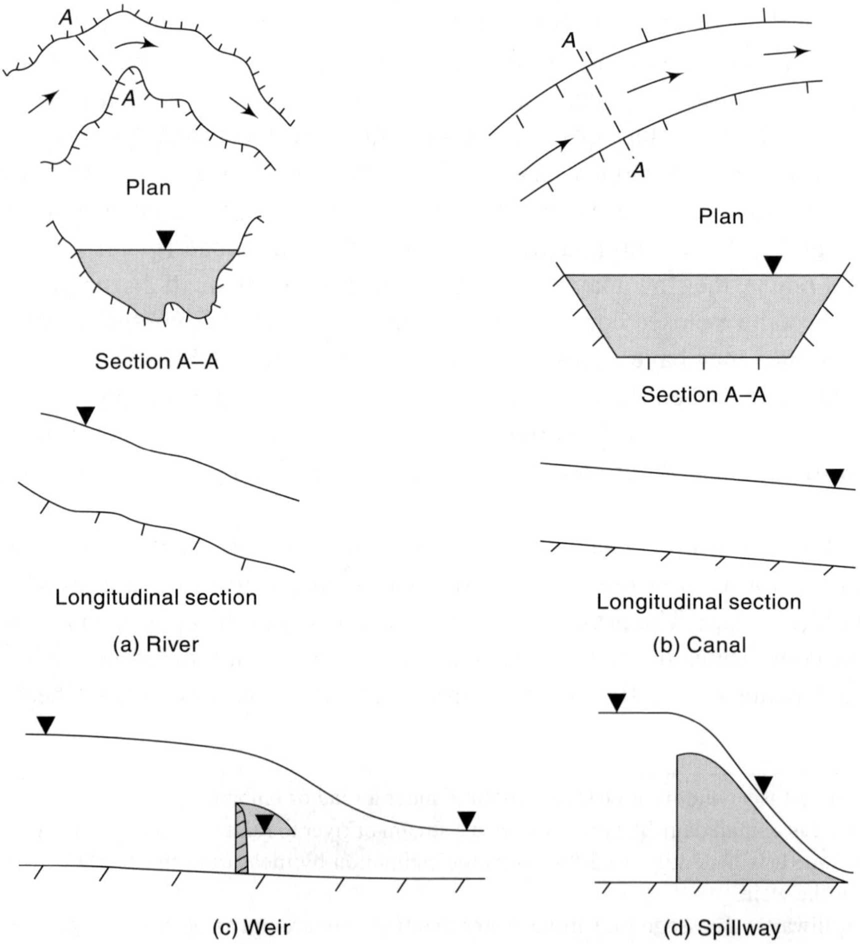

A
A
Plan
Section A–A
Longitudinal section
(a) River
A
A
Plan
Section A–A
Longitudinal section
(b) Canal
(c) Weir
(d) Spillway

Fig. 1.2 Some examples of open channel flow

that the range of variation of boundary roughness, section shape, flow depth, and discharge is much wider for open channel flows.

1.2 Importance of the Study of Open Channel Flows

The study of open channel flows finds numerous applications in civil engineering (and also in some other branches of engineering, e.g., chemical and mechanical). Some of the applications of the principles of open channel flow are discussed in this section.

1.2.1 Measuring the Discharge in a River or Canal

The amount of water carried by a channel per unit time is a very important parameter in various projects. For example, a flood control project would need an accurate estimate of the maximum flow expected in a river. Similarly, an estimate of the discharge in a canal would help to plan the irrigation water allocation for various outlets. One way of measuring the discharge is to measure the flow velocity and multiply it by the flow area. However, as we would see later, since the velocity at a cross section varies from one point to the other, we have to decide the location of measurement (Fig. 1.3). For example, in Fig. 1.3(a), point B appears to be the most logical choice for measuring the velocity due to its central location. Obviously a single velocity measurement may result in large error and it would be preferable to subdivide the channel

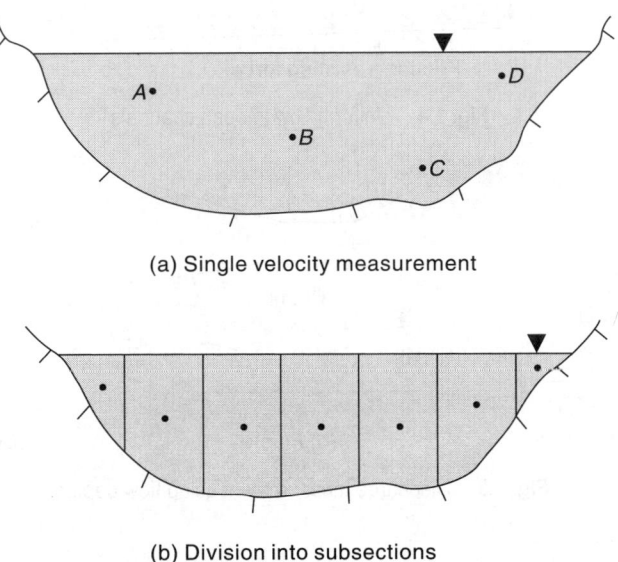

(a) Single velocity measurement

(b) Division into subsections

Fig. 1.3   Discharge measurement in open channels

area into different subareas and perform velocity measurements in each
[Fig. 1.3(b)].

1.2.2 Developing a Relationship between the Depth of Flow and the Discharge in a Channel

Since the depth of flow is much easier to measure than the velocity, it would
be much more efficient to develop a relationship between the water depth[4]
and the discharge. We can then measure the flow depth at any time and obtain
a quick estimate of the discharge at that time. When a large reach of channel
has essentially *uniform* flow, the flow depth can be related to the discharge by
considering a balance of the driving and resisting forces (Fig. 1.4). This would,
however, require an estimate of the channel resistance, which may not be
known with acceptable accuracy. Other techniques of estimating discharge
with the help of measurement of depth, without a need to know the channel
resistance, include the flow under a sluice gate, over a weir, through a flume,
at a drop, etc. (Fig. 1.5). Estimation of the discharge from the measured depths
in these cases requires a proper application of the conservation of mass and
the energy and momentum principles.

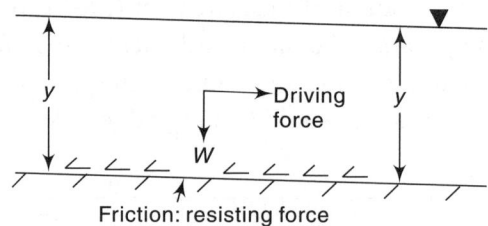

Fig. 1.4 Uniform flow in open channels

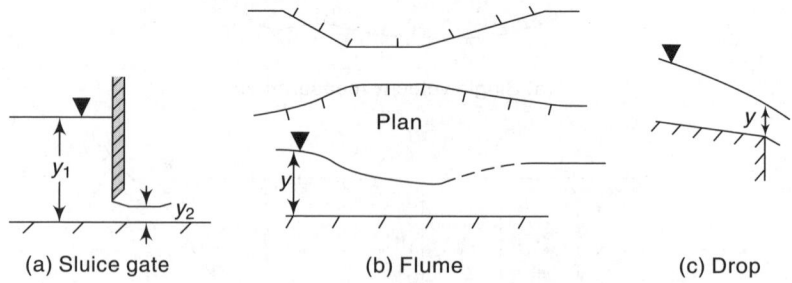

Fig. 1.5 Discharge measurement using flow depths

---

[4]The water depth also varies across a section. Typically the water depth represents the
depth above the deepest point of the section.

1.2.3 Designing a Canal to Carry Given Amount of Water

Design of a canal is largely dependent on the discharge to be carried by it. For example, an irrigation canal has to be designed keeping in mind the water requirements at different locations along its length, which, in turn, depends on the crop pattern and their water requirements. There are aspects of design such as the shape and width of cross section, the bed level and the alignment that have to be decided by the engineer. While the bed level and longitudinal alignment are generally governed by the topography, there could be an infinite number of combinations of channel shape and dimensions which would be able to carry the *design discharge*. One should be able to choose the cross section which would be the optimum[5] under given conditions (Fig. 1.6).

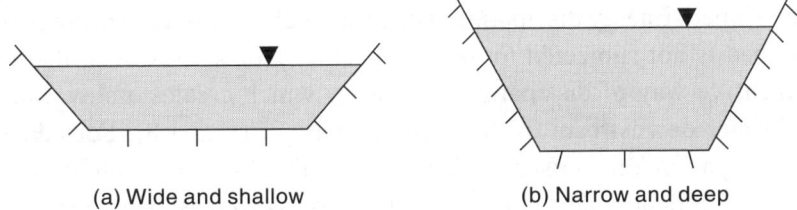

(a) Wide and shallow (b) Narrow and deep

Fig. 1.6 Various designs for the same discharge

1.2.4 Estimating the Area of Submergence due to Construction of a Dam on a River

Any obstruction in a channel affects the flow conditions and results in a change in the flow depth and velocity. Some of these effects may not be critical (for example, a bridge on a river changes the overall flow depth and velocity only marginally)[6] but some other may cause a considerable change. A large dam on a river (or even a small weir across a canal) would cause the water to back-up behind it (Fig. 1.7). Analysis of this phenomenon, known as the backwater effect, has an important bearing on estimating how much area would be submerged under the reservoir created by the dam. Since the backwater effect continues for a long distance (sometimes a few kilometres) upstream of the dam, one should have the tools to compute the variation of flow depth along the river upstream of the dam, in order to be able to estimate the submergence.

---

[5]The optimum, of course, is generally based on economic considerations, e.g., the cost of excavation, filling, lining, etc. However, sometimes social and/or political factors may have to be considered in designing an optimum channel. We assume that optimum refers to the minimum cost.

[6]Near the bridge piers, however, there may be significant change which may adversely affect the foundation. We will not discuss these in this book.

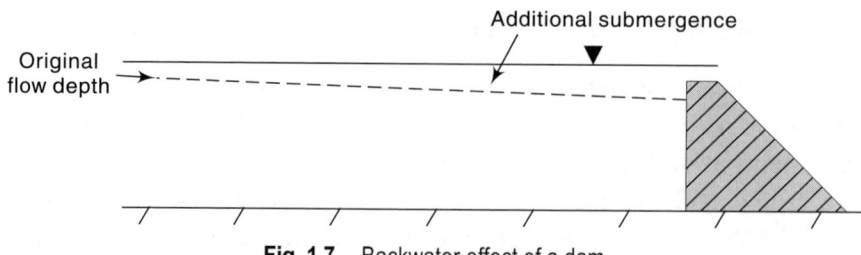

Fig. 1.7 Backwater effect of a dam

1.2.5 Preventing Very High Velocity Flows from Damaging the Channel

Water falling from a high elevation (or flowing through a very steep channel) may have a very large velocity. If left unchecked, this may cause damage to the channel bed. Energy dissipating structures could be used to ensure that the channel bed is not subjected to this high velocity flow. However, there is a less expensive way of dissipating the energy which creates a slow moving pool of water downstream of the high velocity jet (Fig. 1.8). For adequate energy dissipation, one must be able to analyse this flow situation and decide about the water depth in the slow moving portion as well as the length of floor which would still be subjected to the high velocity jet.

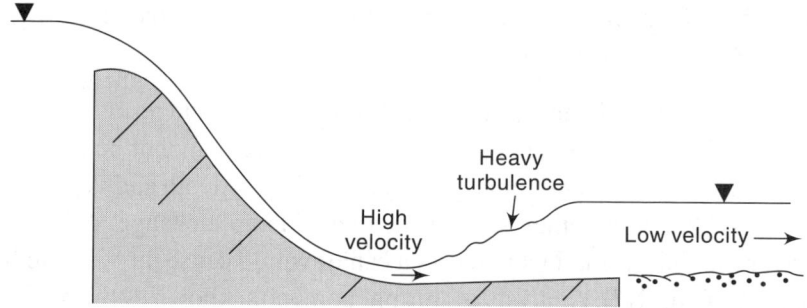

Fig. 1.8 Energy dissipation using a slow moving pool

1.2.6 Estimating the Change in the Flow Conditions due to a Change in the Bed Width or Bed Elevation

A variety of flow situations involve a change in the width of the channel or its bed level. For example, carrying a canal over a river through an aqueduct[7] may be made less expensive if the canal width is reduced (Fig. 1.9). When a

---

[7]An aqueduct is a bridge-like structure supporting a conduit or canal passing over a river or low ground.

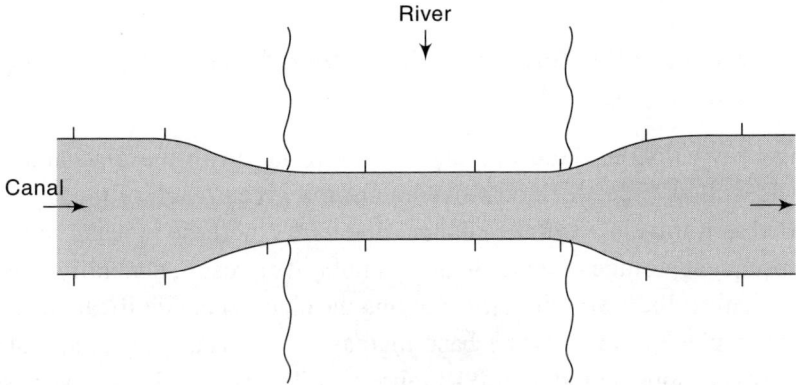

Fig. 1.9   Reduction in width of a canal at an aqueduct

channel having a flow with a uniform velocity and depth encounters such transitions, the flow characteristics undergo a change and we would require to estimate this change in order to design such structures.

1.2.7   Estimating the Change in the Flow Depth due to the Overland Runoff

During periods of intense rain, and for some time after the rain has stopped, the rainwater would flow over the land surface towards the river (Fig 1.10). This would result in an increase in discharge of the river and make the analysis of the flow more complicated due to this spatial variation. Similarly, at low discharges, the evaporation loss may be a significant fraction of the discharge and has to be accounted for. Sometimes, a weir is constructed in the sidewall of a channel to divert water when the water level rises above a fixed level, i.e., the level of the weir crest. Here also we will have the discharge in the channel decreasing in the downstream direction. In order to find the flow depth and velocity (and the diverted discharge), we should have the ability to analyse the flows with varying discharge.

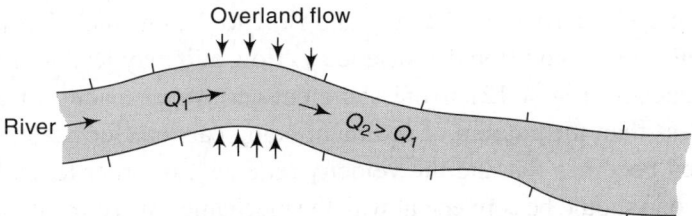

Fig. 1.10   Increasing discharge in the river due to overland flow (Q_1 and Q_2 are discharges at the locations shown.)

1.2.8 Estimating the Time Taken by a Flood Wave to Pass through a Given Length of a River

During heavy floods, there is a significant variation in the discharge with space as well as time. A flood wave entering a given reach of the river gets modified as it travels down the channel due to the increase in water level and the consequent storage of water in the channel. As a result, generally the peak reduces (unless there are tributaries joining the channel or significant overland flow takes place) and the time base increases (because the stored water is released over a long period of time). Figure 1.11 illustrates a flood wave passing through a river. It is important for the design of structures built on the river to know the approximate flood level and discharge at different locations and it requires the capability of incorporating both spatial and temporal variations in flow characteristics.

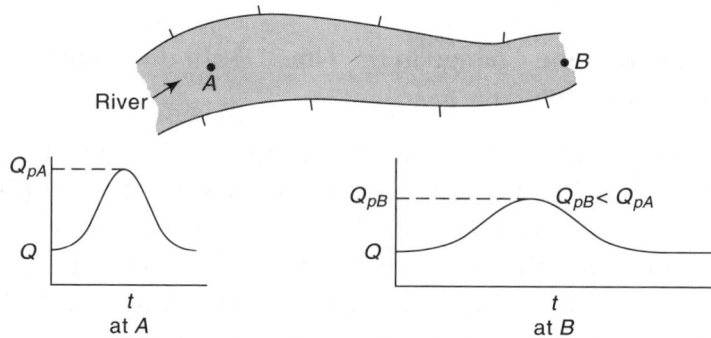

Fig. 1.11 A flood wave passing down a river (Q_p denotes the peak discharge.)

1.2.9 Estimating the Amount of Sediment Carried by a Channel

Most human-made channels have erodible boundaries and the force of the flowing water may sometimes be sufficient to loosen the boundary particles and carry these with it. Even for channels with non-erodible boundaries, sediments may be carried by the overland flow to the channel. The presence of sediments causes additional resistance to flow and may lead to deposition in some reaches (Fig. 1.12) if the sediment-carrying capacity of the flow becomes less than the amount of sediment being carried (for example, if the channel bed becomes flat and the velocity reduces). In order to analyse this behaviour, one should be conversant with the mechanics of sediment transport.[8]

---

[8]However, as we would stress again a little later, the 'mechanics of sediment transport' is itself a separate subject. We will not cover it extensively in this book but will provide enough background for the reader to be able to answer the basic questions related to the sediment-carrying capacity of a channel.

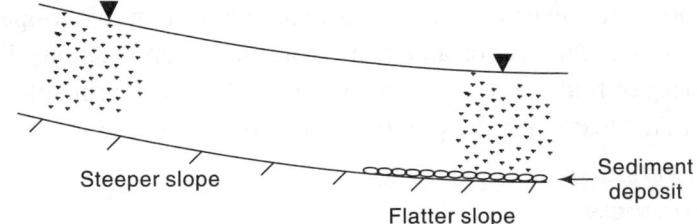

Fig. 1.12 Deposition of sediments in a channel

1.2.10 Studying the Spread of Pollutants in a River

Increasing pollution of rivers by municipal and industrial waste (especially in India, and not much in most of the developed countries) has led to increased awareness of the problem and significant efforts to reverse the damage already done. A number of industries situated near rivers and discharging their waste (with little or no treatment) into them have been closed or shifted to other places to reduce the amount of pollutants. However, the rivers are still heavily polluted. Any effort on improving the riverwater quality would need a thorough understanding of the process of advection and dispersion[9] of pollutants (Fig. 1.13).[10]

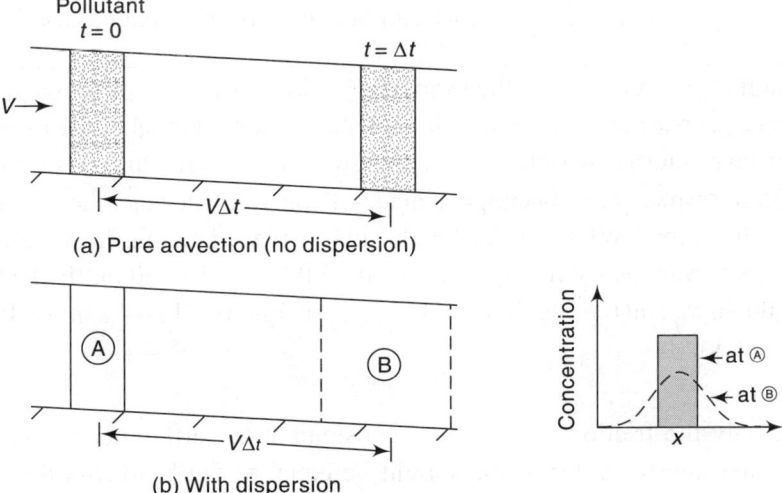

Fig. 1.13 Transport of pollutants through an open channel (*V* is the velocity of flow, *t* is the time, and *x* the distance.)

---

[9]*Advection* refers to the transport of pollutants due to the bulk movement of water and *dispersion* refers to the spreading of the pollutants about their 'mean location' due to concentration gradient or velocity variations. Detailed description is given in Chapter 9.

[10]Again, this topic is too vast to be covered here and is more related to environmental engineering than open channel flow. However, the brief treatment given in this book will enable one to understand some elementary concepts of the process.

This book would expose the reader to all the applications of the open channel flow discussed in this section and many more similar applications. Even if the reader does not find the solution to a problem, the basic principles provided throughout the text would help him or her in finding the solution.

1.3   Overview

A description of any subject is based on some assumptions about the background of the reader. For this book, we assume that the reader is familiar with the basic concepts of fluid mechanics, including fluid properties, forces in stationary and moving fluids, control volume analysis (Reynold's transport theorem), continuity, momentum, and energy equations for closed conduit flows, laminar and turbulent flows in circular pipes, and boundary layer theory. In addition, basic knowledge of mathematical and numerical methods, including differential calculus and solution of differential equations, is assumed. To aid the reader in recalling some of these prerequisites, Appendices A, B, and C list the important formulae and concepts (without detailed description) needed during the study of this book. A reference to these Appendices has been made at appropriate places in the text. The following example illustrates how the reader can take help of the Appendices.

Example 1.1   A culvert in the form of a 5 m long circular pipe (0.3 m internal diameter) carries water from one side of a road embankment to the other. The invert level and crown elevation as shown in Fig. 1.14 are 100.00 m and 100.35 m respectively. The pipe is made of concrete (mean surface roughness height 0.6 mm). What would be the discharge when the headwater and tailwater elevations are respectively (a) 100.60 m and 100.30 m, (b) 100.35 m and 100.30 m, and (c) 100.20 m and 100.15 m? Ignore all losses other than the friction loss.

Solution
Since only the friction loss is to be considered, the difference between the headwater and tailwater levels should be equal to the head loss due to the friction.

For case (a), the culvert behaves as a closed conduit and Darcy–Weisbach equation (*Appendix A*) is used to write

$$h_f = f \frac{L}{D} \frac{V^2}{2g}$$

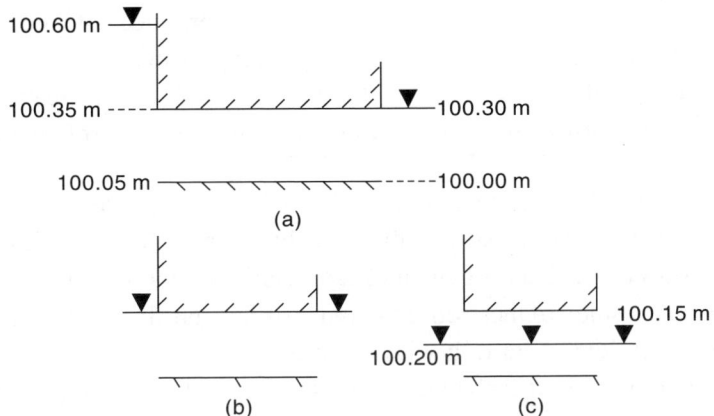

Fig. 1.14 Culvert working as a closed conduit and an open channel

$$\Rightarrow \quad 100.60 - 100.30 = f\,\frac{5}{0.3}\frac{V^2}{2 \times 9.81}$$

$$\Rightarrow \qquad\qquad fV^2 = 0.353 \qquad\qquad\qquad (1.1)$$

The velocity of the flow may be obtained through iterations using the Moody chart (Fig. A.1) or Churchill equation (A.21) or the discharge may be obtained directly using Eq. (A.22). We demonstrate the iterative method using the Moody chart here.

Since $k_s/D = 0.002$, we start with the rough pipe value of $f = 0.023$ from the Moody chart and, using Eq. (1.1), obtain the velocity as 3.92 m/s. The corresponding Reynolds number, Re, is computed as 1.18×10^6 and the friction factor is read from the chart as 0.0235. The next iteration results in a velocity of 3.88 m/s, which is very close to the value at the previous iteration and no further iterations are required. The discharge is therefore, **0.274 m³/s.**

For case (b), the culvert may be considered as either closed conduit or open channel since both the headwater and the tailwater levels are *just touching* the respective crown levels. However, since we do not yet know how to find the discharge for an open channel, we again use the pipe-flow equation with the head loss equal to 0.05 m, and obtain $fV^2 = 0.0589$. Starting with $f = 0.023$, we get $V = 1.60$ m/s and Re $= 4.8 \times 10^5$. From the Moody chart, $f = 0.025$, which results in $V = 1.53$ m/s. Further iterations are not needed and the discharge is **0.108 m³/s.** Note that Eq. (A.22) would directly (without any need of iterations) provide the discharge as 0.274 m³/s in case (a) and 0.111 m³/s in case (b).

Case (c) involves flow with a free surface condition since both the headwater and tailwater levels are below the corresponding crown elevations. At this stage, we are not able to estimate the discharge. By the time the reader finishes the next chapter, however, he or she would be able to solve problems like this.

Now that we have seen our limitations in solving open channel flow problems, we would discuss how to study these flows. We first classify open channel flows on the basis of channel characteristics and flow properties, and then take up the study of these different types of flow one by one with increasing degree of complexity, starting with Chapter 2 with the simplest case where the flow depth, velocity, channel cross section, etc. are not changing with time and location (i.e., uniform flow). Unlike most other books on the subject, we will not describe the basic equations of continuity, momentum, and energy in the initial chapters. We assume that the reader is already familiar with the concepts of these equations in other fields. However, adaptation of these concepts to open channel flow have been introduced as and when needed. For example, the study of uniform flow does not require the direct application of the general momentum or energy equation. Therefore, these equations have not been presented in Chapter 2. The energy equation and the concept of specific energy have been introduced only when they are needed in the chapter on Nonuniform Flow. Also, since a very good book dealing with the historical development of the field of hydraulic engineering is available (Rouse and Ince 1980), we have limited the discussion of historical facts to a minimum.[11] References have not been provided for very old or hard-to-find articles. In such cases, the author's name and the year have been mentioned to put things in the proper historical perspective.

1.4 Classification of Open Channel Flows

Classifying the open channel flow in different categories helps us in determining which parameters are significant and which can be safely ignored while analysing the flow. For example, in flow through closed conduits, we know that if the flow is laminar, the boundary roughness does not influence the frictional losses. Similarly, for very high Reynolds number, the friction factor is not affected by the Reynolds number and is only dependent on the relative roughness of the boundary. Therefore, for free surface flows also, we

---

[11]In fact, at some places we have taken the liberty of ignoring the chronological order of events if it suits our main aim of explaining things in a logical sequence.

would expect some parameters to be important in some flow situations and insignificant in some other situations. We classify the open channel flows based on channel characteristics and flow properties; look at some examples and see how this classification affects the analysis of flow.

1.4.1 Classification Based on Channel Characteristics

Channels can be classified as prismatic and non-prismatic channels, natural and artificial channels, and rigid boundary and mobile boundary channels. Accordingly, the flows through channels can also be classified.

Prismatic and nonprismatic channels

A channel is said to be *prismatic* if it is in the form of a prism, i.e., the cross section and the bed slope do not change along the channel length. If there is a change in cross section and/or slope, the channel is *nonprismatic*. A laboratory flume laid at a constant bed slope and with a uniform cross section is a prismatic channel while a river with varying cross-section and bed slope is a nonprismatic channel (Fig. 1.15). Obviously it is much simpler to analyse the flow in a prismatic channel since the invariance of cross section and bed slope implies that the flow depth, velocity, etc. will be same at every section for a given discharge under normal flow conditions. However, for a nonprismatic channel, flow characteristics may change along the length making the analysis more difficult. Also, the variation in cross section leads to additional losses due to expansion and contraction, which should be accounted for in the analysis. In this book, we will assume that the channel is prismatic, unless specified otherwise (e.g., in Section 3.6.3).

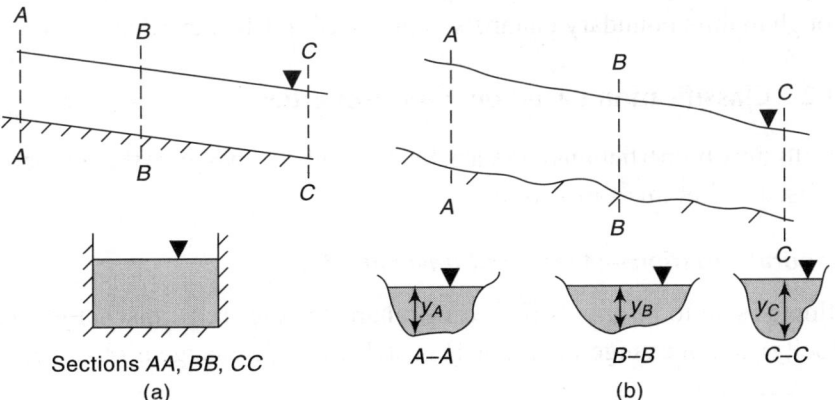

Fig. 1.15 Prismatic and nonprismatic channels

Natural and artificial channels

A river, an estuary,[12] and a land surface during overland runoff, are all examples of natural channels while a laboratory flume, a canal, and a parking lot during overland runoff, represent artificial (or human-made) channels. Generally, natural channels would be nonprismatic while the artificial channels are likely to be prismatic. Also, natural channels typically have an irregular cross section [Fig. 1.15(b)] while artificial channels have regular (e.g., circular, trapezoidal, rectangular) cross sections [Fig. 1.15(a)]. This may lead to ambiguity in defining the flow depth for natural channels. Conventionally, the flow depth is measured as the height of water surface above the deepest point in the cross section.

Rigid boundary and mobile boundary channels

If the material on the bed and sides of a channel is loose and easily movable due to the flow of water,[13] the channel is called a *mobile boundary channel*. Conversely, if the material is not easily movable (e.g., a metal flume, concrete lined canal), the channel is a *rigid boundary channel*. Clearly, analysis of flow through a mobile boundary channel is more complicated than that of flow through a rigid boundary channel, due to the process of sediment erosion and deposition and the resulting additional resistance to flow. These processes may also occur in rigid boundary channels due to sediment inflow from elsewhere (e.g., from the river into a lined canal or from upstream areas to a parking lot). Sediment transport in channels is a subject in its own right and there are various textbooks available on this topic (e.g., Garde and Ranga Raju 2000). In most of this book, therefore, we will consider the channels to be rigid boundary channels carrying no sediments. A brief description of flow through mobile boundary channels is provided in Chapter 8.

1.4.2 Classification Based on Flow Properties

Various flow properties and the classification of flows based on these properties are discussed in this subsection.

Temporal variation—steady and unsteady flows

A flow is said to be *steady* if the flow characteristics (e.g., discharge, depth, velocity) do not change with time [Fig. 1.16(a)]. If a change is observed with

---

[12]An *estuary* is an arm of the sea at the lower end of a river, i.e., where the river joins the sea. The tide meets the river current in the estuary.

[13]Since water is the most common fluid flowing in an open channel, we will assume that the flowing fluid is water, unless otherwise mentioned.

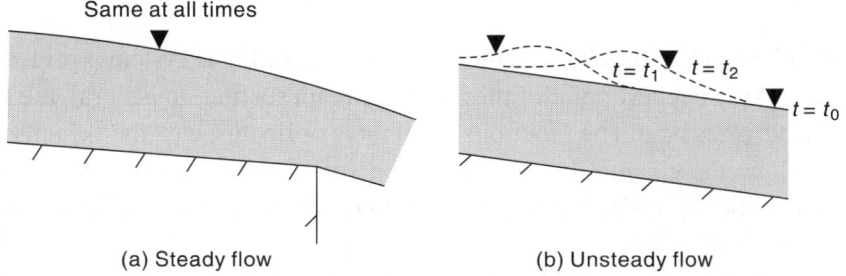

Fig. 1.16   Steady and unsteady flows

time, the flow is *unsteady* [Fig. 1.16(b)]. In general, it is almost impossible to have a strictly steady flow. No matter how good a control exists on the flow conditions, there is bound to be some change with time. Since the analysis of the steady flow is simpler than that of the unsteady flow (due to the absence of temporal derivatives), one may decide to treat a given flow as steady under certain conditions (e.g., discharge/depth variation within, say, 1% of a mean value). Sometimes, it is possible to change the frame of reference and convert an unsteady flow situation into an equivalent steady flow (e.g., in a tidal wave, by moving with the wave).

Spatial variation—one-dimensional, two-dimensional, and three-dimensional flows

Any point in space may be conveniently described through a three-dimensional orthogonal coordinate system. The particular system to be used for a problem depends on the relevant geometry. For example, flow through a circular pipe is best studied in a cylindrical coordinate system. In open channel flows, we would commonly use the Cartesian coordinates (X, Y, Z),[14] with X-axis along the channel bed in the direction of flow (longitudinal distance), Y orthogonal to X upwards (channel depth),[15] and Z orthogonal to both X and Y (channel width) following the right hand rule[16] (see Fig. 2.1). While analysing a

---

[14]Generally, lowercase letters are used to represent the coordinate axes. We use uppercase letters in this book to distinguish the axes from the flow depth (y) and the bed elevation (z).

[15]Note that the depth is not measured vertically but perpendicular to the bed. Generally, the bed slope is very small and the vertical depth and perpendicular depth are almost identical. But for steep channels, such as flow over a spillway, these two depths will differ significantly from each other.

[16]If the thumb, index finger, and middle finger of the right hand are held in a mutually orthogonal position, such that the thumb indicates the X-axis and the index finger the Y-axis, then the middle finger indicates the Z-axis. Thus, if X points in the downstream direction, and Y upwards, Z will point towards right when looking at a cross-section from the upstream side.

particular reach of the channel, the origin of the coordinate system is preferably placed at the deepest point of the most upstream section. In general, the flow characteristics would be functions of all three ordinates, e.g., the velocity will vary along the channel length and also across a cross section. However, to simplify the analysis, we may approximate the actual (three-dimensional) velocity distribution by a simpler profile. For example, if the channel is very wide, we may consider the flow to be independent of Z-location (two-dimensional in the X-Y plane). This assumption would be valid for most of the cross section except very near the sides, where velocity would depend on Z. Similarly, the Y-dependence of velocity may be removed by averaging the velocity across the depth (two-dimensional, X-Z plane). The most commonly used technique for reducing the dimensionality of a problem is the averaging of velocity over the entire cross section. This average velocity is then only a function of X (one-dimensional or 1-D) and the governing equations become simpler[17] without loosing too much of the detail. (In most cases our primary interest is in knowing the variation of flow characteristics along the channel length and not across a section.) In this book, we will adopt the 1-D simplification by using the cross section average velocity [Fig. 1.17(a)]. A brief discussion of the velocity variation in a cross section is provided, however, in Chapter 3 to make the reader aware of the degree of approximation introduced by the 1-D assumption.

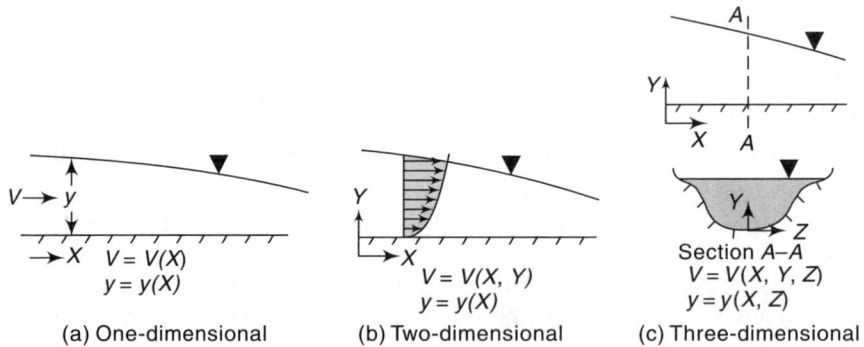

Fig. 1.17 One-, two-, and three-dimensional flows

Spatial variation—uniform and nonuniform flows

Uniform flow indicates that the flow depth, velocity, and discharge do not change in the longitudinal direction. As mentioned for steady flow, it would

---

[17]The averaging process introduces some complications during computation of momentum and energy of flow. These are accounted for by using the so-called *correction factors*, which will be discussed at appropriate places.

be rare to achieve a strictly uniform flow but a number of flow situations may be idealized as uniform flow. For example, flow in a long prismatic laboratory flume may show minor variations in depth, velocity, and discharge with space and time. However, for all practical purposes, it may be analysed as steady uniform flow. Clearly, assumption of uniformity leads to a drastic simplification of analysis since we need to obtain the flow characteristics at a single section only as all characteristics remain same at all sections.

If the flow is not uniform, i.e., the flow depth, velocity, or discharge is varying spatially, it is known as *nonuniform (or varied) flow*. Flow over a weir (depth decreases as we approach the weir) and flow upstream of a dam (depth increases as we approach the dam) are some examples of nonuniform flows (Fig. 1.18). Looking at the temporal and spatial variability together, it should be obvious that unsteady uniform flow is practically non-existent since it would require the flow conditions to be same throughout the channel at some instant and then change uniformly over the entire channel length at the next instant. Therefore, to simplify the classification, the term *unsteady* flow is used to indicate *unsteady nonuniform* flow. Similarly, *uniform* flow implies *steady uniform* flow (since unsteady uniform flow is non-existent) and *nonuniform* flow implies *steady nonuniform* flow (since the unsteady nonuniform flow has already been classified as unsteady flow).

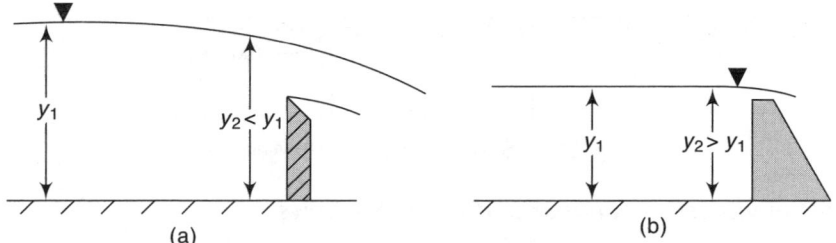

Fig. 1.18   Uniform and nonuniform flows

Nonuniform flow can be further classified into gradually varied flow (GVF)[18] when the depth changes gradually, rapidly varied flow (RVF) when the depth changes significantly over a short distance, and spatially varied flow (SVF)[19] when the discharge changes due to lateral inflow or outflow.

---

[18]The flow is classified as gradually varied if the streamline curvature is so small that pressure distribution at a section may be assumed to be hydrostatic.

[19]For most flow situations, we assume that the discharge is constant along the channel length. Therefore, spatially varied flow could be a classification by itself. Here we include it in nonuniform flow but devote a separate chapter to it (Chapter 6) in keeping with the generally followed practice.

The flow upstream of a dam may be treated as gradually varied [Fig. 1.18(b)], flow over a weir is rapidly varied [Fig. 1.18(a)], and flow in a side channel spillway[20] and a side weir is spatially varied (Fig. 1.19). The computation of water depth (and velocity) at different locations in a GVF is simpler than that in a RVF since the pressure distribution in GVF is assumed to be hydrostatic and is, therefore, a function of the flow depth. For a RVF, the pressure distribution is not hydrostatic and generally experimental observations are utilized to help in the computations. The SVF is further complicated by the variation of discharge along the length.

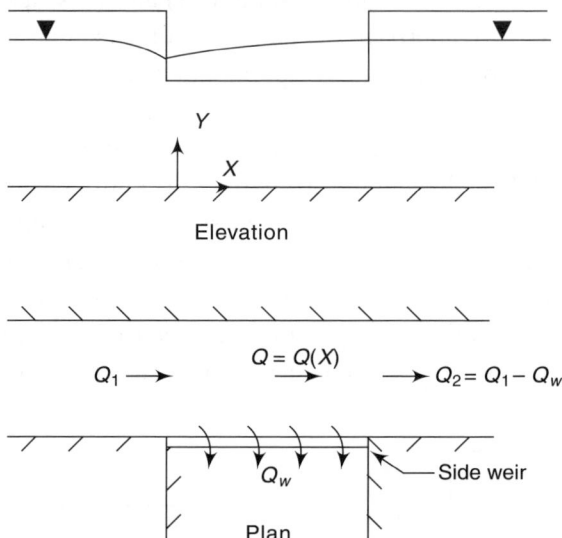

Fig. 1.19   Spatially varied flow at a side weir

Effect of viscosity—laminar and turbulent flows

Similar to the flow through a closed conduit, the flow through an open channel may be classified according to the relative magnitude of inertial and viscous forces. For a circular pipe, the state of flow (laminar or turbulent) depends on the *Reynolds number*

$$\text{Re} = \frac{\rho V L}{\mu} \tag{1.2}$$

where ρ is the mass density, V is the cross section average velocity, L is a characteristic length, and μ is the dynamic viscosity. Generally, the pipe

---

[20]A side channel spillway has its crest parallel to the channel downstream of the spillway. The discharge in the channel increases along its length as additional water spills over the crest.

diameter is used as the characteristic length, and the critical Reynolds number (below which the flow is laminar) is about 2,300.[21] Since open channel flows occur in channels with various cross-sectional shapes, we need to define another characteristic length. For non-circular conduits, the *hydraulic radius*, R,[22] defined as the area of cross section divided by the perimeter, is used as the characteristic length. Considering that the hydraulic radius of a circular pipe is equal to one-fourth of the diameter, the critical Reynolds number, with hydraulic radius as the characteristic length, is about 2,300/4, i.e., 575. For open channels, the hydraulic radius is defined as the area of flow section divided by the wetted perimeter. For a very wide channel, the hydraulic radius may be replaced by the flow depth to obtain a rough idea about the Reynolds number. Since water is the fluid in most open channel flows,[23] we use $\rho = 1{,}000$ kg/m^3 and $\mu = 0.001$ N s/m^2. For a typical velocity value of 1 m/s and flow depth of 1 m, we get Re = 1,000,000. Measurements suggest that the critical value of Re for open channel flows is also about 500. Therefore, the flow through open channels for all practical cases would be turbulent.[24] If we look at very slow flows with velocity of the order of a few centimetres per second, the flow depth would have to be of the order of a few centimetres to sustain laminar flow. Such conditions may exist only during sheet flow of overland runoff. Hence, we will not consider laminar flow in this book.

Effect of gravity—subcritical and supercritical flows

Gravity is the driving force behind flows through open channels. Hence, it stands to reason that the ratio of inertial to gravitational forces will play a major role in open channel flow analysis. Using dimensional analysis with inertial force $\rho V^2 L^2$ and gravitational force $\rho L^3 g$, where V is the average velocity, L is a characteristic length, and g is the gravitational acceleration, we obtain the ratio of these forces as V^2/gL. Following the convention of

---

[21]Under carefully controlled conditions, laminar flow may be obtained for a Reynolds number as high as 40,000! The *lower* critical Reynolds number has been reported variously in the range of 2000 to 2300.

[22]Some books use the notation R_h for the hydraulic radius to distinguish it from the pipe radius. However, in this book, we would use pipe diameter, D_0, and not its radius. Therefore, we use R for hydraulic radius.

[23]Sometimes, we will have seawater as the flowing fluid (estuaries). In laboratory experiments, we may use some other fluid to study, say, the effects of surface tension and viscosity. Throughout this book, we assume that the fluid is water unless mentioned otherwise.

[24]In fact, the Reynolds number, Re, is generally so high that the *friction factor* becomes independent of Re (fully rough zone of flow).

using the first power of velocity, we define a dimensionless number (the *Froude number*) as

$$F_r = \frac{V}{\sqrt{gL}} \tag{1.3}$$

to represent the relative importance of inertial and gravitational forces. It would be shown in Chapter 3 that the characteristic length used in the Froude number is the *hydraulic depth, D,* defined as the ratio of area of flow to the top width of the flow section. Further, it would be shown that

(i) the energy of flow at any section for a given discharge is minimum when $F_r = 1$ (Chapter 3),

(ii) the sum of pressure force and momentum of flow at any section for a given discharge is minimum when $F_r = 1$ (Chapter 4), and

(iii) a small amplitude disturbance created at the free surface in stationary water moves with a velocity of $\sqrt{gD}$ (Chapter 7) implying that the flow velocity is equal to wave velocity when $F_r = 1$.

Thus the condition $F_r = 1$ is quite critical in applying the energy and momentum equations (and also in analysing the spreading of a disturbance) and is known as *critical flow*. Clearly, $F_r > 1$ denotes a velocity larger than the critical velocity (and depth smaller than the critical depth) and is termed *supercritical*, rapid, or shooting flow, and $F_r < 1$ denotes a velocity smaller than the critical velocity and is termed *subcritical*, streaming, or tranquil flow. As F_r becomes larger than (or smaller than) 1, the energy keeps on increasing from its minimum value at $F_r = 1$. For a supercritical flow, therefore,

(i) the energy will increase with a decrease in depth,

(ii) the sum of pressure force and momentum will increase with a decrease in depth, and

(iii) a small disturbance on the surface will not be able to travel upstream[25]

since $V > \sqrt{gD}$.

Conversely, for a subcritical flow,

(i) the energy will increase with an increase in depth,

(ii) the sum of pressure force and momentum will increase with an increase in depth, and

(iii) a small disturbance on the surface will be able to travel upstream.

---

[25]This provides us a rough, but quick, method of estimating whether a flow is supercritical or subcritical. If we drop a stone in a channel and the resulting wave moves both upstream and downstream (of course, the upstream movement will be much slower than the downstream one), the flow would be subcritical.

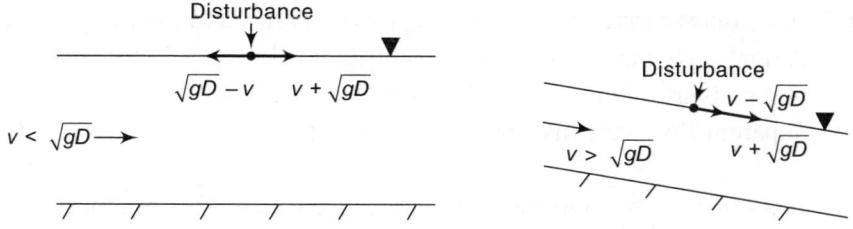

Fig. 1.20 Subcritical and supercritical flows

Most flows in nature are subcritical but the flow over a spillway, a steep laboratory flume, or downstream of a sluice gate, would be supercritical (Fig. 1.20). As we will see in later chapters, the contrasting nature of variation of energy and momentum for subcritical and supercritical flows would be important while analysing gradually varied flow profiles, hydraulic jump,[26] and channel transitions involving change of width and/or bed level.

REFERENCES

Garde, R.J. and Ranga Raju, K.G. (2000). *Mechanics of Sediment Transportation and Alluvial Stream Problems*, 3rd edn, New Age International, New Delhi.

Rouse, H. and Ince, S. (1980). *History of Hydraulics*, reprint. Iowa Institute of Hydraulic Research, University of Iowa, Iowa City.

EXERCISES

1.1 Classify the following flows as uniform, gradually varied, rapidly varied, spatially varied, or unsteady flow:
 (a) Flow in a long laboratory flume
 (b) Flow near the ungated end of a laboratory flume
 (c) Flow upstream of a dam
 (d) Flow over a weir
 (e) An open drain by the side of a road during a rainfall event
 (f) Progress of a tidal wave in an estuary
 (g) Flow in an unlined canal considering evaporation and seepage losses

1.2 For a short laboratory flume, the effect of end conditions causes the flow to be nonuniform. What would you do to achieve a nearly uniform flow?

---

[26]A hydraulic jump is formed when a supercritical flow meets a subcritical flow.

1.3 An estuary carries water at a velocity of 1.5 m/s and flow depth of 2 m. A tidal wave moves upstream at a constant velocity of 2 m/s. How would you convert this into a steady state flow situation? What would be the apparent flow velocity upstream of the wave after the conversion?

[3.5 m/s]

1.4 In a sugar factory, molasses (dynamic viscosity $= 7 \, Ns/m^2$, mass density $= 1500 \, kg/m^3$) is carried away in a 2 m wide rectangular open channel at a flow depth of 1 m and volumetric flow rate of 0.5 m³/s. Is the flow laminar or turbulent? What would be the speed of a small disturbance created on the surface? What is the Froude number?   [3.13 m/s, 0.08]

2 Uniform Flow

2.1 Introduction

As described in Chapter 1, the term *uniform flow* refers to the condition in which flow parameters (depth, velocity, discharge) are virtually same at all cross sections, i.e., there is no spatial variation. Although it is rare to find completely uniform flows in nature, many flow situations may be approximated as uniform flows. For example, flow in a long reach of a prismatic channel may be considered uniform. For other situations, where flow is considerably nonuniform, e.g., upstream of a small weir or large dam, it would be helpful to know what would have been the depth (and velocity) of flow had there been no disturbance to the flow. Thus analysis of this normally occurring flow becomes important for nonuniform flow conditions also. The analysis may involve determination of the uniform flow depth (or normal depth) for a given discharge or the estimation of discharge for a given flow depth. We start with the uniform flow in a prismatic channel of simple cross-section and boundary characteristics and then move on to composite boundary channels (which have different types of boundaries comprising the perimeter) and compound channels (which have a distinct central portion flanked by shallower sections). Finally, we discuss the concept of the most efficient section and look at the design of a channel to carry a given discharge under uniform flow conditions.

2.2 Flow Through Prismatic Channels

Most laboratory flumes have regular (mostly rectangular) cross sections and most natural channels have irregular cross sections. However, because of their large width, compared to the flow depth, natural channels may sometimes be approximated as having a rectangular or trapezoidal cross section. Since the trapezoidal section encompasses the rectangular and triangular sections as special cases with zero side slopes and zero bottom width, respectively, we would generally analyse the trapezoidal section. The other commonly encountered section is a circular section (not flowing full), e.g., storm water drains. Moreover, as we will see a little later, two different normal depths may carry the same discharge for circular conduits flowing partly full. Therefore, the analysis of circular open channels is also described in detail. In this section, we discuss uniform flow in a prismatic channel of an arbitrary cross section.

Figure 2.1 shows the longitudinal and cross sections of a prismatic channel. As mentioned in Chapter 1, the orthogonal coordinate system with X-axis along the direction of flow, Y-axis along the flow depth, and Z-axis along the channel width, will be followed throughout this book.[1] The vertical direction is represented by the z-axis, positive upward.[2]

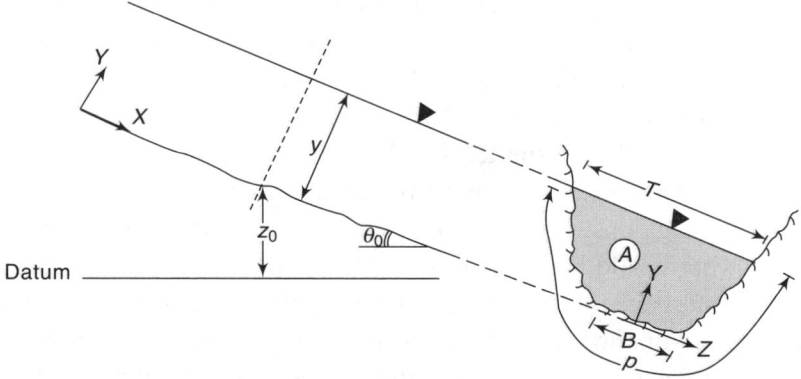

Fig. 2.1   The longitudinal and cross sections of a prismatic channel

Some important geometric elements shown in Fig. 2.1 are discussed below.

Bed slope (S_0)

If θ_0 denotes the angle of the general bed profile from the horizontal, the bed slope ($= \tan \theta_0$) is generally expressed as, say, 1 in 1000, indicating a drop of 1 m in 1000 m *horizontal* length. For such small slopes, it will be nearly equal to a drop of 1 m in 1000 m length *along the channel bed* (the exact value being 1 m in 1000.0005 m!). However, for increasingly larger bed slopes, the

---

[1]For an irregular cross section, X-axis will be taken along the line joining the deepest points of two consecutive sections. When there are bends in the channel, the X-axis will also be bent as it follows the flow direction. As mentioned in Chapter 1, Y-axis is orthogonal to the X-axis and for a steep bed slope, the flow depth y will be different from the vertical depth (which is easier to measure). However, for most practical applications, the channel slope is small and y may be taken equal to the vertical depth.

[2]It may be better to take the z-axis pointing downward since gravity acts in that direction. However, we will follow the usual convention of measuring the vertical distance above a datum.

two values will become more different (e.g., a 1 m drop in 10 m horizontal length is equivalent to a 0.995 m drop in 10 m length along the channel bed). Since most cases of practical interest involve small slope, we assume that $S_0 = \sin \theta_0 \approx \tan \theta_0$. For steep channels, e.g., flow over a spillway, we express the bed slope as $S_0 = \sin \theta_0 = -(dz_0/dX)$, z_0 being the elevation of the channel bottom above an arbitrary horizontal datum.

Flow depth (y)

The depth of flow, y, is the depth in the Y-direction, i.e., perpendicular to the bed. Typically the laboratory and field measurements of flow depth provide the *vertical* depth at a point. For generally encountered bed-slopes, the two depths would be very close (e.g., for a slope of 1 vertical to 10 horizontal, which is quite steep for typical open channel flows, y would be 99.5% of the vertical depth). Therefore, we would assume that y is same as the vertical depth. For very steep channels, we would use the correction factor of $\cos \theta$ to obtain y from the vertical depth. In most of the open channel flow problems, flow depth is the primary variable of interest. Once the depth is known, the velocity can be readily obtained since discharge is generally known.

Bed width or bottom width (B)

This is the channel width at the bottommost point of the cross section. For a natural channel, it may not be possible to define a bottom width due to irregularity of the cross-section. However, we may approximate the shape by, say, a trapezoid and obtain a bottom width. For a triangular channel, the bottom width will be zero.

Top width (T)

This is the width of the channel cross section at the water surface. Even for natural channels, it is well defined. For a rectangular channel, the top width will be same as the bottom width. For other section shapes, the top width will depend on the flow depth. It is an important parameter since it determines the rate of increase of flow area with change in flow depth. The top width generally increases with increase in flow depth but in some cases, e.g., a circular channel flowing more than half full, it may decrease with increase in depth.

Wetted perimeter (P)

The channel perimeter in contact with the flowing fluid is known as the *wetted perimeter*. It plays an important role in open channel flow since the resistance to flow due to the boundary shear is directly proportional to the wetted perimeter. Clearly, P will increase with an increase in the flow depth.

Flow area (A)

The area occupied by the flowing fluid in any cross section is called the *flow area*. This area is in the Y-Z plane but for most practical cases, it may be taken in a vertical plane. As gravity is the driving force in open channel flows, the flow area plays a vital role since the weight flux of fluid is directly proportional to A.

The velocity of flow will vary over a cross section from zero at the boundary to the maximum near the water surface. In general, it will have all three components (X, Y, Z) but the longitudinal component is predominant. We perform a one-dimensional analysis since generally we are not interested in the variation of velocity over a section. The quantity of interest is the average velocity over the cross section, V, defined as

$$V = \frac{\int_A v_X \,(Y, Z)\, dA}{A} = \frac{Q}{A} \tag{2.1}$$

where v_X is the velocity component in the X direction and Q is the discharge at that section. In general, V will be a function of X since the flow depth and, therefore, area would change along the channel length. However, for uniform flow, V is the same at all sections.

We now state the problem to be solved for uniform flow in a prismatic channel: Given the discharge, bed slope, and the nature of the boundary (e.g., the grain size for natural channels or the material comprising the boundary, since it affects the resistance to flow), what will be the flow depth under uniform flow conditions? Once the flow depth is known, the velocity can be obtained after computing the area of flow. The three basic equations, continuity, momentum, and energy, can be used to answer this question. In this case, however, application of the continuity equation does not help us as it just tells us that depth, velocity, and discharge[3] are the same at every section, which we already know! Since there is no spatial variation of flow properties, the momentum equation simplifies to a force balance equation stating that the driving force due to gravity and the resisting force due to boundary shear should be equal and opposite. This helps us in relating the average shear stress

---

[3]Strictly speaking, the mass of the fluid should be conserved. However, since we are dealing with water at close to atmospheric pressure and temperature, the fluid is assumed to be incompressible and continuity equation may be thought of as a volume conservation equation.

at the boundary with the flow depth and other geometric parameters. Since the velocity and pressure distributions are same at every section in a uniform flow, the kinetic energy and flow work would also be same.[4] The energy equation then states that the energy loss (which here indicates the conversion of mechanical energy into heat energy) between any two sections is due to the drop in elevation along a streamline. It implies that the slope of the *energy line* (also called friction slope, S_f, since most of the losses are due to friction at the boundary) is equal to the bed slope, S_0.

The application of the continuity, momentum, and energy equations, will thus provide us with relationships between the average boundary shear and the flow characteristics and between the energy loss and the bed slope. These, by themselves, are not sufficient to determine the normal depth. Therefore, we make use of the extensive studies done on closed-conduit flow regarding the shear stress distribution and energy loss and correlate the two to help us obtain the uniform flow depth. We assume that the reader is familiar with the basic theory of closed conduit flow. A brief description, nonetheless, is provided in Appendix A.

Figure 2.2 shows the longitudinal section of an open channel with a control volume bounded by two sections, 1 and 2, ΔX apart. The weight of water within the control volume is given by

$$W = \rho g A \Delta X \tag{2.2}$$

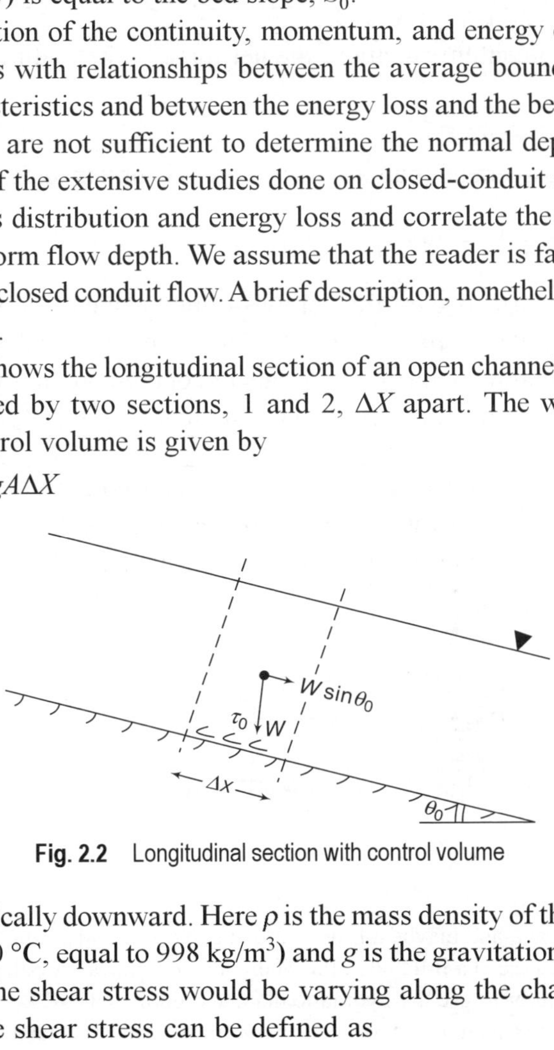

Fig. 2.2   Longitudinal section with control volume

and it acts vertically downward. Here ρ is the mass density of the flowing fluid (for water at 20 °C, equal to 998 kg/m^3) and g is the gravitational acceleration (9.81 m/s^2). The shear stress would be varying along the channel perimeter and an average shear stress can be defined as

---

[4]It would be shown in later chapters that the kinetic energy (and momentum) would be larger than that based on the average velocity. We need not concern ourselves with this as the kinetic energy is same at all sections.

$$\overline{\tau} = \frac{\int_0^P \tau_{nX}(t)\,dt}{P} \tag{2.3}$$

in which t and n[5] denote the tangential and normal directions respectively for a length element along the channel boundary. For the one-dimensional analysis, the actual variation of shear stress is not needed, and only the average shear stress, $\overline{\tau}$, is required. However, for the sake of completeness, typical shear stress distributions on the boundary of channels of different cross-sectional shapes are shown in Appendix B.

From the momentum equation, the net force in the X-direction should be zero. Therefore,

$$W \sin \theta - \overline{\tau} P \Delta X = 0 \tag{2.4}$$

from which the average boundary shear stress is obtained as

$$\overline{\tau} = \rho g \frac{A}{P} S_0 = \rho g R S_0 \tag{2.5}$$

The quantity A/P occurs frequently in open channel flow computations and is called the *hydraulic radius*,[6] R. The correlation of this parameter for the open channel flows with the pipe diameter for pressure flow through circular pipes enables us to obtain a resistance equation.

Example 2.1   A 5 m wide rectangular channel is laid at a longitudinal slope of 1 in 4000 and carries water at a uniform flow depth of 1.5 m. Find the hydraulic radius and the average boundary shear stress.

Solution
We have
Area of flow, $A = 5 \times 1.5$ m^2 = 7.5 m^2
Wetted perimeter, $P = 5 + 2 \times 1.5$ m = 8 m

$\therefore$ Hydraulic radius, $R = \dfrac{A}{P}$ = **0.9375 m**

---

[5]A purely temporary, and maybe a little confusing, use of t and n. In general t will represent time and n, as seen a little later, the Manning's roughness coefficient.

[6]Some authors prefer to use the hydraulic diameter, defined as $4A/P$, rather than the hydraulic radius. The advantage of using the hydraulic diameter is that it is equal to the pipe diameter for pressure flow through circular pipes. Therefore, all equations and charts developed for circular pipes could be directly used for other cases by replacing the pipe diameter with the hydraulic diameter. However, we will follow the more commonly used notation and use the hydraulic radius for the open channel flow.

Also

average boundary shear, $\bar{\tau} = \rho g R S_0 = 1000 \times 9.81 \times 0.9375 \times \dfrac{1}{4000}$ N/m^2

$$= \textbf{2.3 N/m}^2$$

2.3 Comparison of Open Channel Flow with Closed Conduit Flow

For pressure flow through a circular pipe, the average shear stress (in fact, the shear stress at a cross section is same at all points on the boundary due to radial symmetry) is given by [see Eq. (A.17) in Appendix A]

$$\bar{\tau} = C_f \frac{\rho V^2}{2} = f \frac{\rho V^2}{8} \tag{2.6}$$

where V is the cross section average velocity, C_f is the skin friction coefficient, and f is the Darcy–Weisbach friction factor.

The energy slope is given by the well-known Darcy–Weisbach equation

$$S_f = \frac{h_f}{L} = \frac{f V^2}{2 g D_0} = \frac{f V^2}{8 g R} \tag{2.7}$$

where D_0 is the diameter of the pipe and we have used the fact that the hydraulic radius for a circular pipe is equal to $(\pi D_0^2/4)/\pi D_0$, i.e., $D_0/4$. The use of hydraulic radius enables us to translate the information available for pressure flow through circular pipes into the flow of fluids through open channels. For example, the conventional friction factor diagram (Moody's chart, Fig. A.1 in Appendix A) for a circular pipe is presented in terms of a Reynolds number based on the pipe diameter and a relative roughness also based on the pipe diameter. We assume that a similar (but not same)[7] relationship holds good for free-surface flows provided we replace the pipe diameter by 4 times the hydraulic radius for the open channel flow. This enables us to arrive at resistance equations for open channel flows.

2.4 Resistance Equations

A comparison of Eqs (2.5) and (2.6) leads to a *resistance equation* for open channel flows as

---

[7]There are differences between the Moody's chart for pressure flow through circular pipes and the corresponding chart for free-surface flow through circular, triangular, and rectangular channels (ASCE 1963, Yen 2002).

$$V = \sqrt{\frac{8\,gRS_0}{f}} = C\sqrt{RS_0} \tag{2.8}$$

in which $C\left(= \sqrt{8g/f}\right)$ is a coefficient known as the Chezy C since this form of resistance equation was first proposed by a French engineer, A. Chezy, in the 1770s. It should be noted that the recommendation by Chezy was not based on theoretical analysis described in the previous section. He obtained the equation on the basis of measurements in French rivers and found that C was different for different channels. Also note that the same equation may be obtained by using Eq. (2.7) and the fact that for uniform flow the friction slope is equal to the bed slope. For SI units, a typical range of variation of the value of C is 40 to 80 $m^{0.5}/s$. We will, however, not discuss the Chezy's equation further since we believe that the Manning's equation described later in this section is a better alternative.

Since C is inversely proportional to the square root of the friction factor, and the friction factor depends on the *relative roughness*,[8] it seems reasonable to conclude that C would be a function of the hydraulic radius (and, consequently, the flow depth) as well as the boundary roughness. One would, therefore, expect that C not only changes from one channel to another, but would also be different at different depths *for the same channel*. One way to develop a resistance coefficient which is independent of the flow depth is to incorporate the variation of the friction factor, f, into the Chezy's equation.

If we assume that the combination of relative roughness and Reynolds number is such that the flow may be treated as fully rough, the friction factor becomes independent of the Reynolds number and is given by the Karman–Nikuradse formula [see Eq. (A.19) in Appendix A]

$$\frac{1}{\sqrt{f}} = -2.0 \log\left(\frac{k_s}{3.715\,D_0}\right) = 2.0 \log\left(\frac{R}{k_s}\right) + 2.34 \tag{2.9}$$

in which k_s is the equivalent sand-grain roughness. Figure 2.3 shows a double-logarithmic plot of this equation over a wide range of R/k_s values. Also shown on this figure is a straight line with a slope of 1/6 (indicating a power-law relationship, $1/\sqrt{f} = C_1(R/k_s)^{1/6}$, between the variables). It is seen that for

---

[8]We know that it is true for pressure flow through pipes and assume that a qualitatively similar behaviour would be observed in free-surface flows also. Further, it should be noted that the friction factor also depends on the Reynolds number. However, most open channel flow conditions result in a very large Reynolds number and the flow may be assumed to be in the fully rough (or fully turbulent) zone.

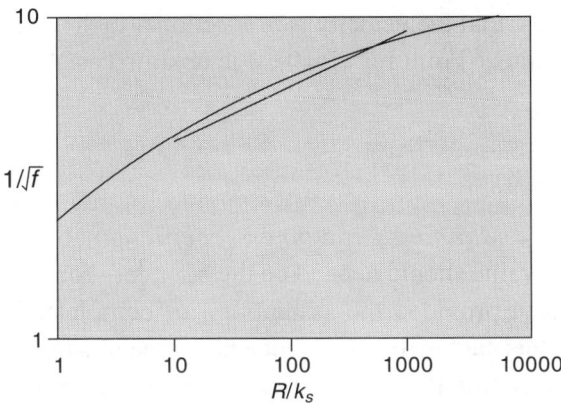

Fig. 2.3 Variation of the friction factor with relative roughness for fully rough flows

intermediate R/k_s values, the one-sixth power law is a reasonable approximation of the logarithmic relationship. We may, of course, use a different power and get a better fit in different ranges. For example, for small values of R/k_s, a slope of 1/3 is better, while for large values of R/k_s, a slope of 1/10 provides the best fit. However, considering the approximations involved in extrapolating the pipe-flow friction factor to the open channel flows and various other assumptions made in the analysis, 1/6 is considered a *good value* to work with. (This is also supported by a number of observations indicating that the velocity is proportional to the two-thirds power of the hydraulic radius. Incidentally, a power of 1/7 fits the logarithmic relationship more closely than 1/6, with the coefficient, C_1, equal to 3.2.) The value of the coefficient, C_1, in the power-law relationship is not critical for our purpose since we will resort to empiricism to obtain the resistance (or roughness) coefficient. For the fit shown in Fig. 2.3, i.e., for the range of R/k_s between 10 and 1000, C_1 is close to 2.8.

Using the power-law relationship, we may write the Chezy's coefficient as

$$C = \sqrt{\frac{8g}{f}} = \frac{C_1 \sqrt{8g}}{k_s^{1/6}} R^{1/6} = \frac{R^{1/6}}{n} \tag{2.10}$$

and modify the resistance equation as

$$V = \frac{1}{n} R^{2/3} S_0^{1/2} \tag{2.11}$$

This is the widely-used *Manning's equation* named after the Irish engineer, R. Manning, who concluded in the 1880s, based on analysis of a large set of experimental data, that the velocity is proportional to the two-thirds power of the hydraulic radius.[9] From Eq. (2.10), n is obtained as

$$n = \frac{k_s^{1/6}}{C_1 \sqrt{8g}} \qquad (2.12)$$

and is called the *Manning's roughness coefficient*, or more commonly, *Manning's n*. Its value should depend on the boundary roughness only and not on the flow depth provided the assumption of completely turbulent rough flow is valid. Other factors that influence the value of Manning's n are cross-sectional and alignment irregularities, vegetation, obstruction, etc. The roughness coefficient, its units, values, and various influencing factors are described in the next section.

Example 2.2   For the channel in Example 2.1, the boundary is made of concrete (k_s = 1 mm). Assuming the flow to be fully rough, obtain the friction factor from the Moody's chart. What are the corresponding values of the Chezy's constant and Manning's n? What is the flow velocity?

Solution
The hydraulic radius has been obtained in Example 2.1 as 0.9375 m.
The relative roughness = $k_s/4R$ = 2.7E–04 and corresponding f = 0.0145.
From Eq. (2.10),

$$C = \sqrt{8g/f} = \textbf{73.6 m}^{\textbf{1/2}}\textbf{/s} \text{ and } n = R^{1/6}/C = \textbf{0.013}$$

The velocity from Eq. (2.11) or Eq. (2.8) is obtained as V = **1.12 m/s**.

2.5   Manning's Roughness Coefficient

A cursory glance at Eq. (2.11) shows that n should not be a nondimensional quantity in order to have a dimensionally homogeneous equation. It may be seen from this equation, as well as from Eq. (2.12), that the units of n should be $m^{-1/3}s$. However, attaching a dimension to n implies that the *numerical value* of n would be different for different systems of units. For example, if n is 1 $m^{-1/3}s$ (an unrealistically high value, as we will see later), in the FPS system, it would be equal to 0.67 ($=3.28^{-1/3}$) $ft^{-1/3}s$. It would necessitate using different sets of tables of n for different systems of units. To avoid this, we

---

[9]A French engineer, Gauckler, was probably the first one to suggest the use of the exponent 2/3 in the 1860s. Manning found the exponent to be 4/7 for some channels but settled on 2/3 as the more widely applicable value.

believe that the most straightforward methodology is to treat the 1 in the numerator as a dimensional constant having units of $m^{1/3}s^{-1}$ and n as dimensionless.[10] In the FPS system, it should be replaced by 1.49 ($=3.28^{1/3}$) $ft^{1/3}s^{-1}$. The n value, of course, would then be the same for all systems of units.

The value of n typically ranges from about 0.01 for smooth metal flumes in the laboratory to as large as 0.2 for natural channels with irregular section and presence of bushes, trees, etc. on the boundary. Extensive tables of n values for different boundaries and figures showing channels with widely different n values are available (e.g., Chow 1959, CFRS 2007, LMNO 2007, USGS 2007) to help one choose an appropriate value of n. Table 2.1 shows the typical values of n for some of the commonly encountered materials.

Table 2.1   Values of Manning's n for different surface materials

Material	n
Glass, PVC	0.010
Brass	0.011
Finished Concrete, Smooth Steel	0.012
Wood	0.012–0.013
Cast Iron	0.013
Brickwork	0.015
Asphalt	0.016
Natural Streams	0.030–0.040
Floodplains	0.040–0.150

Example 2.3 illustrates the use of Manning's equation.

Example 2.3   For the culvert mentioned in Example 1.1, use the Manning's equation (take n for concrete as 0.012) to obtain the discharge when the pipe is flowing just full? What is the discharge for case (c) when the pipe flows half full?

---

[10]There has been some discussion in the engineering community about the dimensions of n and efforts have been made to make the Manning's equation dimensionally homogeneous (e.g., Yen 1992, Swamee 1994, Yen 2002). However, until a dimensionally homogeneous form, with the associated values of n, is widely accepted, we will continue to use the conventional form. Similarly, there have been suggestions that the friction factor form of the resistance equation [Eq. (2.8)] should be preferred to the Manning's equation. Although we feel that it would be more appropriate due to its applicability in all conditions (not only in fully turbulent conditions), we prefer to use the Manning's equation in this text.

Solution

When the pipe is flowing just full, the flow area and wetted perimeter are taken as the entire pipe area and perimeter, respectively. The hydraulic radius is, therefore, one-fourth of the diameter, i.e., $R = 0.075$ m. The bed slope of the pipe is 5 cm in 5 m, i.e., 0.01.

From Manning's equation, therefore, the velocity is obtained as

$$V = \frac{1}{0.012}\, 0.075^{2/3}0.01^{1/2} \text{ m/s} = 1.48 \text{ m/s}$$

and the discharge as **0.105 m³/s** (very close to the value, 0.108 m³/s, obtained treating it as a closed conduit flow in Example 1.1).

When the pipe is flowing half full, the hydraulic radius remains same since both the area and the perimeter become half their *full values*. Therefore, the velocity remains same and the discharge is half[11] of that obtained under full conditions, i.e., **0.052 m³/s**.

Although the boundary roughness is the primary factor influencing the roughness coefficient, there are several other factors which affect it. The *n* values have to be modified if additional resistance to flow is expected due to irregularities in the channel shape, alignment, etc. Some of the factors that influence the value of *n* are described next.

Boundary roughness

Equation (2.12) shows that the value of *n* is proportional to one-sixth power of the average height of the surface roughness, k_s. Using this equation, *with a value of C_1 equal to 2.8*, and the median diameter[12] of the boundary material, d_{50}, as k_s, we obtain

$$n = \frac{d_{50}^{1/6}}{24.8} \tag{2.13}$$

with d_{50} in metres (for the more commonly used units of grain-diameter in mm, the constant in the denominator changes to 78.4. However, we will not

---

[11]It just happens to be true for a circular channel that the discharge at *half full* condition is half of that at the *just full* condition. For a circular channel, the discharge when the flow depth is, say, 75% of the diameter, will not be 75% of the full discharge. Similarly, for a trapezoidal channel, the discharge becomes more than twice when the depth is increased by 100%.

[12]Since the boundaries of natural channels comprise particles of varying sizes, the median diameter, i.e., the diameter for which 50% (by weight) particles are finer, is commonly used to represent the average particle size.

use mm for the diameter). Equation (2.13) is known as the Strickler equation after the Swiss engineer, A. Strickler, who obtained an empirical equation of the same form in 1920s, based on the observations on gravel-bed streams in Switzerland. Many studies conducted since then have confirmed this relationship. The coefficient in the denominator, however, has been reported to vary from about 21 to 26. For coarse-grained boundaries, armouring[13] tends to coarsen the boundary and the median size may not be a suitable indicator of the flow resistance. For such conditions, d_{90} should be used as an indicator of k_s and the coefficient is close to 26. For the culvert mentioned in Example 1.1, the mean height of roughness elements was given as 0.6 mm for concrete. Using this value, we obtain a rough estimate of the Manning's n as $(0.0006)^{1/6}/24.8 = 0.012$, which is same as that given in Table 2.1 and used in Example 2.3. Also note that the small power (1/6) of the particle size implies that the roughness coefficient is not very sensitive to small errors in measurement of the grain-size. For example, increasing the grain-size to twice its value will increase n by only about 12%.

Cross section and alignment irregularities

In natural channels, the cross section shape and size may vary considerably from one location to the other. Similarly, meandering of rivers may introduce large deviation from a straight alignment. These will result in additional resistance to flow and are typically accounted for by increasing the n value by an amount dependent on the nature of these variations. For example, a gradual change in cross section may not need any modification of n but a sudden change may cause an increase of the order of 0.005. Similarly, channel bends with large radius of curvature may not influence the roughness coefficient but a pronounced curvature may cause an increase of the order of 0.002. Excessive meandering may lead to an increase in n of about 25–30%.

Vegetation and flow depth

The presence of vegetation on the channel bed and sides leads to additional resistance and increases the value of n. However, for large depth of flow and flexible vegetation, bending over of the vegetation tends to reduce the resistance and results in a lower n. For most streams, n decreases with increase in flow depth because the effect of channel bed irregularities is more pronounced at lower depths. However, if the floodplains are significantly rougher than the

---

[13]Armouring refers to the formation of a coarser layer on the surface when the channel boundary consists of particles of different sizes. It is described in Chapter 8 on Mobile Bed Channels.

main channel, the *n* value will increase as the depth becomes larger and flow spills onto the floodplains. For most natural channels, therefore, the *n* value will be lowest at a flow depth approaching the bankfull stage.

Sediment load and obstructions

Presence of obstructions, e.g., bridge piers, and transport of sediment load, either in suspension or near the bed, increases the *n* value since it causes additional head loss. (Details of sediment transport process are provided in Chapter 8.)

Except for the boundary roughness, the effect of other parameters on the value of *n* is difficult to quantify. The selection of a proper value of *n* for a channel is, therefore, more of an art than a science. Field experience, aided by tables and figures, is the best guide for this purpose. For the rest of this chapter, we would assume that a proper value of *n* has been obtained. This *n* value may be different for different segments of the boundary, e.g., a compound channel with a main channel and floodplains or a composite channel with metal bed with glass sides. We will, however, first discuss the case of a uniform roughness over a simple cross section and look at the application of the Manning's equation to compute the uniform flow parameters.

2.6 Normal Depth and its Computation

The Manning's equation relates the velocity of flow and, therefore, discharge, with the flow depth and the bed slope under uniform flow conditions. For a channel of specified bed slope and cross-section, a given discharge will, therefore, flow at a uniform depth, which satisfies the Manning's equation. Conversely, if we measure the flow depth under uniform flow, we should be able to compute the discharge from the Manning's equation. The computation of discharge for a known flow depth is rather straightforward (as seen in Example 2.3) but finding out the flow depth for a given discharge involves the solution of a nonlinear equation since both the area and the hydraulic radius are, generally, nonlinear functions of depth. Other possibilities include computation of the roughness coefficient or bed slope when all other parameters are known or designing a channel (cross section shape and flow depth) for a given discharge, roughness, and bed slope. (Sometimes, the bed slope may also have to be designed.)

Rewriting the Manning's equation in terms of discharge, we have

$$Q = \frac{1}{n} AR^{2/3} S_0^{1/2} \tag{2.14}$$

For a given channel cross section, the term $AR^{2/3}$ will depend only on flow depth and is known as the *section factor for uniform flow*, Z_u.[14] (An implicit assumption here is that n is independent of the flow depth. In some cases, n may have to be treated as being dependent on the depth. Then, *conveyance* of the channel, K, defined as $K = AR^{2/3}/n = Q/S_0^{1/2}$, would be a better parameter to use. In this book, we ignore the dependence of n on depth and use section factor only.) The section factor is given by

$$Z_u = AR^{2/3} = \frac{A^{5/3}}{P^{2/3}} \left(= \frac{nQ}{S_0^{1/2}} \right) \tag{2.15}$$

and is, therefore, known for given values of n, Q, and S_0. As previously mentioned, we analyse the flow through trapezoidal channels and obtain the rectangular and triangular channel solutions as special cases. Figure 2.4 shows the cross section of a trapezoidal channel with bottom width B and side slopes of m horizontal : 1 vertical. (We assume that both the sides have same slope. A more general treatment may include different side slopes of m_1 and m_2 but will not be significantly different from what is described here.) The flow depth under uniform flow conditions, i.e., the *normal depth*, is denoted by y_n.

The section factor is then given by

$$Z_u = \frac{(By_n + my_n^2)^{5/3}}{(B + 2\sqrt{1+m^2}\, y_n)^{2/3}} \tag{2.16}$$

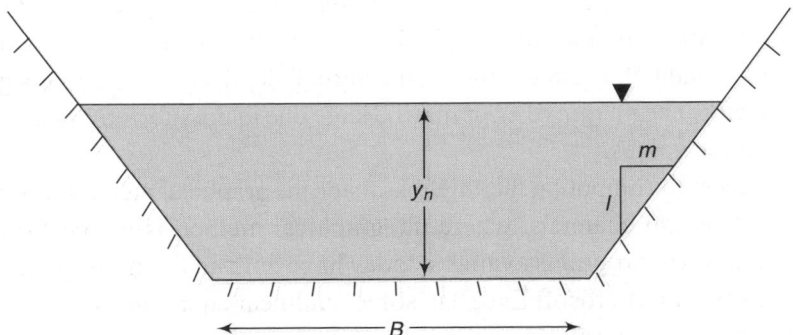

Fig. 2.4   Cross section of a trapezoidal channel

---

[14]Commonly Z is used as the symbol for the section factor. Since the Z-axis is used infrequently in this text, we decide to use the same symbol (with subscript u denoting uniform flow) and believe it will not cause confusion.

and the discharge corresponding to a given normal depth can be easily computed. Putting $m = 0$ in Eq. (2.16), we obtain the expression for rectangular channels and putting $B = 0$, we get the section factor for triangular channels. To obtain the normal depth for a given discharge, the section factor is computed from Eq. (2.15) using the known values of n, Q, and S_0. However, since Eq. (2.16) is nonlinear, computation of the normal depth for a known value of section factor is not straightforward. [For triangular channels $(Z_u \propto y_n^{8/3})$ and for very wide rectangular channels $(Z_u \propto y_n^{5/3})$, Eq. (2.16) can be directly solved for the normal depth. A rectangular (or trapezoidal) channel may be treated as wide when the bottom width is more than, say, 20 times the flow depth.] Earlier when computers were not so easily available, graphical methods were commonly used to get the normal depth for a given section factor. To reduce the number of variables, Eq. (2.16) may be nondimensionalized as

$$\frac{Z_u}{B^{8/3}} = \frac{\left(\dfrac{y_n}{B} + m\dfrac{y_n^2}{B^2}\right)^{5/3}}{\left(1 + 2\sqrt{1+m^2}\,\dfrac{y_n}{B}\right)^{2/3}} \tag{2.17}$$

to make the dimensionless section factor, $Z_u/B^{8/3}$, a function of the dimensionless depth, y_n/B, and the side slope m. (Note that the division by the bottom width renders the analysis inapplicable for triangular channels. However, for triangular channels, we do not need to use the graphical method since the normal depth can be directly computed for any section factor.) Figure 2.5 shows the plot of Eq. (2.17) with the $m = 0$ curve representing a rectangular channel. From the figure, the normal depth can be read for a given section factor. The margin of error may be large due to the subjectivity involved in reading the values.

Easy access to computing facilities has made the graphical methods obsolete, except for natural channels, where the graphical method is the best option. Most spreadsheet programs available today have built-in functions (e.g., Goal Seek or Solver in Microsoft Excel) to solve nonlinear equations, which can be used to obtain the normal depth for a given discharge. However, it may be cumbersome if a large number of such computations have to be performed. Also, sometimes, the computation of the normal depth may be required in a computer program for, say, flow profile computations under nonuniform flow conditions. In such cases, an iterative solution of Eq. (2.16) may be programmed using the fixed-point iteration or the Newton iteration scheme (see Appendix C). These schemes start with an initial guess and iteratively improve upon it

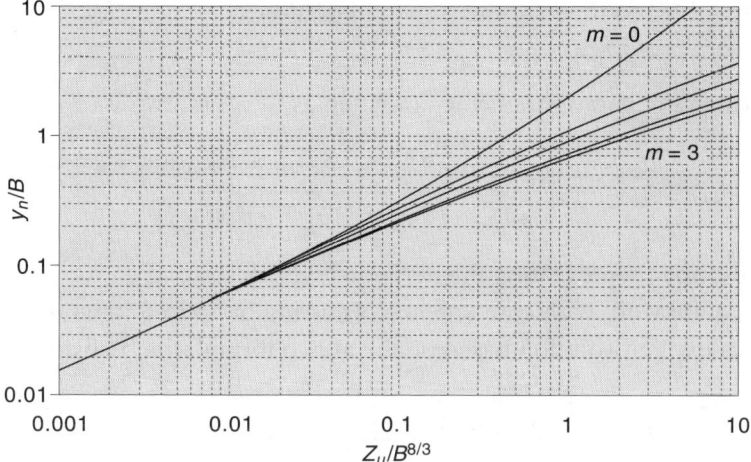

Fig. 2.5   Variation of nondimensional section factor with nondimensional depth for trapezoidal channels, $m = 0$ (topmost line), 0.5, 1, 2, and 3 (bottom line)

till an acceptable accuracy is achieved. The number of iterations depends on the initial guess but has been found to be small (generally smaller than 10). We briefly describe a formulation[15] with Newton iteration scheme.

For a trapezoidal channel using the nondimensional depth[16] and discharge[17] as (Terzidis 2005)

$$\eta = \frac{my_n}{B} \quad \text{and} \quad \Theta = (1 + m^2)^{0.2} \left(\frac{mnQ}{B^{8/3} S_0^{1/2}} \right)^{0.6} \tag{2.18}$$

the Manning's equation may be written as the following nonlinear equation:

$$f(\eta) = \eta^2 + \eta - \Theta \, g_\eta^{0.4} = 0 \tag{2.19}$$

in which, for convenience, we have defined the function g_η as

$$g_\eta = 2\eta + \frac{m}{\sqrt{1 + m^2}} \tag{2.20}$$

---

[15]There are many alternative ways of writing Eq. (2.16) in the form $f(x) = 0$. We use this particular form since it will be subsequently used to derive an approximate, but very accurate, explicit solution.

[16]Note that the definition of nondimensional depth implies that the formulation is not valid for rectangular channels. A similar formulation for rectangular channels, using y_n/B as nondimensional depth, was proposed by Terzidis and is given later in this section.

[17]For the FPS system, there would be an additional factor of 1.486. However, in this book, we will deal only with the SI units and not include this factor.

The derivative required for the Newton iteration is

$$f'(\eta) = 2\eta + 1 - 0.8 \Theta g_{\eta}^{-0.6}$$

using which the iterative scheme is written as

$$\eta_{k+1} = \eta_k - \frac{f(\eta_k)}{f'(\eta_k)} = \frac{\eta_k^2 g_{\eta_k}^{0.6} - 0.8 \Theta \eta_k + \Theta g_{\eta_k}}{(2\eta_k + 1) g_{\eta_k}^{0.6} - 0.8 \Theta} \tag{2.21}$$

with subscript k representing the iteration level. A starting guess, η_0, equal to 0 works quite well, with convergence occurring in up to 5 iterations in most cases. An explicit solution based on the above formulation is listed below:[18]

$$\eta_0 = -0.5 + \sqrt{0.25 + \Theta \sqrt{\frac{m}{\sqrt{1 + m^2}} - 1 + \sqrt{1 + 4\Theta}}}$$

$$g_{\eta_0} = 2\eta_0 + \frac{m}{\sqrt{1 + m^2}} \tag{2.22}$$

$$\eta = \frac{\eta_0^2 g_{\eta_0}^{0.6} - 0.8 \Theta \eta_0 + \Theta g_{\eta_0}}{(2\eta_0 + 1) g_{\eta_0}^{0.6} - 0.8 \Theta}$$

Use of the graphical, iterative, and explicit methods for finding the normal depth in a trapezoidal channel is illustrated in Example 2.4.

Example 2.4   A trapezoidal channel with bottom width of 5 m and side slopes 1H:1V is laid at a longitudinal slope of 1 in 1000. The channel is lined with brickwork and carries a discharge of 20 m^3/s. What would be the normal depth of flow?

Solution
Since the channel has brick lining, the value of n may be taken as 0.015 (Table 2.1). (We make the generally reasonable assumption that n is independent of flow depth. Variations up to 30% have, however, been reported for some channel shapes.)
The section factor is then obtained as

$$Z_u = \frac{nQ}{S_0^{1/2}} = \frac{0.015 \times 20}{0.001^{1/2}} \text{ m}^{8/3} = 9.49 \ m^{8/3}$$

---

[18]Slight modification has been made in the equation for η_0 listed by Terzidis, where a variable exponent was used. This exponent was related to the nondimensional discharge. We believe that Eq. (2.22) is sufficiently accurate for all practical problems (the maximum error over a wide range of parameters was found to be less than 0.01%).

(a) *Using the graphical method*:

The nondimensional section factor, $\dfrac{Z_u}{B^{8/3}} = \dfrac{9.49}{5^{8/3}} = 0.13$

From Fig. 2.5, using the curve for $m = 1$, we get the nondimensional depth, $y_n/B = 0.3$.
Therefore, the normal depth, $y_n = 0.3 \times 5$ m $= \mathbf{1.5}$ **m**.[19]

(b) *Using the Newton's iterative scheme*:

Nondimensional discharge [Eq. (2.18)], $\Theta = (1 + 1)^{0.2}\left(\dfrac{1 \times 0.015 \times 20}{5^{8/3}\,0.001^{1/2}}\right)^{0.6}$

$$= 0.3374$$

The computations are shown in the table below:

Table 2.2 Computation of normal depth using the iterative scheme

Iteration no., k	η_k	g_{η_k} [Eq. (2.20)]	η_{k+1} [Eq. (2.21)]
0	0.0000	0.7071	0.4399
1	0.4399	1.587	0.3040
2	0.3040	1.315	0.2895
3	0.2895	1.286	0.2894
4	0.2894	1.286	0.2894

The normal depth is, therefore, obtained as

$$y_n = 0.2894 \times \frac{B}{m} = \mathbf{1.447}\ \mathbf{m}^{20}$$

(c) *Using the explicit solution*:
From Eq. (2.22), for $\Theta = 0.3374$ and $m = 1$, we get $\eta_0 = 0.2910$, $g_{\eta_0} = 1.289$, and finally, $\eta = 0.2894$ (same as that obtained in b after four iterations) giving a normal depth of **1.447 m**.

Terzidis (2005) has also provided explicit approximations for the normal depth in a rectangular channel, which are listed here after slight modifications. Defining the nondimensional parameters for this case as

$$\eta = \frac{y_n}{B} \quad \text{and} \quad \Theta = \left(\frac{nQ}{B^{8/3} S_0^{1/2}}\right)^{0.6} \tag{2.23}$$

---

[19]If we compute the discharge for this depth, we obtain a value of 21.3 m³/s, which has a 6.5% error!

[20]Considering the approximations and uncertainties involved with the Manning's equation, it appears to be an overkill to report the depth to an accuracy of mm! However, in this book, we mostly use four significant digits.

the explicit equation is written as

$$\eta_0 = \Theta(1 + 1.2\Theta)^{0.7826} \tag{2.24a}$$

$$\eta = \frac{(1.2\eta_0 + 1)\Theta}{(2\eta_0 + 1)^{0.6} - 0.8\Theta} \tag{2.24b}$$

If the channel in Example 2.4 is rectangular (other values remaining same), we get $\Theta = 0.2937$ from Eq. (2.23), $\eta_0 = 0.3720$ from Eq. (2.24a), and $\eta = 0.3658$ from Eq. (2.24b), i.e., a normal depth of **1.829 m**.[21]

For a natural channel, the width is generally much larger compared to the depth and it may be approximated as a wide rectangular channel. However, sometimes, this assumption may not be valid and we would need to account for the irregular variation of the boundary. The graphical method would work best in such cases, by plotting the cross section, obtaining the wetted perimeter and flow area for different flow depths, and preparing a section factor plot which could be used to obtain the normal depth for a given discharge. Example 2.5 illustrates the methodology.

Example 2.5 Cross sections of a natural channel ($n = 0.02$) have been obtained at two locations which are at a longitudinal distance of 1 km from each other (Fig. 2.6). How can a section factor plot be prepared to obtain the normal depth of flow for any discharge?

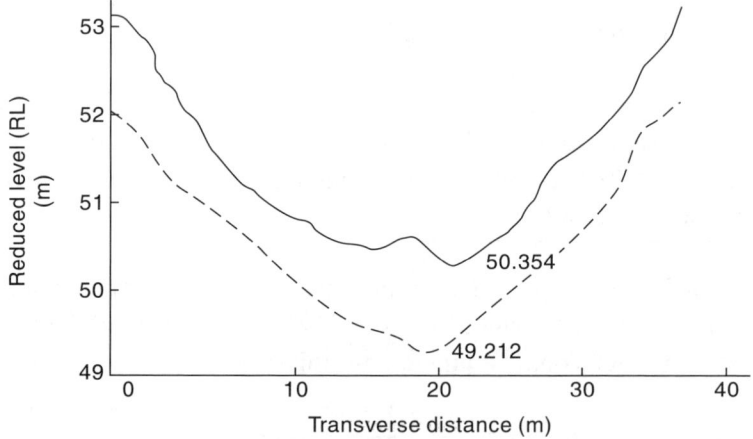

Fig. 2.6 Channel cross sections at two locations 1 km apart (solid line – upstream section, dashed line – downstream section)

---

[21]One might as well use the fixed point iteration scheme (Appendix C) to derive an explicit approximation. One such expression is $\eta = [1 + 2.134\Theta(1 + 2.4\Theta^{1.47})^{0.4536}]^{0.4}$. For this example, we get $\eta = 0.3656$.

Solution

Here we will discuss only the methodology. Since the computations involve plotting the section on a graph paper and *counting* the area, we leave it to the reader (Exercise 2.4) to prepare the section factor plot.

Since only two sections are given, and they are not identical, we will have to decide on how to compute the section factor representing both sections. In absence of any other information about the channel between these two sections, we assume that the bed level follows a constant slope, which may be obtained from the deepest bed level at these two sections.[22] The bed slope, therefore, may be taken as $(50.354 - 49.212)/1000$, i.e., 1.142×10^{-3}. To compute the area and perimeter corresponding to a flow depth, we may compute these separately for the two sections and then take the average or we may plot the two sections together (taking care to bring the deepest points together) and draw an *average* section. The latter technique would reduce the computations but may not be applicable if the *shapes* of the two sections are very different. Once the perimeter and area have been computed for different flow depths, the section factor plot can be easily prepared and the normal depth can be obtained for a given discharge.

From Fig. 2.5, it is seen that the section factor of a trapezoidal channel increases monotonically with flow depth. This is true for most commonly encountered cross sections. From Eq. (2.15), it is clear, since the area of flow always increases with depth, that if the hydraulic radius increases with increase in the flow depth, the section factor would also increase. (The converse, however, is not true. Even if the hydraulic radius decreases with the depth, the section factor may increase due to the increase in area of flow). By differentiating R with respect to y, it can be shown that dR/dy would be positive if the top width is greater than the hydraulic radius, i.e., the product of T and P is more than A. This condition will be satisfied for channels where the top width is continuously increasing with the flow depth. However, for channels where the top width may become smaller with increase in the flow depth, e.g., a circular channel flowing more than half full or a trapezoidal channel with sides sloping inwards, it may happen that the section factor decreases with increase in the flow depth. This implies that at two different flow depths we may obtain the same section factor and, therefore, the same discharge. The

---

[22]One should be careful, however, in making these assumptions. Sometimes, a downstream section may have a higher elevation at its deepest point than that at an upstream section due to local topography. It does not mean that the channel has an *adverse* slope.

occurrence of multiple normal depths is analysed with the help of a circular channel in the next section.

2.7 Multiple Normal Depths: Circular Channel

For a circular channel (Fig. 2.7), various geometric parameters (nondimensionalized through their values when the pipe is flowing just full) are listed below:

$$\frac{A}{A_f} = \frac{D_0^2 \, (\theta - \sin \theta \cos \theta)/4}{\pi D_0^2 /4} = \frac{\theta - \sin \theta \cos \theta}{\pi} \qquad (2.25a)$$

$$\frac{P}{P_f} = \frac{D_0 \theta}{\pi D_0} = \frac{\theta}{\pi} \qquad (2.25b)$$

$$\frac{R}{R_f} = \frac{D_0 \, (\theta - \sin \theta \cos \theta)/4\theta}{D_0 /4} = \frac{\theta - \sin \theta \cos \theta}{\theta} \qquad (2.25c)$$

in which the angle θ is related to the nondimensional depth, y/D_0, as

$\theta = \arccos \left(1 - \dfrac{2 y}{D_0} \right)$ and the subscript f represents the values when $y = D_0$

(i.e., $\theta = \pi$).

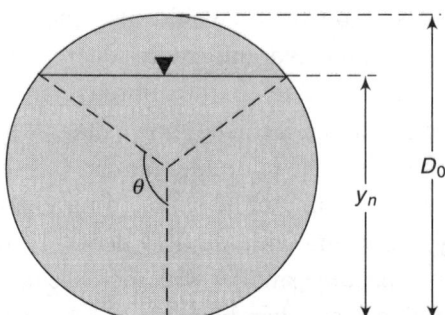

Fig. 2.7   Cross-sectional elements of a circular channel

Figure 2.8 shows a plot of these elements and the decrease in the hydraulic radius for y/D_0 greater than about 0.8 is clearly seen. From Manning's equation, it may be noted that the velocity reaches a maximum when the flow occurs at a depth equal to about $0.82D_0$. (In Exercise 2.5, the reader will find more precise values of this and some other significant elements.) Also note that the

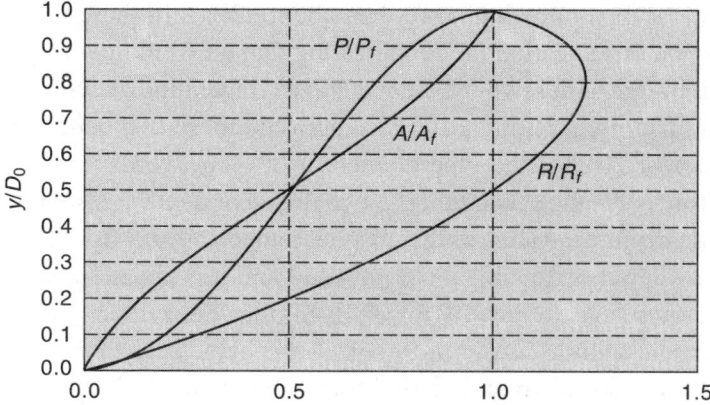

Fig. 2.8 Geometric elements of a circular channel

rate of increase of area with flow depth is minimum when the flow depth is very small or when it is approaching the pipe diameter, while this rate is maximum when the flow depth is close to the half-full stage. This is consistent with the fact that the top width, *T*, represents the rate of change of area, *dA/dy*, and is maximum at the half-full stage.

For discharge computations, it is the section factor which is of interest to us. Figure 2.9 shows a plot of the section factor in a nondimensional form defined as

$$\frac{Z_u}{Z_{uf}} = \frac{(\theta - \sin \theta \cos \theta)^{5/3}}{\pi \theta^{2/3}}$$

in which Z_{uf} is the section factor at full depth and is equal to $\pi D_0^{8/3}/8 \times 2^{1/3}$. Note that the ratio Z_u/Z_{uf} will be same as the ratio Q/Q_f, where Q_f is the discharge when the pipe is flowing just full.

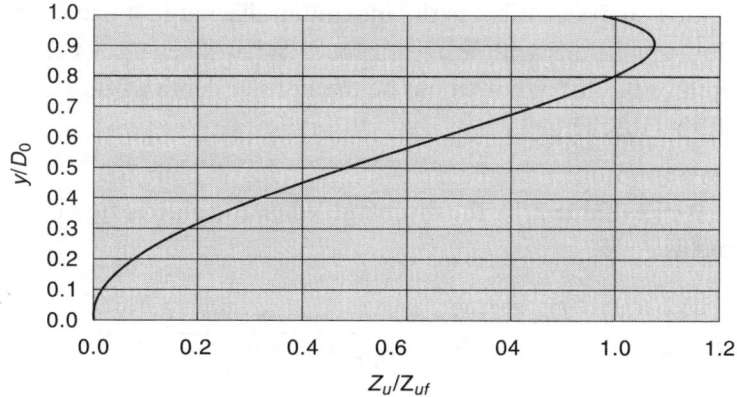

Fig. 2.9 Section factor of a circular channel

Clearly, for pipes flowing more than about 82% full, there would be two depths for the same section factor indicating that uniform flow is possible for a given discharge under two different depths. (Depending on where we are on the section factor curve, the difference between these two depths may be as high as almost 20% of the pipe diameter.) In other words, the driving and resisting forces balance each other at two different depths. Also note that the discharge-carrying capacity of the channel under uniform flow conditions is maximum when it is flowing about 94% full and it is about 7.5% larger than the discharge when the pipe is just full. Similar behaviour is observed for other channels with a *closing top*, e.g., a trapezoidal channel with inward sloping sides (also called an inverted triangular channel). The occurrence of multiple normal depths becomes a critical factor when designing a circular pipe to carry a given discharge. Providing a diameter that will result in the normal depth being more than $0.82D_0$ is generally avoided because a slight disturbance on the free surface may cause the depth to 'jump' to the other possible normal depth resulting in the possibility of pressure flow. To compute the normal depth for a given discharge, one can use Fig. 2.9 to obtain the nondimensional normal depth for the computed value of Z_u/Z_{uf}, which is equal to $8 \times 2^{1/3} nQ/\pi D_0^{8/3} S_0^{1/2}$. An explicit approximation could be derived similar to Swamee (1994) using a nondimensional discharge, $\Theta = nQ/D_0^{8/3} S_0^{1/2}$ as,

$$\frac{y_n}{D_0} = 1.560 \Theta^{0.4666}[1 - 0.5650(0.3353 - \Theta)^{0.4971}] \tag{2.26}$$

which is valid for normal depth up to $0.94D_0$ (at which the maximum nondimensional discharge of 0.3353 occurs) and is accurate to within 1% of the actual value.

Example 2.6   A circular concrete pipe (diameter $= 1$ m) is laid at a longitudinal slope of 1 in 5000. What is the maximum discharge it can carry under uniform flow conditions with a free surface? What discharge is carried at a normal depth of 0.8 m? What would be the normal depth of flow if the pipe carries half this discharge?

Solution
Taking $n = 0.012$ (Table 2.1), the discharge when the pipe is flowing *just full* is computed as

$$Q_f = \frac{1}{n} \frac{\pi D_0^2}{4} \left(\frac{D_0}{4}\right)^{2/3} S_0^{1/2} = \frac{\pi}{0.012 \times 4^{5/3} \times 5000^{1/2}} \text{ m}^3/\text{s}$$

$$= 0.3673 \text{ m}^3/\text{s}$$

From Fig. 2.9, the maximum discharge is carried when the normal depth is equal to about 94% of the diameter and its ratio with the pipe-full discharge is about 1.075. Hence, the maximum discharge under free-surface flow would be **0.3948 m³/s**.

At a normal depth of 0.8 m, from Fig. 2.9,

$$Q = 0.98 \, Q_f = \textbf{0.3600 m}^3\textbf{/s}$$

and for a discharge of half this value, from Fig. 2.9,

$$Z_u/Z_{uf} = 0.49 \text{ and } y_n = \textbf{0.49 m}$$

We can also use Eqs (2.25) and (2.26) to obtain these values. The maximum discharge is

$$\frac{0.3353 D_0^{8/3} S_0^{1/2}}{n} = \textbf{0.3952 m}^3\textbf{/s}$$

For a normal depth of 0.8 m, we get

$$\theta = 2.214 \text{ rad}, \, A = 0.6736 \text{ m}^2, \, R = 0.3042 \text{ m}, \text{ and } Q = \textbf{0.3591 m}^3\textbf{/s}$$

For a discharge equal to half of this value, the dimensionless discharge,

$$\Theta = \frac{nQ}{D_0^{8/3} S_0^{1/2}} = \frac{0.012 \times 0.5 \times 0.3591}{5000^{-1/2}} = 0.1524$$

and, from Eq. (2.26), $y_n = \textbf{0.4910 m}$

The other type of channel with multiplicity of normal depths is the inverted triangular channel, i.e., a trapezoidal channel with negative side slopes. For such channels, a section factor graph similar to Fig. 2.5, shown in Fig. 2.10, demonstrates the possibility of multiple normal depths for some discharge values. These channels are, however, not very common and will not be discussed further.

As mentioned earlier, we have assumed n to be independent of depth. It may show considerable variation with depth in some cases. For example, a circular pipe flowing half full may have a value of n almost 20% higher than that when it is flowing just full. The dependence on depth may also occur due to the channel boundary comprising materials with different roughness characteristics or when there are floodplains adjacent to a main channel and these have very different roughness. The length of a particular material in contact with water will typically depend on depth. Therefore, the *effective roughness* of the channel as a whole would be a function of flow depth. It implies additional complications in the computation of the normal depth for such channels as discussed in the next section.

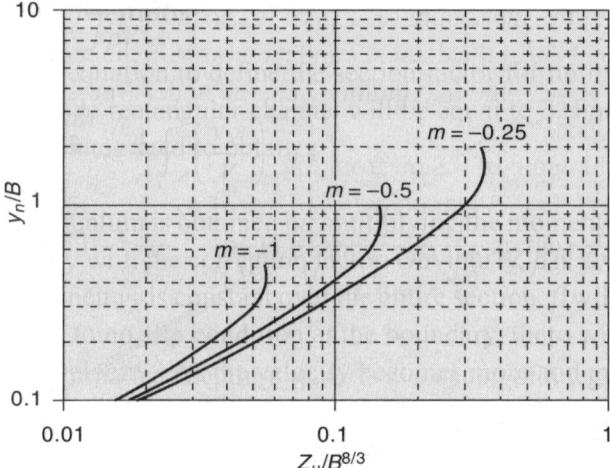

Fig. 2.10 Variation of nondimensional section factor with nondimensional depth for inverted triangular channels for $m = -0.25$, -0.5, and -1

2.8 Normal Depth in a Composite Channel

The cross section of a composite channel[23] is shown in Fig. 2.11. The boundary consists of N different types of materials (e.g., glass sides and metal bottom of a laboratory flume or coarse bed and smooth sides of a natural channel) with the corresponding wetted perimeter being P_i ($i = 1$ to N). We first look at the determination of discharge for a given normal depth and then discuss the more complicated problem of determining the normal depth for a given discharge.

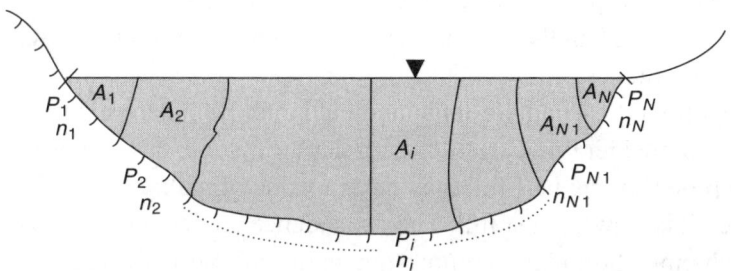

Fig. 2.11 Cross section of a composite channel

In the 1930's, Horton proposed a method which divides a channel area into N arbitrary segments such that the ith boundary segment is attached to an area

---

[23]A more descriptive term would be *channel with composite roughness*. For convenience, we use the shorter term.

A_i. [For this method, it does not matter how these sub-areas are located since the final form for the effective roughness does not involve these areas. Physically, we may think of these as the areas of influence of different roughness lengths (see Exercise 2.7).] It is assumed that each sub-area has the same mean velocity, which is equal to the mean velocity through the entire section. Therefore, application of Manning's equation to each subsection and to the entire area results in

$$\frac{1}{n_i}\frac{A_i^{2/3}}{P_i^{2/3}}S_0^{1/2} = \frac{1}{n_e}\frac{A^{2/3}}{P^{2/3}}S_0^{1/2} \quad \text{for } i = 1 \text{ to } N \tag{2.27}$$

where n_e represents the effective or equivalent value of the roughness coefficient for the composite section and the longitudinal slope has been assumed to be the same for each subsection. The fraction of area associated with each subsection is obtained as

$$\frac{A_i}{A} = \frac{P_i n_i^{3/2}}{P n_e^{3/2}} \tag{2.28}$$

and, since the sum of all these fractions should be equal to 1, we get

$$n_e = \left(\frac{\sum_{i=1}^{N} P_i n_i^{3/2}}{P}\right)^{2/3} \tag{2.29}$$

The discharge for a given normal depth can then be obtained as

$$Q = \frac{1}{n_e} A R^{2/3} S_0^{1/2} \tag{2.30}$$

We may interpret Eq. (2.29) as the 3/2 power of effective n being equal to the weighted arithmetic mean of the 3/2 power of individual roughness elements, with the weight equal to the ratio of the perimeter of that element to the total perimeter. Various other formulae are available for computing the value of n_e. These are based on different assumptions and lead to expressions similar[24] to Eq. (2.29). For example, one of the earliest such methods, proposed by a Russian engineer, N.N. Pavlovskii, in 1931, equated the total resisting force with the resisting force developed in each subsection and resulted in

---

[24] All of these could be interpreted as weighted arithmetic/harmonic mean of some power of individual n values. Although it should be obvious, we should mention that the equivalent roughness coefficient will always lie between the minimum and the maximum value of roughness coefficients for the segments on the wetted perimeter.

$$n_e = \left(\frac{\sum\limits_{i=1}^{N} P_i n_i^2}{P} \right)^{1/2} \tag{2.31}$$

In 1960, a German engineer, K. Felkel, equated the sub-area discharges and arrived at a weighted harmonic mean

$$n_e = \left(\frac{\sum\limits_{i=1}^{N} P_i n_i^{-1}}{P} \right)^{-1} = \frac{P}{\sum\limits_{i=1}^{N} \dfrac{P_i}{n_i}} \tag{2.32}$$

Yen (1991) equated the sub-area shear velocities and suggested a simple weighted mean as

$$n_e = \frac{\sum\limits_{i=1}^{N} P_i n_i}{P} \tag{2.33}$$

While all these expressions require only the perimeter of different roughness elements [although we do not need it, the individual sub-areas may be computed using Eq. (2.28) after obtaining the value of n_e], there are some others that require the area or the hydraulic radius of the sub-areas. For these, two commonly used options for the method of subdivision are: vertical lines through the junction of two different types of boundary elements (which will obviously not work for a rectangular channel with different materials on the bottom and the sides) and bisectors of angles at the junctions (shown as lines 1 and 2 in Fig. 2.12). We only list here one such expression, proposed by G.K. Lotter in 1933, based on equating the sub-area discharges:

$$n_e = \frac{PR^{5/3}}{\sum\limits_{i=1}^{N} \dfrac{P_i R_i^{5/3}}{n_i}} \tag{2.34}$$

i.e., the weighted harmonic mean with weights of $(P_i/P) \times (R_i/R)^{5/3}$. More expressions are listed by Yen (2002) and there is no distinct advantage of using one over the others. We believe that any of the expressions listed here could be used since (i) there is no particular reason for considering the assumptions made in a method to be better than those in other methods, (ii) the results obtained are not greatly different from one another, and (iii) the

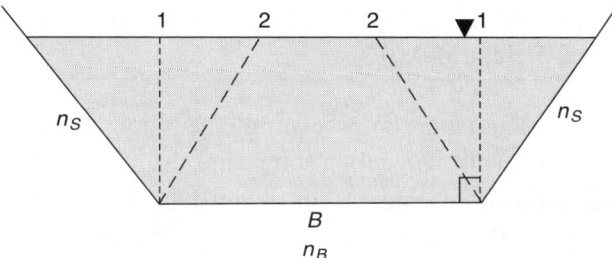

Fig. 2.12   Commonly used methods for division into sub-areas

uncertainty involved in the Manning's equation and the n value probably over-shadows the approximations used in finding the effective roughness.

Example 2.7   A trapezoidal channel (bed width = 5 m, side slopes 3H:1V) is laid at a longitudinal slope of 1 in 2000. The bed is lined with concrete and the sides with asphalt. What would be the equivalent roughness coefficient when the flow depth is 2 m?

Solution

For concrete, n = 0.012 and for asphalt, n = 0.016 (Table 2.1). The corresponding perimeters are 5 m and $2\sqrt{1+3^2}$ × 2 m or 12.65 m.

Using Horton's approach [Eq. (2.29)],

$$n_e = \left(\frac{5 \times 0.012^{3/2} + 12.65 \times 0.016^{3/2}}{5 + 12.65} \right)^{2/3}$$

$$= \mathbf{0.0149}$$

Note that the value is closer to that of asphalt because of larger perimeter compared to concrete.

Similarly, using Pavlovskii method [Eq. (2.31)], n_e = 0.0150, using Felkel method [Eq. (2.32)], n_e = 0.0146, and using Yen's approach [Eq. (2.33)], n_e = 0.0149.

For Lotter's method [Eq. (2.34)], assuming the subdivision of area by vertical lines through the junctions of the bed and the sides, the sub-areas are obtained as 10 m^2 and 12 m^2 respectively corresponding to the bed and the sides. (The perimeter is assumed to be the same as before and the dividing lines are not counted in it.) The relevant hydraulic radii are 2 m and 0.9486 m respectively, and n_e = 0.0125. If we subdivide the area by bisectors of the angles between the bed and the sides, the area corresponding to the bed is 9.35 m^2 and that corresponding to the sides is 12.65 m^2 giving hydraulic radii of 1.87 m and 1.00 m respectively, and resulting in n_e = **0.013**. Therefore, while most of the methods result in a value close to 0.015, Lotter's method results in an equivalent roughness coefficient of 0.013.

As one would realize from the preceding discussion, the computation of normal depth for a given discharge for a composite channel is quite complicated because of the additional nonlinearity introduced due to the dependence of n on flow depth. The total perimeter, P, and some of the individual perimeters, P_i, will be functions of flow depth.[25] The graphical and numerical methods discussed in Section 2.6 would work equally well and are not repeated here. For example, for a rectangular channel with different materials on the bottom and the sides, using Yen's expression [Eq. (2.33)], we get the following nonlinear equation:

$$Q = \frac{P}{\sum_{i=1}^{2} n_i P_i} \frac{A^{5/3}}{P^{2/3}} S_0^{1/2} \quad \Rightarrow \quad \frac{(B + 2\,y_n)^{1/3}\,(By_n)^{5/3}}{Bn_B + 2\,y_n\,n_s} = \frac{Q}{S_0^{1/2}} \tag{2.35}$$

in which n_B is the n value for the bottom and n_S, for the sides. (Note that the last term in Eq. (2.35) is the conveyance of the channel.) The nondimensional form may be written as

$$\frac{(1 + 2\eta)^{1/3}\,\eta^{5/3}}{1 + 2\dfrac{n_S}{n_B}\,\eta} = \frac{n_B Q}{B^{8/3}\,S_0^{1/2}} \tag{2.36}$$

A plot of the nondimensional section factor with nondimensional depth can be prepared for the given value of the roughness ratio, n_S/n_B, for use in obtaining the normal depth. An iterative scheme can also be developed along the lines of Eq. (2.21).

Although we have mentioned the channels with a main channel section and adjacent floodplains as being composite channels because the roughness characteristics of these two sub-areas are quite different,[26] there is an additional complication even when the roughness may be assumed to be same. This happens because of the use of the entire area and wetted perimeter in computing the hydraulic radius. As the flow spills onto the floodplains, there is a sudden increase in the wetted perimeter since the floodplains are much wider than the

---

[25]For a rectangular channel with metal bed and glass sides, for example, the perimeter corresponding to the metal boundary will not depend on the depth.

[26]A recent study (Yang et al. 2005b) has concluded that the application of the equivalent roughness formula discussed in this section may give poor results for a compound section with a main channel and floodplains.

main channel. Though there is an increase in area also, it may so happen that the hydraulic radius is smaller than that computed at the bankfull stage, i.e., when the flow depth is just enough to fill the central channel without spilling onto the floodplains. Thus there is a (physically untenable) possibility that the application of the Manning's equation to the whole section, when the depth is just above the bankfull stage, would produce a discharge smaller than the bankfull discharge. Various methods to get around this problem have been proposed and some of them are discussed in the next section.

2.9 Flow in a Compound Channel

The term *compound channel*, or *two-stage channel*, generally represents a channel with a central main channel flanked by floodplains on either side. The flow mostly takes place in the main channel but spills over to the floodplains at higher discharges (Fig. 2.13).

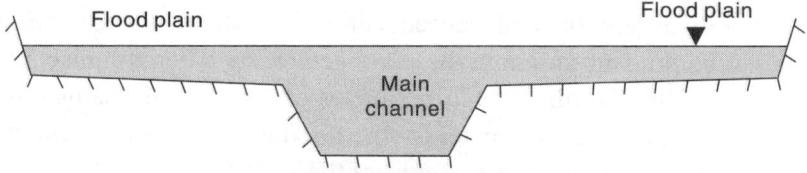

Fig. 2.13 Cross section of a compound channel

The simplest method of computing the discharge through a compound channel is the *single-channel method*, in which the entire channel is treated as one. As discussed in the previous section, it may result in an underprediction of discharge (sometimes, by as much as 50%) when the flow depth is just above the bankfull stage. This happens because the hydraulic radius of the entire section is not a very good indicator of the channel resistance as the velocities in the main channel and in the floodplains are significantly different.

Example 2.8   For the channel in Example 2.4, consider the bankfull depth to be 1.447 m and the floodplains to be 50 m wide on either side with vertical sides. Assuming the Manning's *n* for the floodplain to be same as the main channel (0.015), find the discharge when the depth of flow on the floodplains is equal to 10 cm.

Solution
As seen in Example 2.4, the bankfull discharge is 20 m³/s. For a flow depth of 1.547 m, i.e., water depth of 0.1 m on the floodplain, using the single channel method, we get

$A = 19.83 \text{ m}^2$, $P = 109.3$ m, $R = 0.1814$ m, and $Q = \mathbf{13.40 \text{ m}^3/\text{s}}$

This implies that the discharge has *reduced* from a bankfull value of 20 m³/s by more than 30% on *increasing* the flow depth by 10 cm.

To avoid this underprediction of discharge, the divided-channel method uses subdivisions of the channel through vertical, horizontal, or inclined lines at the junction of the main channel and floodplains (Fig. 2.14). Pezzinga (1994) examined four different methods of subdivisions—vertical, horizontal, diagonal, and bisector—shown by points *V*, *H*, *D*, and *B* in Fig. 2.14. *B* is obtained by bisecting the angle formed by the main channel and floodplain boundaries and *D* is the central point of the water surface in the main channel. Pezzinga concluded that the diagonal and bisector subdivisions worked better than the vertical and horizontal ones. The differences were, however, not very significant. Therefore, in Example 2.9, we use a vertical subdivision. Another choice we have to make is whether to consider these dividing lines as part of the perimeter or not. While these are not actual *boundaries*, there do exist a large velocity differential and, consequently, shear stress at these lines, and it would not be prudent to ignore these altogether. As a compromise, we may include these lines in the wetted perimeter of the central channel and not consider them as wetted perimeter for the floodplains. (Since the floodplains are generally very wide and the lengths of these artificial boundaries are very small, it generally does not make much of a difference even when these lines are included in the perimeter.)

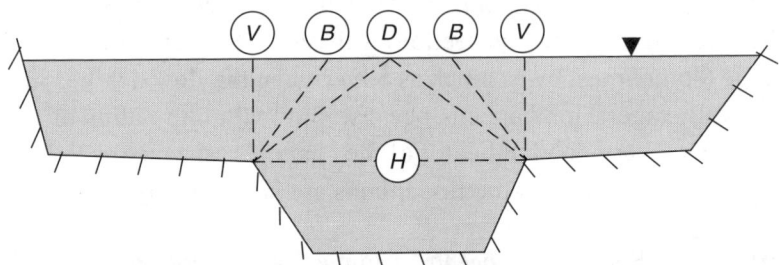

Fig. 2.14 Subdivisions of a compound channel section

Example 2.9 Repeat Example 2.8 using the divided-channel method.

Solution

Dividing the channel into three subsections using vertical lines, and taking advantage of the symmetry, we compute the discharge in the main channel and (one side of) floodplains as

Main channel: $A = 9.829 \text{ m}^2$, $P = 9.293$ m, $R = 1.058$ m, $V = 2.189$ m/s, and $Q = 21.51 \text{ m}^3/\text{s}$

Floodplain: $A = 5$ m^2, $P = 50.1$ m, $R = 0.0998$ m, $V = 0.4536$ m/s, and $Q = 2.268$ m^3/s

The total discharge $= 21.51 + 2 \times 2.268$ m^3/s $= \mathbf{26.05}$ **m**^{3}**/s**

Also note the large difference in the velocity in the main channel and that in the floodplains. Had we considered the floodplains to be rougher than the main channel, this difference would be still larger.

Since the momentum transfer due to difference in velocities at the subdividing lines is not considered in the divided channel method, it may cause an overestimation of discharge by as much as 30% for shallow floodplain flows (the so-called *kinematic effect*, see van Prooijen, et al. 2005). To account for the momentum transfer, an apparent shear stress may be applied at the *shear layer* between the main channel flow and the floodplain flow. Several formulae have been developed to represent this shear stress but they pertain to specific experimental conditions and there is no universally applicable formula. We mention a few of the recently developed methods but do not describe them in detail since a little more work is needed before widespread practical application of these methods becomes a reality.

The coherence method (Ackers 1991) empirically adjusts (based on the depth of flow on the floodplains) the discharge calculated for the main channel and the floodplains in the divided channel method, and requires assumptions about the geometry of the channel. The lateral distribution method (Shiono and Knight 1991) calculates the discharge through an integration of a two-dimensional momentum equation and accounts for eddy viscosity and secondary currents.[27] Use of this method involves a tedious parameter estimation process. In the exchange discharge model (Bousmar and Zech 1999), the interactions between the main channel and the floodplains are accounted for by considering a momentum transfer proportional to the product of the velocity gradient at the interface and the mass exchanged through this interface due to turbulence. The computations are, however, quite involved and require a numerical solution. Seckin (2004) has recommended the use of the exchange discharge model and coherence method over the single- and divided channel methods. Another study (Yang, et al. 2005a) compared different alternatives of evaluating the discharge for compound channels and concluded that the coherence method provided the most accurate predictions of the actual

---

[27]Secondary currents represent velocity components in the transverse (*Y-Z*) plane. These are driven by turbulence, cause circulation from the boundary towards the centre, and are responsible for the maximum velocity occurring not at the surface but slightly below it.

discharge. We believe that for most practical compound channel problems, the accuracy provided by the divided channel nethod is sufficient. For very small depths over the floodplains, knowing that the single channel method underpredicts the discharge and the divided channel method overpredicts it, it may be a good idea to compute the discharge using both these methods and then make a judicious choice about the likely discharge.

In all our discussions so far, we have taken the flow depth (or discharge) as the variable and have considered the other cross section parameters constant. If we relax this constraint and assume that the channel width (or side slope or diameter, as the case may be) is also variable, we will get infinitely many combinations of the depth and width (or other parameters) that will provide the same discharge. In other words, for a rectangular channel laid at a specified longitudinal slope and with specified boundary roughness, there would be infinite combinations of channel width and flow depth that would result in the same discharge under uniform flow conditions. '*Which of these combinations is the best*' is the question we try to answer in the next section.

2.10   Efficient Sections

For the channel of Example 2.4, assuming that the bottom width is also variable, we obtain various combinations of width and depth resulting in the same section factor, i.e., the same discharge. Figure 2.15 shows a plot of these combinations (note that for a flow depth of about 1.5 m, the bottom width is 5 m, as per the conditions obtained in Example 2.4). Also shown are the corresponding flow area and wetted perimeter at different flow depths. It is seen that the area and perimeter are not monotonic functions of flow depth. There is a depth between 2 and 2.5 m at which the area is minimum. Also, there is a depth in the same range at which the perimeter attains a minimum value (from the figure, it is difficult to say whether both these depths are same or different). This gives us a scheme to define the *most efficient section* as the one having a minimum flow area or a minimum wetted perimeter for a given discharge, roughness, and bed slope. (Also, since discharge is specified, minimum flow area implies maximum velocity, and therefore, maximum value of hydraulic radius.) We will now look at how to obtain the conditions for efficient sections and show that the minimum area and perimeter occur at the same flow depth.[28]

---

[28]However, we may note from Fig. 2.15 that, at least for this case, there is not much change in area and perimeter over a wide range of depth (and width) values. Therefore, the *cost difference* between the most efficient section and a slightly different section may not be very large.

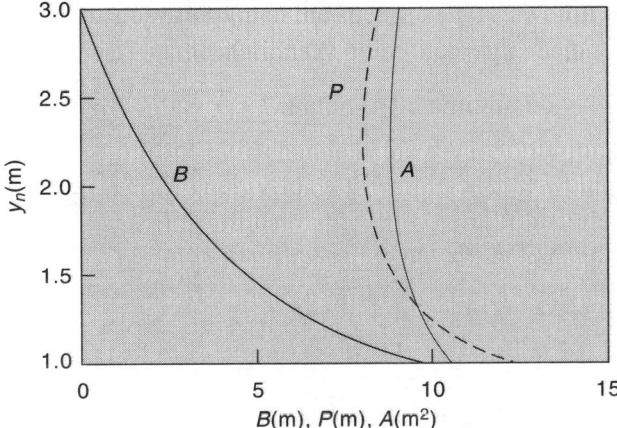

Fig. 2.15 Various combinations of channel width and flow depth providing the same discharge

We may treat the problem of finding the most efficient section as one of the constrained optimization, in which the area (or perimeter) is to be minimized subject to the constraint on discharge (Monadjemi 1994). Assuming that the area and wetted perimeter are functions of two independent variables (e.g., bed width and flow depth for rectangular channels, flow depth and channel diameter for circular channels, flow depth and side slopes for triangular channels)[29], we formulate the problem as follows:

Minimize $A(x_1, x_2)$

$$\text{subject to } \phi(x_1, x_2) = \frac{nQ}{S_0^{1/2}} - \frac{A^{5/3}}{P^{2/3}} = 0$$

where x_1 and x_2 are the two independent variables (e.g., width and depth for a rectangular channel). Using the method of Lagrange multiplier, this problem is converted into the unconstrained problem of minimization of the modified objective function $A(x_1, x_2) + \lambda\phi(x_1, x_2)$, which is now a function of three variables x_1, x_2, and λ. Note that the constraint is automatically satisfied by the condition that the partial derivative of the modified objective function with respect to λ should vanish. The other two conditions are as follows:

$$\frac{\partial(A + \lambda\phi)}{\partial x_1} = 0 \tag{2.37a}$$

---

[29]To fit a trapezoidal channel within this framework, we may assume that either the side slope or the bed width is held constant. However, as will be shown a little later, we may generalize the formulation to the case where area and perimeter are functions of three independent variables.

$$\frac{\partial(A + \lambda\phi)}{\partial x_2} = 0 \tag{2.37b}$$

from which λ may be eliminated to obtain

$$\frac{\partial A}{\partial x_1}\frac{\partial \phi}{\partial x_2} = \frac{\partial A}{\partial x_2}\frac{\partial \phi}{\partial x_1} \tag{2.38}$$

From the definition of ϕ, we have

$$\frac{\partial \phi}{\partial x_1} = -\frac{5\,A^{2/3}}{3\,P^{2/3}}\frac{\partial A}{\partial x_1} + \frac{2\,A^{5/3}}{3\,P^{5/3}}\frac{\partial P}{\partial x_1} \tag{2.39a}$$

$$\frac{\partial \phi}{\partial x_2} = -\frac{5\,A^{2/3}}{3\,P^{2/3}}\frac{\partial A}{\partial x_2} + \frac{2\,A^{5/3}}{3\,P^{5/3}}\frac{\partial P}{\partial x_2} \tag{2.39b}$$

Substitution in Eq. (2.38) results in

$$\frac{\partial A}{\partial x_1}\frac{\partial P}{\partial x_2} = \frac{\partial A}{\partial x_2}\frac{\partial P}{\partial x_1} \tag{2.40}$$

(It can be easily shown that the same equations are obtained if we use Chezy's equation to define ϕ.)

For example, in a rectangular channel, $A = By_n$, $P = B + 2y_n$, and using $x_1 \equiv B$ and $x_2 \equiv y_n$, Eq. (2.40) is written as $y_n \times 2 = B \times 1$, indicating that a rectangular channel is most efficient when the flow depth is half of the bed width (the hydraulic radius turns out to be half of the flow depth). By using this ratio, the actual values of these parameters may be obtained from the Manning's equation.

Similarly, for a triangular channel with side slope mH:1V, $A = my_n^2$, $P = 2\sqrt{1+m^2}\ y_n$, and from Eq. (2.40),

$$y_n^2 \times 2\sqrt{1+m^2} = 2my_n \times \frac{2\,my_n}{\sqrt{1+m^2}}$$

implying that a triangular channel is most efficient when its side slope is 1:1, i.e., the half bottom angle is 45°. The hydraulic radius in this case is the depth divided by $2^{3/2}$.

For a trapezoidal channel with *fixed* side slopes of mH:1V, we get

$$y_n \times 2\sqrt{1+m^2} = (B + 2my_n) \times 1$$

This indicates that for the most efficient trapezoidal channel of a given side slope, m,[30]

$$B = 2\left(\sqrt{1+m^2} - m\right) y_n \tag{2.41}$$

and the hydraulic radius is half of the flow depth.

The problem of minimization of the wetted perimeter can be formulated on similar lines and it is easily seen that it would result in the same equation, i.e., Eq. (2.40). From Manning's equation also, for a given discharge, roughness, and bed slope, if A is minimum, P has to be minimum, since $A^{5/3}/P^{2/3}$ is constant. For cases where the area and perimeter are functions of three independent variables, the equivalent of Eq. (2.40) is given as

$$\frac{\partial A}{\partial x_1}\frac{\partial P}{\partial x_2} = \frac{\partial A}{\partial x_2}\frac{\partial P}{\partial x_1} \tag{2.42a}$$

$$\frac{\partial A}{\partial x_1}\frac{\partial P}{\partial x_3} = \frac{\partial A}{\partial x_3}\frac{\partial P}{\partial x_1} \tag{2.42b}$$

For a trapezoidal channel, using the bed width, side slope, and flow depth as variable, we get

$$y_n \times \frac{2my_n}{\sqrt{1+m^2}} = y_n^2 \times 1 \tag{2.43a}$$

$$y_n \times 2\sqrt{1+m^2} = (B + 2my_n) \times 1 \tag{2.43b}$$

From the first of these, we get $m = 1/\sqrt{3}$ (an angle of 60°) and the second is same as Eq. (2.41).

Example 2.10   A trapezoidal channel is to have a longitudinal slope of 1 in 1000 and is to be lined with brickwork. It has to carry a discharge of 20 m³/s and the slope stability requires the side slopes to be limited to an angle of 30°. What should be the bed width for the most efficient section?

Solution

The side slope, $m = \cot 30° = \sqrt{3}$.
From Eq. (2.41), $B = 0.536\, y_n$, resulting in
$$A = 2.268 y_n^2, \quad P = 4.536\, y_n, \quad \text{and } R = 0.5\, y_n$$

---

[30]If we keep y_n fixed and treat B and m as the variables, x_1 and x_2, we will obtain $m = 1/\sqrt{3}$. Monadjemi (1994) mentioned that there is no solution when B is treated as constant but Swamee (1995) extended the methodology to this condition also.

From the Manning's equation (with $n = 0.015$), therefore,

$$Q = \frac{1}{n} 2.268 y_n^2 (0.5 \, y_n)^{2/3} S_0^{1/2} = \frac{1.429 \, S_0^{1/2}}{n} \, y_n^{8/3}$$

which results in

$$y_n = 2.034 \text{ m and } B = \textbf{1.090 m}$$

The flow area for this channel is equal to 9.381 m^2 and the wetted perimeter is 9.225 m and all other combinations of bed width and flow depth that carry the same discharge would have a larger area and perimeter. For example, if a bed width of 2 m is provided, the normal depth would be 1.824 m and the flow area and wetted perimeter would be 9.410 m^2 and 9.296 m respectively. These values are higher, though not very significantly, than the optimum values. Similarly, a bed width of 0.5 m would result in a normal depth of 2.189 m, area of 9.394 m^2 and perimeter of 9.256 m, again larger but not significantly different.

Since the cost of excavation is roughly proportional to the area and the cost of lining is directly proportional to the perimeter, the hydraulically most efficient section would also be the most economic section. However, earthwork cost typically increases with depth, water loss through evaporation increases with increase in the top width, and provision of freeboard also complicates the cost analysis. We will not discuss these aspects of the *most economic section*. Swamee, et al. (2000) discusses the most economic section considering a number of factors, e.g., depth-dependent earthwork cost, lining cost, and cost of water lost through seepage and evaporation.

As seen from Fig. 2.15, although the bottom width and flow depth vary considerably in the vicinity of the optimum combination, the area and wetted perimeter do not change much. Same observation has also been made in Example 2.10. The cost of the most efficient section, therefore, may not be significantly different from that of the one which has a little larger or smaller width. For a rectangular section, it is easy to show (Exercise 2.11) that providing a width 20% larger than the optimum value will cause the flow depth to be 16% smaller but the flow area would be only about 1% larger than the optimum area and the perimeter will be only 2% larger than the optimum perimeter. Similarly, providing a 20% smaller width results in 26% larger flow depth, 1% larger area, and 3% larger perimeter. It can, therefore, be concluded that section dimensions *close to* the optimum values would also work equally well. Thus, in addition to economy of the section, we should also consider a number of other factors while deciding on the cross section, longitudinal slope, lining material, etc. Some of the important factors are described in the next section.

2.11 Design of Channels

Various elements of design of a channel to carry a specified discharge are: longitudinal slope, lining material,[31] and shape and dimensions of the section. The guiding principles for designing a channel are the economy and practicality. The economy should consider not only the capital cost but also the operation and maintenance cost. For example, providing a large bed slope will result in higher velocities and, therefore, smaller cross-sectional area. It will reduce the excavation cost but may increase the cost of maintenance of the lining material since the higher velocity will tend to damage the lining. Similarly, for a channel in a hydropower project, providing a very flat bed slope leads to lower velocities and will increase the capital cost (due to larger area), and the maintenance cost (since lower velocities are conducive to deposition of sediments and plant growth, which will have to be periodically removed). On the other hand, it will result in higher power output due to availability of larger head. Sometimes, the most economical sections may not be practical. For example, a semi-circular section is the most hydraulically efficient section (Exercise 2.10) but is difficult to construct. Even a trapezoidal section is not very practical since sharp corners may lead to zones of low velocity and rapid deposition or plant growth. In this section, we look at how some of these considerations affect various elements of design.

Longitudinal slope

The longitudinal slope is typically governed by the topography. An irrigation canal, for example, is designed to follow a ridge and the bed slope will be decided accordingly. We may decide to choose the longitudinal profile in such a way as to balance the excavation and filling (cut and fill) so as to avoid large scale transport of material. Sometimes, we may provide a flatter slope than that dictated by the topography as it will lead to availability of larger *head* at the delivery point. However, a minimum permissible velocity, known as the nonsitting velocity, must be maintained in the channel to avoid silt deposition. The nonsilting velocity is generally taken to be about 0.6 to 1.2 m/s.

Lining

Lining prevents erosion of the boundary material and also reduces the seepage

---

[31]In this section, we consider the channel to be a rigid bed channel (mobile bed channels are discussed in Chapter 8). Therefore, the boundary is considered either lined using brick, stone, concrete, plastic, metal, etc. or unlined but passing through non-erodible rock formation.

loss. The material of the lining is decided based on availability, cost, and the primary purpose of the lining. For example, if stones are easily and cheaply available nearby, stone lining would be preferred. If water scarcity exists, the irrigation canal should aim at minimizing the seepage loss and a concrete lining should be provided. Sometimes, brick-over-plastic lining is an inexpensive way of significantly reducing the seepage loss. For clear water flow, linings can withstand high velocity flows but when water carries sand or gravel particles, the maximum permissible velocity can be reduced. For concrete lined channels, the maximum permissible velocity can be 2 to 3 m/s while channels excavated in hard rock strata may easily withstand a velocity as high as 4 m/s. Sometimes, velocities as high as 5–10 m/s may be permitted if the channel is fully lined with soil-cement, gabion,[32] or concrete.

Shape and dimensions of the section

Once the longitudinal slope and the lining material are decided, the method described in the previous section can be used to obtain the most efficient section for the given discharge. For this, we first decide the shape of the section. For smaller discharges (up to about 50 m³/s), a triangular section (modified at the bottom to avoid the sharp corner, reduce the perimeter, and increase the hydraulic radius) may be used (Exercise 2.12). For higher discharges, a triangular section would result in very large flow depths and a trapezoidal channel with rounded corners is preferred (Fig. 2.16 and Exercise 2.11). The geometric elements of a rounded trapezoidal channel may be written as

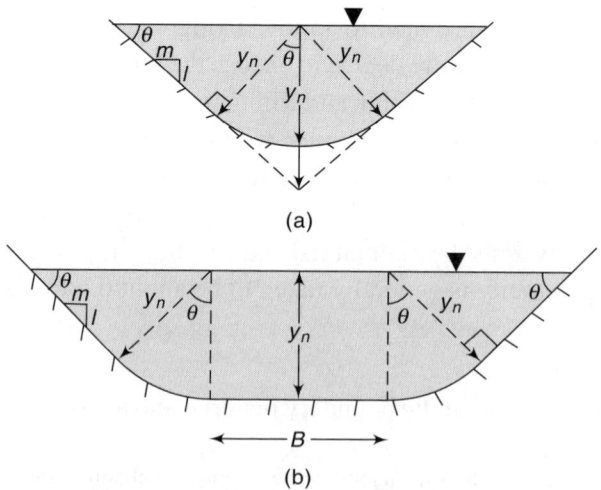

(a)

(b)

Fig. 2.16 Modified triangular and trapezoidal sections

---

[32]A gabion is a cage of wickerwork or iron filled with earth or stones and is used for erosion control.

$$P = B + 2my + 2\theta y = B + 2y\left(m + \tan^{-1}\frac{1}{m}\right) = B + 2My \qquad (2.44)$$

$$A = By + my^2 + \theta y^2 = By + My^2$$

in which M [$= m + \arctan(m)$] is introduced to simplify the expressions. For rounded triangular channels, same expressions would work with $B = 0$.

The normal depth for any given discharge can be obtained by following a methodology similar to that for sharp-cornered trapezoidal sections. Note that m is replaced by M in the expression for area of flow and, in the expression for the perimeter, $\sqrt{1+m^2}$ is replaced by M. The Manning's equation is written in terms of the nondimensional variables as

$$\Theta\left(=\frac{M^{5/3}nQ}{B^{8/3}S_0^{1/2}}\right) = \frac{(\eta+\eta^2)^{5/3}}{(1+2\eta)^{2/3}} \qquad (2.45)$$

in which Θ is nondimensional discharge and η is dimensionless normal depth, My_n/B. Figure 2.17 shows a plot of these parameters, which can be used to obtain the normal depth for any discharge for a known bed width and side slope. For triangular section, Manning's equation will directly give the normal depth as

$$y_n = 1.19\left(\frac{nQ}{MS_0^{1/2}}\right)^{3/8}$$

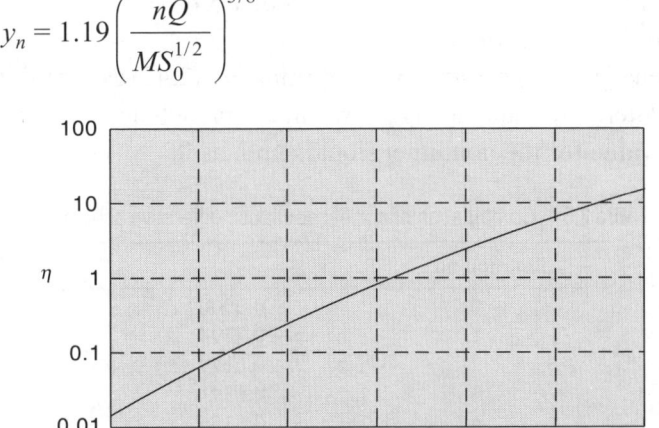

Fig. 2.17 Variation of nondimensional discharge with nondimensional depth for rounded trapezoidal channels

The graphical or iterative solution can be avoided by using the following explicit approximation, which is accurate to within 1%:

$$\eta = \left(\Theta^{-1.58} + \frac{0.74}{\Theta}\right)^{-0.3794} \qquad (2.46)$$

The following example illustrates the computation of normal depth in a rounded trapezoidal channel.

Example 2.11 A trapezoidal channel with rounded bottom has a bed width of 4 m and side slopes of 1H:1V. It is laid at a longitudinal slope of 1 in 1000, is lined with brickwork, and carries a discharge of 20 m³/s. What would be the normal depth of flow?

Solution

As in Example 2.4, the value of n is taken as 0.015. Since $m = 1$, $M = 1 + \pi/4$ = 1.785. The dimensionless discharge is then obtained as

$$\Theta = \frac{M^{5/3} nQ}{B^{8/3} S_0^{1/2}} = \frac{1.785^{-5/3} \times 0.015 \times 20}{4^{8/3} \times 0.001^{1/2}} = 0.6183$$

(a) *Using the graphical method*:

From Fig. 2.16, we get the nondimensional depth, $\dfrac{My_n}{B} = 0.65$.

Therefore, the normal depth, $y_n = 0.65 \times \dfrac{4}{1.785}\, m = \mathbf{1.457\ m}$

[Note that in Example 2.4, a similar trapezoidal channel (with sharp corners and a bed width of 5 m) had a normal depth of 1.5 m.]

(b) *Using the fixed-point iterative scheme*:

Nondimensional discharge, $\Theta = 0.6183$. Writing Eq. (2.45) as $\eta = \Theta^{3/5}(1 + 2\eta)^{2/5}/(1 + \eta)$, the iterative sequence is shown in the table below, starting with an initial guess value for the nondimensional depth as 0.

Table 2.3 Computation of normal depth using Iterative scheme

Iteration no., k	η_k
0	0.0000
1	0.7494
2	0.6179
3	0.6391
4	0.6356
5	0.6361
6	0.6360

The normal depth is, therefore, obtained as

$$y_n = 0.6360 \times \frac{B}{M} = \mathbf{1.425\ m}$$

(c) *Using the explicit solution*:

From Eq. (2.46), for $\Theta = 0.6183$, we get $\eta = 0.6332$ (almost same as that obtained in method (b) after six iterations) giving a normal depth of **1.419 m**.

In Example 2.11, we assumed that the side slopes were known. Generally, the side slopes are chosen based on stability and may vary from an almost vertical side in hard rocks to a slope of about 2H:1V for channels through loose sand and 3H:1V for channels through sandy loam. The procedure described in the previous section may then be used to obtain the most efficient section for the chosen side slope and channel shape. Sometimes, however, empirical curves are used to decide the bottom width and flow depth as a function of the design discharge. The velocity obtained for the designed section should be compared with the minimum and maximum permissible velocities. If it does not satisfy the requirements,[33] the section may have to be modified.

A freeboard[34] has to be added to prevent the possibility of waves and fluctuations in discharge causing overflow of water from the channel sides. This is particularly important in cases where this water may damage the channel substructure. There is no universal rule for the freeboard, and values ranging from about 5 to 30% of the flow depth have been provided in most channels. Typical freeboards range from about 0.5 m at low discharges (about 1 m³/s) to about 1 m for high discharges (about 100 m³/s). Sometimes, the freeboard is computed as 'the velocity head + about 15 cm' subjected to a minimum of about 30 cm. Because of the infrequent occurrence of the depth increasing above the design depth, the lining does not have to continue through the freeboard. However, it is advisable to have some freeboard for the lining also. A value of one-third (at low discharges) to one-half of the bank freeboard is considered sufficient as the lining freeboard.

Example 2.12 A concrete lined canal passes through loose sand and is to be designed to carry a discharge of 100 m³/s. Topography of the area suggests a general slope of about 1 in 4000. The velocity has to be maintained in the range of 1-2 m/s to ensure nonsilting and nonscouring conditions. Design the section.

---

[33]Sometimes, a Froude number [see Eq. (1.3)] criterion is also specified restricting the Froude number to about 0.5. For higher Froude numbers, the water surface becomes wavy and large disturbances are observed at the bends. This may cause spilling if enough freeboard is not provided.

[34]*Freeboard* is the margin of safety provided above the design water level in the channel up to the bank height (or lining height).

Solution

Since the discharge is large, we will choose a modified trapezoidal section with rounded corners (Fig. 2.16). For channels passing through loose sand, a side slope of 2H:1V may be used. A bed slope of 1 in 4000 is provided based on the topography. To achieve a minimum area, we would use the maximum permissible velocity of 2 m/s, which results in a flow area of 50 m^2 and hydraulic radius of 1.87 m (from the Manning's equation with $n = 0.012$) giving a wetted perimeter of 26.7 m. From Eq. (2.44), the area and perimeter for a modified trapezoidal section are written as

$$A = By_n + [m + \tan^{-1}(1/m)]y_n^2 = By_n + 2.464y_n^2$$
$$P = B + [2m + \tan^{-1}(1/m)]y_n = B + 4.928y_n$$

we have two equations which can be solved for the bed width and normal depth. From the second equation, we get $B = 26.7 - 4.928y_n$ and substitution into the first results in a quadratic equation, which gives $y_n = 2.41$ m and $B = 14.8$ m (a second solution of the quadratic equation is 8.43 m, which results in negative width). The width and depth appear to be reasonable and there is no need of modifications. Had the general ground slope been very steep, we would have obtained a smaller depth and larger width (e.g., with a bed slope of 1 in 2000, we get a bed width of about 40 m, which is not desirable as it takes up a huge area). In that case, we may provide a bed slope flatter than that suggested by the topography. Similarly, for flat terrains, the required flow depth may become very large. In that case, we may have to design with a velocity smaller than the maximum permissible velocity of 2 m/s (the velocity should not, however, go below the nonsilting velocity of about 0.6 m/s).

A freeboard of 1 m is sufficient with lining carried up to 0.5 m above the design depth. Thus a section with bed width of 15 m and total depth of 3.5 m should be provided with lining on the sides to a height of 3 m.

In designing of channels, we assume the design discharge as a constant value. However, once we have designed and built the channel, it is likely that the discharge would not be same throughout the year. For example, an irrigation canal may have a smaller discharge during periods of low demand. Thus the actual discharge in the channel would vary over a wide range causing a change in the Froude number [see (Eq. 1.3)]. Since Froude number is an important parameter in open channel flow, and significantly affects the propagation of small disturbances created in the flow, it may be desirable to keep it more or less uniform even though the discharge varies. For this purpose, a generalized Froude number (similar to the generalized slope concept proposed by Jones and Tripathy in 1965) may be defined as follows.

Writing the Froude number as

$$F_r = \frac{V}{\sqrt{gD}} = \frac{V\sqrt{T}}{\sqrt{gA}} = \frac{1}{n}\frac{R^{2/3} S_0^{1/2} T^{1/2}}{g^{1/2} A^{1/2}} \Rightarrow \frac{ng^{1/2} F_r}{S_0^{1/2}} = \frac{T^{1/2} A^{1/6}}{P^{2/3}} \quad (2.47)$$

we can use appropriate nondimensionalization to define a generalized Froude number. For example, for rectangular or trapezoidal channels, the bed width, B, can be used as the normalizing variable to obtain a generalized Froude number, F_*, as

$$F_* = \frac{ng^{1/2} F_r}{B^{1/6} S_0^{1/2}} = \frac{(1+2m\eta)^{1/2} (\eta + m\eta^2)^{1/6}}{(1+2\sqrt{1+m^2}\,\eta)^{2/3}} \quad (2.48)$$

in which η is the nondimensional depth, y/B. Figure 2.18 shows a plot of the generalized Froude number for rectangular ($m = 0$) and trapezoidal channels. It is seen that for a given channel, the Froude number may vary over a wide range as the discharge is changed and the normal depth to bed width ratio changes accordingly. However, it is interesting to note that for $m = 0.5$ (and, to some extent, for $m = 0.25$), the range of variation of Froude number is quite small over a wide range of y/B values (0.2 to 2). Thus if the discharge is likely to vary over a wide range, we may be able to maintain the Froude number as reasonably constant by adopting a side slope of 0.5H:1V or 0.25H:1V (if permitted by the site conditions) and designing the channel dimensions accordingly. Another noticeable feature is the presence of multiple depths which provide the same generalized Froude numbers for $m = 0$ and 0.25 (and, in fact, for any m less than 0.466, see Exercise 2.14).

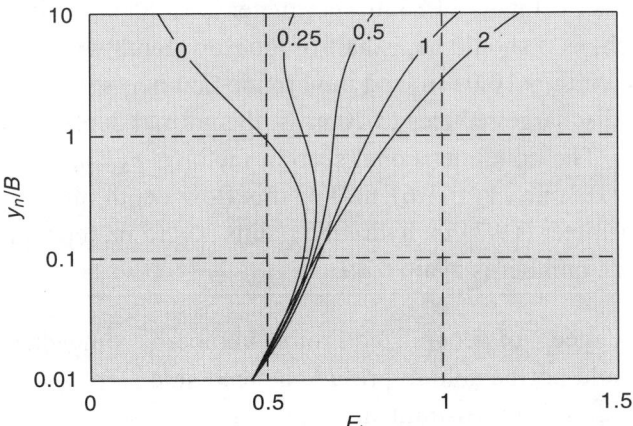

Fig. 2.18 Variation of generalized Froude number with nondimensional depth for rectangular and trapezoidal channels (Numbers on the curve represent the side slope, m.)

Example 2.13 demonstrates the design of a channel with an almost constant Froude number over a wide range of discharge values.

Example 2.13   A concrete lined canal is to be designed to carry a discharge of 100 m^3/s, although the discharge during some periods may be as low as 20 m^3/s. Design the section and the bed slope such that the Froude number is nearly 0.5 for the entire discharge range.

Solution

Since the Froude number is to be kept constant, we would use a side slope of 0.5H:1V (assuming that such steep slopes can be provided).

From Fig. 2.18, the width-to-depth ratio should then be between 0.2 to 2 to maintain a constant Froude number and the generalized Froude number, F_*, would be about 0.7.

From Eq. (2.41), the most efficient section for this side slope ($m = 0.5$) will have $B = 1.25y$, i.e., $y/B = 0.8$.

With $n = 0.012$ and $F_r = 0.5$, Eq. (2.48) results in $B^{1/6}S_0^{1/2} = 0.0268$. The Manning's equation can now be used to obtain the bed width as

$$100 = \frac{1}{0.012} \frac{(B \times 0.8B + 0.5 \times 0.64B^2)^{5/3}}{(B + 2\sqrt{1.25} \times 0.8B)^{2/3}} S_0^{1/2}$$

$$= \frac{1}{0.012} 0.6096 B^{8/3} \frac{0.0268}{B^{1/6}}$$

from which, **$B = 5.58$ m**.

The bed slope and flow depth are computed as 4.05×10^{-4} and 4.46 m respectively. The area of flow is 34.83 m^2, perimeter is 15.55 m, hydraulic radius is 2.24 m (which, as it should be, is half of the flow depth within the round-off margin), top width is 10.04 m, and the Froude number is 0.49.

When the discharge reduces to 20 m^3/s, the normal depth can be obtained from Fig. 2.5. The nondimensional section factor is $nQ/S_0^{1/2}B^{8/3} = 0.122$ for which $y_n/B = 0.3$, i.e., $y_n = 1.67$ m. For this flow depth, the area of flow is 10.71 m^2, perimeter is 9.31 m, hydraulic radius is 1.15 m, top width is 7.25 m, and the Froude number is again 0.49.

While the concept of generalized Froude number is somewhat useful, generally, we would not be able to provide a steep side slope. Hence, practical utility of Fig. 2.18 is rather limited.

We have focussed our attention on design of rigid boundary channels. For unlined channels (except those through non-erodible material), the size of particles on the channel boundary will also affect the design since the velocity should not be large enough to remove these particles. Also, the seepage loss through the boundary may be a significant fraction of the channel discharge and may be utilized to provide a reduced section as we move downstream in the channel. Chapter 8 describes the process of flow and sediment transport through mobile bed channels and the methodology for designing such channels.

SUMMARY

This chapter has provided an overview of open channel flows under uniform conditions in which the flow depth, velocity, and discharge, stay the same at all sections. Resistance equations relating the flow velocity to the bed slope, boundary roughness, and the hydraulic radius, has been developed. The resistance equation has been applied to simple cross sections to estimate the discharge for a given normal flow depth and the normal depth for a given discharge. Occurrence of multiple normal depths has been demonstrated through analysis of a circular channel. More complex sections, viz., the composite section and the compound section, has then been analysed to illustrate various difficulties and the methods to overcome those. Finally, the design aspects has been enumerated, including the selection of bed slope, side slope, cross section shape, and the efficient section.

Some advanced topics, which have been briefly stated, may interest the reader. Some of the areas which may require additional research are dimensionally homogeneous resistance equation; explicit estimation of normal depth for circular channels, selection of the best formula for equivalent roughness coefficient in a composite channel, a good way of subdividing a compound channel or to account for the shear stress at the main channel— floodplain boundaries, and a comprehensive design of the most economical section. Recent references pertaining to some of these areas are provided here. The website of the book maintains an updated list of the references which the reader may find of interest.

The design of a channel is typically based on the assumption of uniform flow conditions occurring throughout the channel. However, in practice, there are a number of situations where the flow will not be uniform. For example, we may have to construct a weir across the channel which will raise the water level on the upstream side. This would require raising the bank levels, at least in the immediate vicinity of the weir. In order to compute this change in depth, and consequently velocity, of flow, we would have to use equations other than the resistance equation. In the next chapter, we will address the issue of computation of the water surface profiles by introducing the energy equation and the concept of specific energy.

REFERENCES

Ackers, P. (1991): 'Hydraulic design of straight compound channels', *Rep. No. SR28*, HR Wallingford, Wallingford, UK.

ASCE (1963): 'Task force report: Friction factor in open channels', *J. of Hydr. Div.*, ASCE, **89**(2), pp. 97–143.

Bousmar, D. and Zech, Y. (1999): 'Momentum transfer for practical flow computation in compound channels', *J. of Hydr. Eng.*, **125** (7), pp. 696–706.

CFRS (2007): "Manning's *n* values", *http://www.fsl.orst.edu/geowater/FX3/help/8_Hydraulic_Reference/Mannings_n_Tables.htm* (last accessed on Mar. 20, 2007), Corvallis Forestry Research Community.

Chow, V. T. (1959): *Open-Channel Hydraulics*, McGraw-Hill, New York.

LMNO Engineering (2007): 'Fluid flow calculation website', *www.lmnoeng.com/manning.htm* (last accessed on Feb. 12, 2007).

Monadjemi, P. (1994): 'General formulation of best hydraulic channel section', *J. of Irrig. & Dra. Eng.*, **120**(1), pp. 27–35.

Pezzinga, G. (1994): 'Velocity Distribution in Compound Channel Flows by Numerical Modeling', *J. of Hydr. Eng.*, **120**(10), pp. 1176–98.

Seckin, G. (2004): 'A comparison of one-dimensional methods for estimating discharge capacity of straight compound channels', *Cana. J. of Civ. Eng.*, **31**(4), pp. 619–631.

Shiono, K. and Knight, D.W. (1991): 'Turbulent open channel flows with variable depth across the channel', *J. of Fluid Mech.*, **222**, pp. 617–46.

Swamee, P.K. (1994): 'Normal-depth equations for irrigation canals', *J. of Irrig. & Dra. Eng.*, **120**(5), pp. 942–948.

Swamee, P.K. (1995): 'Discussion on general formulation of best hydraulic channel section by P. Monadjemi', *J. of Irrig. & Dra. Eng.*, **121**(2), p. 222.

Swamee, P.K., Mishra, G.C., and Chahar, B.R. (2000): 'Comprehensive design of minimum cost irrigation canal sections', *J. of Irrig. & Dra. Eng.*, **126**(5), pp. 322–27.

Terzidis, G.A. (2005): 'Explicit method to calculate the normal depth of trapezoidal open channel'.

USGS (2007): 'Surface water field techniques', *wwwrcamnl.wr.usgs.gov/sws/fieldmethods/Indirects/nvalues/index.htm* (last accessed on Feb. 12, 2007).

van Prooijen, B.C., Battjes, J.A., and Uijttewaal , W.S.J.(2005): 'Momentum exchange in straight uniform compound channel flow', *J. of Hydr. Engg.*, **131**(3), pp. 175–83.

Yang, K.J., Cao, S.Y., Liu, X.N., Zhang, Z.X., and Shuili, X. (2005a): 'Comparison of methods for calculating flow capacity of channels with compound cross section', *J. of Hydr. Eng.* (China), **36**(5), pp. 563–68.

Yang, K.J., Cao, S.Y., Liu, X.N., and Shuili, X. (2005b): 'Analysis on methods for predicting composite roughness of river channel with compound cross section', *J. of Hydr. Eng.* (China), **36**(7), pp. 780–86.

Yen, B.C. (ed.), 1991: 'Hydraulic resistance in open channels', in *Channel Flow Resistance: Centennial of Manning's Formula*, Water Resource Publications, Highlands Ranch, Colorado, pp. 1–135.

Yen, B.C. (1992): 'Dimensionally homogeneous Manning's formula', *J. of Hydr. Eng.*, **118**(9), pp. 1326–32.

Yen, B.C. (2002): 'Open channel flow resistance', *J. of Hydr. Eng.*, **128**(1), pp. 20–39.

EXERCISES

2.1 The channel in Exercise 1.4 is concrete lined and is laid at a slope of 1 in 1000. What would be the velocity of flow using Manning's equation? Why is this velocity different from that in Exercise 1.4? Develop a resistance equation for this flow. What would be the depth of flow if the same discharge of water flows through the channel under uniform flow conditions? [1.66 m/s, 0.27 m]

2.2 The Manning's equation is applicable to completely turbulent and fully rough flows only although it has been found to work well in other conditions also. Using the criterion for rough and smooth boundaries [Eq. (A.11) in Appendix A] and using an Strickler type equation [Eq. (2.12)], obtain the critical value of the parameter $n^6 \sqrt{RS_0}$ beyond which the Manning's equation should be applicable. [9.6×10^{-14}]

2.3 For a horizontal channel $S_0 = 0$ and for a channel with its bed rising in the direction of flow (adverse slope) S_0 is negative. Obviously the Manning's equation cannot be used to obtain the normal depth for a given discharge. What does this signify and why?

2.4 For the problem of Example 2.5, prepare the section factor plot and obtain the normal depth for a discharge of 100 m^3/s. [2.4 m]

2.5 For uniform flow in circular channels, the maximum velocity would be obtained when the hydraulic radius is maximum, whether we use Chezy's equation or Manning's equation. Use Eq. (2.25) to obtain the value of θ (and y_n/D_0) for which this will occur. Using the Manning's equation obtain the depth for maximum discharge and its ratio with the pipe-full discharge.

Note: The condition for maximum discharge will be different for Chezy's and Manning's equations.   [129° (0.81), 0.94 D_0, 1.076]

2.6 The trapezoidal channel of Example 2.4 was designed for a discharge of 20 m³/s. Due to water withdrawal, seepage, and evaporation, the discharge at a downstream section is estimated to be only 15 m³/s. To reduce the seepage beyond this point, the bottom is lined with concrete but the sides are kept brick-lined. Find the depth of flow assuming uniform flow conditions. Use Eq. (2.33) for the equivalent roughness.   [1.149 m]

2.7 A 1 m wide rectangular laboratory flume has steel bed and glass sides and is laid at a slope of 1 in 500. What would be the discharge when uniform flow occurs at a flow depth of 0.5 m. Use Eq. (2.29) for effective roughness and also obtain the areas corresponding to each roughness type using Eq. (2.28). Compare these areas with those obtained using the bisector method. What would be the discharge if Eq. (2.34) is used to compute the effective roughness?   [0.805 m³/s, 0.813 m³/s]

2.8 For the compound channel of Example 2.8, find the discharge when the depth in the main channel is 1.6 m, using both the single and the divided channel methods. For the divided channel method, try all four methods of dividing the channel shown in Fig. 2.14.

[20.22 m³/s, vertical: 32 m³/s, horizontal: 23.6 m³/s]

2.9 A channel is designed to carry a discharge of 20 m³/s, with Manning's $n = 0.015$ and bed slope of 1 in 1000 (for trapezoidal channel, side slope, $m = 1/\sqrt{3}$). Find the channel dimensions of the most efficient section if the channel is (a) trapezoidal, (b) rectangular, and (c) triangular.

[(a) $B = 2.60$ m, $y = 2.25$ m; (b) 4.26 m, 2.13 m; (c) $y = 3.02$ m]

2.10 Show that the most efficient circular section will have a depth equal to half the diameter. For the channel in the previous exercise, find the dimensions of the efficient (semi)circular channel. Compare the flow area and wetted perimeter of all sections obtained in Exercises 2.9 and 2.10 and see which section is the best overall.   [$D_0 = 4.67$ m]

2.11 The most efficient rectangular channel has a bed width equal to twice the flow depth. To carry the same discharge, if the bed width is made larger than the optimum width, the flow depth will become smaller and vice-versa. Obtain an equation relating the percentage change in flow depth, perimeter, and area to that in the bed width compared to their optimum values. Using this equation, verify the values mentioned at the end of Section 2.10.

$$\left[(1-\delta y)^{2.5}\,(1+\delta B)^{2.5}-1-\frac{\delta B}{2}+\frac{\delta y}{2}=0\right]$$

2.12 Figure 2.16 shows the modified trapezoidal section generally used for a channel with larger discharges. If the side slopes are fixed at 1H:1V (i.e., $\theta=45°$), what would be the relationship between the bottom width and the depth for the hydraulically efficient section? If the discharge is 100 m³/s, Manning's n is 0.012, and the bed slope is 1 in 4000, what would be the normal depth? [$B=0$, 4.85 m]

2.13 Design an asphalt lined canal passing through loose sand to carry a discharge of 40 m³/s. The bed slope has been selected as 1 in 5000 based on the topography and the efficient modified triangular section has to be used (try to optimize the side slope first).[If slope = 2H:1V, $y=3.54$ m]

2.14 Figure 2.18 shows that the generalized Froude number achieves a maximum value for the side slopes smaller than 0.5H:1V. Obtain an expression relating the value of y/B to the side slope m at the extreme values of the generalized Froude number and show why $m=0.466$ may be treated as a critical value of the side slope.

$$\left[(-4m^2\,\sqrt{1+m^2}\,\left(\frac{y}{B}\right)^3+(4m\,\sqrt{1+m^2}-10\,m^2)\left(\frac{y}{B}\right)^2+\right.$$
$$\left.(6\sqrt{1+m^2}-10\,m)\frac{y}{B}-1=0\right]$$

3 Nonuniform Flow: Gradually Varied

3.1 Introduction: Examples and Importance

Uniform flow, discussed in the previous chapter, occurs over long undisturbed reaches of prismatic channels. However, there are several *disturbances* that may cause the flow depth (and velocity) to change from one section to the other. For example, the channel slope may change, a small weir or a large dam may be constructed across the channel, or a sluice gate[1] may be built to control the downstream flow. While some of these situations would result in a rapid variation in depth (e.g., flow near a weir), others would cause a slow variation (e.g., flow upstream of a large dam). Figure 3.1 shows a possible variation of flow depth in a long channel which ends in a free overfall. (Later in the chapter, we will see that there are various possible water surface profiles depending on the length and slope of the channel.) The regions of uniform flow, gradually varied flow (GVF), and rapidly varied flow (RVF) are demarcated in the figure by two lines, the exact locations of which are rather subjective. For example, uniform flow has a constant depth (equal to the normal depth) and GVF has slowly varying depth. It is quite difficult to judge the exact location where one flow finishes and the other starts since the water surface is tangential to the normal depth. Similarly, the dividing line between GVF and RVF will shift based on our judgement of the location where the slow variation of depth ends and the fast variation begins. Clearly, some empiricism will be involved in this determination. For example, we may say that the uniform flow extends up to the point where the depth is 99% of the normal depth. On the other hand, we may say that the flow will be rapidly varied from the edge (or brink) to a distance equal to, say, six times the flow depth at the end. In this chapter, we discuss the GVF and in the next, the rapidly varied flow.

Our prime interest in solving a GVF problem is the variation of flow depth. When a dam is constructed on a river, additional land is submerged due to the

---

[1] A sluice is a water channel that is controlled at its upstream end by a gate, called the sluice gate. We would use the term sluice gate to represent a gate anywhere in the channel, not only at the upstream end.

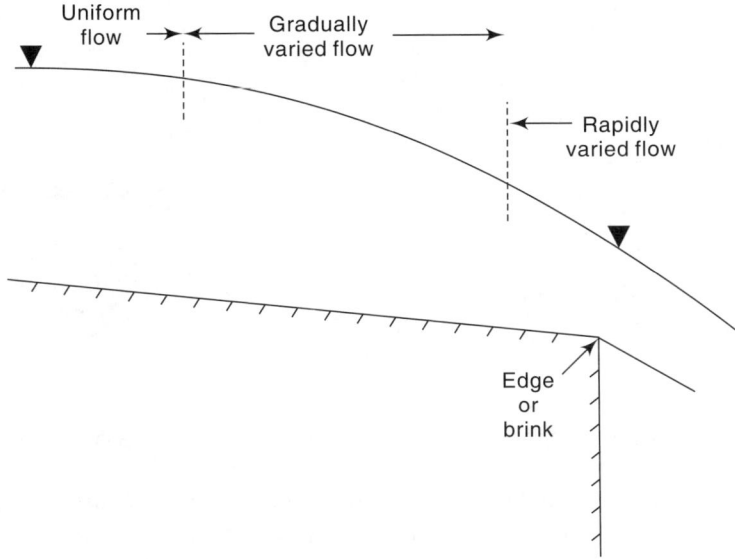

Fig. 3.1   Flow in a channel ending in a free overfall

formation of the reservoir. The submergence is, however, not limited to the area just upstream of the dam since the *backwater*[2] effect is felt for a considerable distance (sometimes, of the order of a few kilometres). Similarly, if the bed slope of a channel becomes smaller after a certain section, the uniform flow depth increases (section factor being inversely proportional to the square root of the bed slope) and creates a backwater effect upstream of that point. The additional area submerged due to the backwater effect may be obtained by computing the flow depth at various locations and collating this information with the topographic map of the area.

3.2   Applications of Continuity, Momentum, and Energy Equations to GVF

We would first look at the applications of the continuity, momentum, and energy equations to GVF to obtain an equation relating the flow depth variation to some channel and flow properties. Various techniques of solving this equation would then be discussed.

3.2.1   Application of Continuity Equation to GVF

Figure 3.2 shows the longitudinal section of an open channel along with a

---

[2]Backwater refers to the water backed up in its course by an obstruction (or, sometimes, by an opposing current).

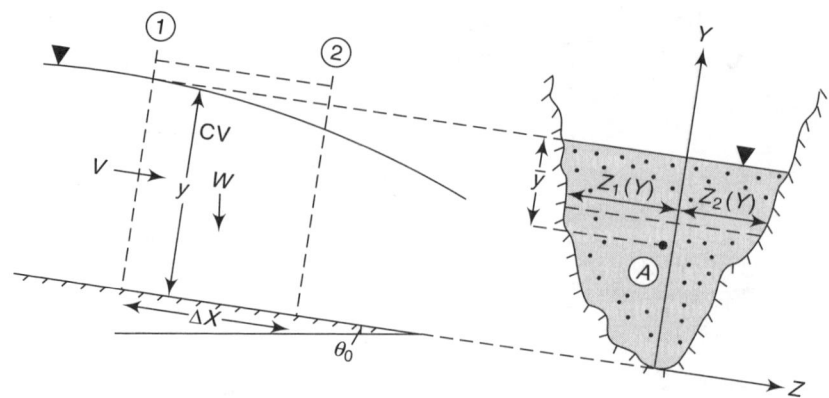

Fig. 3.2 Longitudinal section of a channel with gradually varied flow

control volume (CV) and the cross sections at both ends of this CV. As stated earlier, we perform a one-dimensional analysis by using the cross-sectional average velocity [Eq. (2.1)]. Since the flow is steady and the discharge is not changing spatially,[3] the continuity equation is written as (in terms of volume, since water is incompressible)

$$VA - \left(VA + \frac{\partial (VA)}{\partial X} \Delta X \right) = 0 \implies VA = \text{constant} (= Q) \qquad (3.1)$$

Inflow rate	Outflow rate	Rate of change
into CV	from CV	within CV

3.2.2 Application of Momentum Equation to GVF

Application of the X-momentum equation over the CV shown in Fig. 3.2 requires information about the forces acting on the CV and the net efflux of momentum across the control surface (CS) [see Eq. (A.3), Appendix A]. The forces acting on the CV are due to the pressure at the cross sections, shear stress at the boundary, weight of water within the CV, and air resistance at the water surface. The air resistance is generally insignificant and is ignored. The component of weight in the X-direction is given by

$$F_W = W \sin \theta_0 = \rho g A \Delta X S_0 \qquad (3.2)$$

Estimation of the other two forces, pressure and shear, require some assumptions to be made about the distribution of pressure over a cross section and shear stress on the boundary. Appendix B shows theoretical pressure

---

[3]Here we consider that there are no lateral inflows or outflows. Chapter 6 deals with such cases.

distributions for open channels with steep bed slope and/or curved streamlines. One of the basic assumptions of GVF is the *small curvature of streamlines*, which allows us to ignore the acceleration normal to the flow direction and enables us to write the pressure at any point in the flow field as[4]

$$p = \rho g(y - Y) \cos \theta_0 \qquad (3.3)$$

in which y is the flow depth (see Section B.2, Appendix B). The pressure force over a section is, then,

$$F_p = \rho g \cos \theta_0 \int_0^y \int_{z_1(Y)}^{z_2(Y)} (y - Y) dZ \, dY \qquad (3.4)$$

in which Z_1 and Z_2 represent the variation of channel width at different heights from the bed (see Fig. 3.2). We see that the integral in Eq. (3.4) is the moment of the area of cross section, A, about the water surface and may be written as $A \bar{y}$, $\bar{y}$ being the distance (along Y-axis) of the centroid of the cross-sectional area from the water surface. In most cases, the channel bed slope will be very small and cos θ_0 may be taken as unity. We restate the fact that Eq. (3.4) has been obtained by assuming that the curvature of streamlines is small and write it as

$$F_p = \rho g A \bar{y} \cos \theta_0 \qquad (3.5)$$

The distribution of shear stress on the boundary for a *uniform flow* is shown in Appendix B. As discussed in Chapter 2, since we use the one-dimensional analysis, we need to obtain an *average shear stress* on the boundary. For nonuniform flow, some studies (e.g., Graf and Song 1995, Dey and Lambert 2005) have been conducted regarding the shear stress distribution on the boundary and its comparison with that in uniform flow, but it may be a while before these are used in practice. The commonly used assumption is that the average shear stress at the boundary is given by an expression similar to that used for uniform flow [Eq. (2.5)] with the bed slope replaced by a friction slope or *shear slope*.[5] This slope at a given section with flow depth y is taken equal to the shear slope for a uniform flow (or, the bed slope, since for uniform

---

[4]Since most open channel flow problems have atmospheric pressure at the free surface, we write the pressure in terms of gage pressure and not absolute pressure. Since it is the pressure difference which is important, it does not matter whether we use absolute pressure or gage pressure.

[5]Shear slope is not a commonly used term. Since the shear stress is due to friction, we may use the term *friction slope*. However, to distinguish it from the *energy slope*, discussed later in this section, we use the terms shear slope and energy slope and ultimately employ the friction slope to represent both of these.

flow conditions, the shear slope and bed slope are same) at that section taking place with the same flow depth, y. This assumption, although not proven rigorously, has been found to work well in practice as shown by comparison of the observed and computed GVF water surface profiles. It is likely to work better for GVF with decreasing depth (increasing velocity) because expanding flows (decreasing velocity) may have large non-frictional losses due to eddies. Use of this assumption leads to an average shear stress

$$\bar{\tau} = \rho gRS_\tau \tag{3.6}$$

with the shear slope, S_τ, given by the Manning's equation,

$$Q = \frac{1}{n} AR^{2/3}S_\tau^{1/2} \implies S_\tau = \frac{n^2 Q^2}{A^2 R^{4/3}} \tag{3.7}$$

where the area and hydraulic radius are computed for the flow depth, y. The force due to boundary shear on the CV is then obtained as

$$F_\tau = \bar{\tau} P\Delta X = \rho gAS_\tau \Delta X \tag{3.8}$$

Example 3.1   A 1 m wide rectangular channel is laid at a bed slope of 1 in 1600, has a Manning's n of 0.01, and ends in a free overfall. For a discharge of 0.25 m³/s, the flow depth at a section 1 m upstream of the brink was found to be 0.25 m and the flow was gradually varied. Find the friction slope and compare it with the bed slope.

Solution
The friction slope is obtained as [Eq. (3.7)]

$$S_\tau = \frac{n^2 Q^2 P^{4/3}}{A^{10/3}} = \frac{0.01^2 \, 0.25^2 \, (1 + 2\times 0.25)^{4/3}}{0.25^{10/3}}$$

which is equivalent to **1 in 917**, more than the bed slope of 1 in 1600. The normal depth for this case is 0.3037 m.

Once all the forces are known, the net force in the X-direction can be computed and set equal to the net momentum efflux. For computing the momentum flux at a section, we need to know the velocity distribution over a cross-section. Appendix B shows a few typical velocity profiles in channels of various shapes. Again, since we are using a one-dimensional analysis, it would suffice to integrate the momentum flux over a differential area to obtain the momentum flux over the entire area. The momentum efflux in the X-direction is, therefore, obtained as

$$M_X = \rho \int_A v_X \, (\vec{V} \cdot dA\hat{n}) \tag{3.9}$$

in which $\vec{V}$ is the velocity vector, dA is an elemental area in the cross section, and $\hat{n}$ is the area vector (i.e., a unit vector normal to the area pointing *outward* from the control surface). Due to the assumption of *negligible curvature of streamlines*, $\hat{n}$ is taken to be in the positive X-direction for the outflow section and in the negative X-direction for the inflow section. The magnitude of the momentum flux at any section is then written as

$$\rho \int_A v_X^2 dA \tag{3.10}$$

which is more than the momentum obtained by using the average velocity [Eq. (2.1)], i.e., $\rho V^2 A$. This can be shown mathematically by writing v_X as a sum of the mean velocity V and the deviation v'.[6] A physics-based explanation is that when we replace the actual velocity profile by the average velocity profile (which is defined so as to make the mass in both profiles equal), we subtract some mass (shown by $M1$ in Fig. 3.3) and add an equal mass ($M2$). However, $M1$ has a much higher average velocity (V_1) compared to $M2$ (V_2). Therefore, the net effect of the average velocity profile is a reduction in momentum compared to the actual momentum.

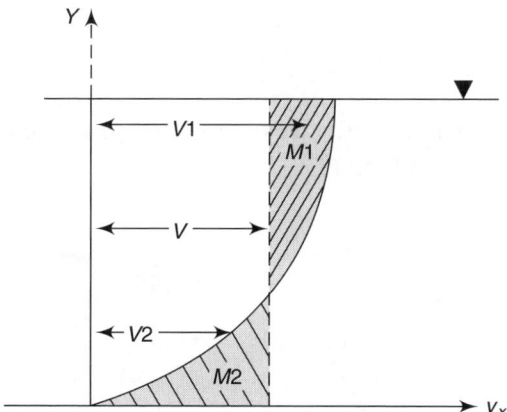

Fig. 3.3 Actual and average velocity profiles and the associated momentum flux

One way to evaluate Eq. (3.10) in a one-dimensional framework is to use a different average velocity, say V_M, such that $\rho V_M^2 A$ is equal to the value given by Eq. (3.10). However, a more convenient method is based on the *momentum correction factor* (also called the Boussinesq coefficient after the French mathematician J. Boussinesq who proposed it in 1877). The momentum correction factor, β, is defined through

---

[6] $\int_A v_X^2 dA = \int_A (V + v')^2 \ dA = V^2 A + \int_A v'^2 \ dA \geq V^2 A$ (since the areal average of deviations vanishes)

$$\rho \int_A v_x^2 \, dA = \beta \rho V^2 A$$

$$\Rightarrow \quad \beta = \frac{\int_A v_x^2 \, dA}{V^2 A} = \frac{1}{A} \int_A \left(\frac{v_X}{V} \right)^2 \, dA \qquad (3.11)$$

i.e., the actual momentum is obtained by multiplying the momentum based on the average velocity with the correction factor. Obviously, β is always greater than unity. (Only when v_X is constant over the entire section, β will be equal to one. However, due to no slip condition at the boundary, there will always be some variation in velocity.) As the velocity becomes more and more uniform over a section, β will come closer to one. For most open channel flow conditions, the value of β is found to be less than 1.15 and is, generally, taken as unity.

Example 3.2 In a rectangular channel, the velocity may be assumed to vary as one-seventh power of the distance from the boundary (Fig. 3.4) and is given by

$$v_X(Y, Z) = v_{\max} \left(\frac{Y}{y} \right)^{1/7} \left(1 - \frac{2\,|Z|}{B} \right)^{1/7}$$

where $v_{\max}$ is the maximum velocity over the cross section. What would be the momentum correction factor?

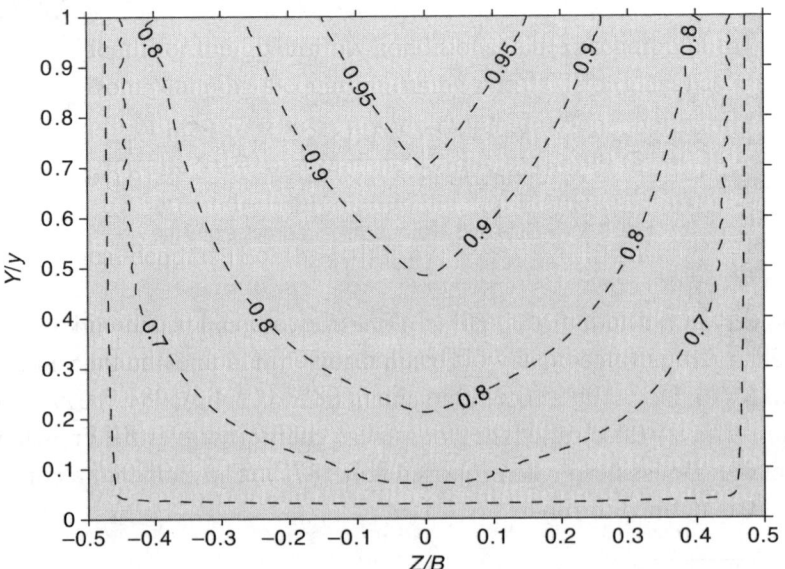

Fig. 3.4 Velocity variation following the one-seventh power law

Solution

Taking advantage of symmetry, we find the cross section average velocity as

$$V = \frac{\int_A v_X \, dA}{A} = \frac{\int_0^{B/2} \int_0^y v_{max} \left(\frac{Y}{y}\right)^{1/7} \left(1 - \frac{2Z}{B}\right)^{1/7} dYdZ}{By/2} = \frac{49}{64} v_{max}$$

Using Eq. (3.11), the momentum correction factor is obtained as

$$\beta = \frac{1}{A} \int_A \left(\frac{v_X}{V}\right)^2 dA = \frac{\int_0^{B/2} \int_0^y \left(\frac{64}{49}\right)^2 \left(\frac{Y}{y}\right)^{2/7} \left(1 - \frac{2Z}{B}\right)^{2/7} dYdZ}{By/2}$$

$$= 1.032$$

The correction factor is quite close to unity and is, therefore, ignored for most practical cases.

The net momentum efflux across the control surface is obtained as

$$M_X = \beta\rho\left\{\left[V^2 A + \frac{\partial(V^2 A)}{\partial X}\Delta X\right] - V^2 A\right\} = \beta\rho\Delta X Q^2 \frac{\partial(1/A)}{\partial X} \qquad (3.12)$$

Note that β has been assumed to be constant indicating that the velocity profiles at all sections are similar (but not same)[7] and we have used the continuity equation given by Eq. (3.1) to eliminate the velocity. Application of the momentum equation to the CV results in

$$F_W + F_{p1} - F_{p2} - F_\tau = M_X$$

That is

$$\rho g A \Delta X S_0 + \rho g A \bar{y} \cos \theta_0 - \rho g \cos \theta_0 \left[A\bar{y} + \frac{\partial(A\bar{y})}{\partial X}\Delta X\right] - \rho g A \Delta X S_\tau$$

$$= \beta\rho\Delta X Q^2 \frac{\partial(1/A)}{\partial X}$$

which may be simplified as

$$S_0 - S_\tau = \frac{\cos \theta_0}{A} \frac{\partial(A\bar{y})}{\partial X} + \frac{\beta Q^2}{gA} \frac{\partial(1/A)}{\partial X} \qquad (3.13)$$

---

[7]Equation (3.11) indicates that β depends only on the ratio v_X/V and not the actual values of v_X and V.

(note that the bed slope is treated as constant). To evaluate the two partial derivatives, we make use of the fact that at any given section, area and the depth to its centroid are functions of the flow depth only, and the flow depth, being the depth to the water surface from the deepest point of the section, is only a function of X (and not Y and Z). The partial differentiation thus becomes ordinary differentiation and application of the chain rule leads to

$$\frac{\partial (A\bar{y})}{\partial X} = \frac{d(A\bar{y})}{dy}\frac{dy}{dX}$$

$$\frac{\partial (1/A)}{\partial X} = -\frac{1}{A^2}\frac{dA}{dy}\frac{dy}{dX}$$

The derivatives of area and the moment of area with respect to the flow depth are obtained by taking a small increment in flow depth, Δy, (see Fig. 3.5) as follows

$$\frac{dA}{dy} = \lim_{\Delta y \to 0} \frac{(A + T\Delta y) - A}{\Delta y} = T \qquad (3.14a)$$

$$\frac{d(A\bar{y})}{dy} = \lim_{\Delta y \to 0} \frac{\left[A(\bar{y} + \Delta y) + T\Delta y\frac{\Delta y}{2}\right] - A\bar{y}}{\Delta y} = A \qquad (3.14b)$$

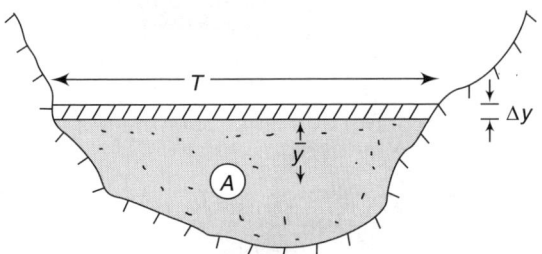

Fig. 3.5 Change in the area and the moment of area with change in flow depth

Note that $d(A\bar{y})/dy$ may also be evaluated by using the Leibniz integral rule[8] but we find it simpler to do it from the first principles. Using these expressions for the derivative, Eq. (3.13) may be written as

---

[8]Leibniz integral rule states that $\dfrac{d}{dt}\displaystyle\int_{a(t)}^{b(t)} f(x,t)\,dx = \int_{a(t)}^{b(t)}\dfrac{\partial f(x,t)}{\partial t}\,dx + f[b(t),t]\dfrac{db(t)}{dt}$

$-f[a(t),t]\dfrac{da(t)}{dt}$. In the present case, $A\bar{y}$ involves an integral [Eq. (3.4)] whose limits are functions of y.

$$\frac{dy}{dX} = \frac{S_0 - S_\tau}{\cos\theta_0 - \dfrac{\beta Q^2 T}{gA^3}} \tag{3.15}$$

This equation relates the rate of change of the flow depth in the longitudinal direction to the flow and channel properties and is called the *gradually varied flow equation* (since it is derived from the equation of motion, it is also called the *dynamic equation* of GVF, and sometimes, it is called the *differential equation* of GVF). In the next subsection, we derive a similar equation by applying the mechanical energy equation.

3.2.3   Application of Energy Equation to GVF

The mechanical energy equation [see Eq. (A.7)] applied to the CV shown in Fig. 3.2 results in the following equation:

$$\iint_{A_1} \left(\frac{p}{\rho} + gz + \frac{v_X^2}{2} \right) \rho v_X \, dA$$

$$= \iint_{A_2} \left(\frac{p}{\rho} + gz + \frac{v_X^2}{2} \right) \rho v_X \, dA + \rho g Q h_L \tag{3.16}$$

where h_L is the head loss (i.e., loss of mechanical energy[9] per unit weight). It should be noted that we have neglected the kinetic energy due to the transverse components of velocity, v_Y and v_Z. This may be interpreted as these components being very small compared to the longitudinal component or their contribution to the kinetic energy being equal at both sections. From Fig. 3.6, it is observed that for the element dA, the pressure, p, and elevation from datum, z, are given by

$$p = \rho g(y - Y)\cos\theta_0, \quad z = z_0 + Y\cos\theta_0$$

$$\Rightarrow \qquad \frac{p}{\rho} + gz = g(z_0 + y\cos\theta_0) \tag{3.17}$$

indicating that the sum of the flow work and specific potential energy is constant over the cross section and may be taken out of the integral.

Equation (3.16) is, therefore, written as

$$\rho g(z_0 + y\cos\theta)Q + \iint_{A_1} \rho \frac{v_X^3}{2} \, dA$$

---

[9]In other words, the mechanical energy that gets converted into thermal energy due to viscous action.

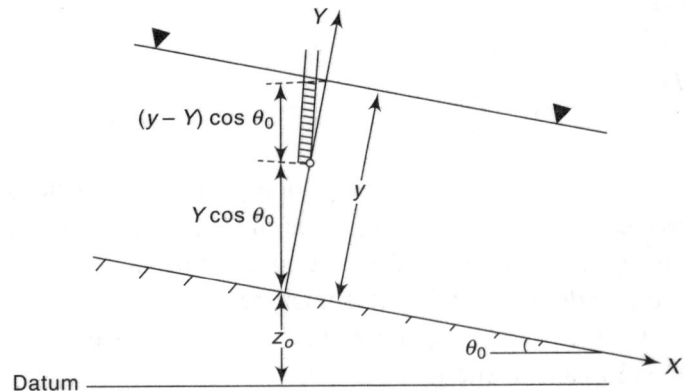

Fig. 3.6   Variation of the pressure and elevation along the Y-axis

$$= \rho g \left[z_0 + \frac{dz_0}{dX} \Delta X + \left(y + \frac{dy}{dX} \Delta X \right) \cos \theta_0 \right] Q$$

$$+ \iint_{A_2} \rho \frac{v_X^3}{2} \, dA + \rho g Q h_L \tag{3.18}$$

We now define an *energy correction factor* (or Coriolis coefficient, named after the French mathematician G. Coriolis who proposed it in 1836) on similar lines as the momentum correction factor [Eq. (3.11)] as

$$\alpha = \frac{1}{A} \iint_A \left(\frac{v_X}{V} \right)^3 dA \tag{3.19}$$

which represents the ratio of the actual kinetic energy over the cross-section to the kinetic energy computed on the basis of average velocity.[10] The integrals in Eq. (3.18) are then written as

$$\iint_A \rho \frac{v_X^3}{2} \, dA = \alpha \frac{\rho A V^3}{2} = \frac{\alpha \rho Q V^2}{2} \tag{3.20}$$

and Eq. (3.18) may be written in terms of *heads* as

$$z_0 + y \cos \theta_0 + \alpha \frac{V^2}{2g} = z_0 + \frac{dz_0}{dX} \Delta X$$

$$+ \left(y + \frac{dy}{dX} \Delta X \right) \cos \theta_0 + \frac{\alpha}{2g} \left(V^2 + \frac{dV^2}{dX} \Delta X \right) + h_L \tag{3.21}$$

---

[10]α is always greater than unity and, for most open channel flows, is less than 1.3. Higher values are obtained in channels with highly nonuniform velocity distribution over the cross-section (e.g., in compound channels). We would, generally, take it as unity, unless otherwise mentioned. For the one-seventh power law distribution used in Example 3.2, the energy correction factor may be easily computed as 1.092.

On the left hand side, the various terms represent the *elevation head*, *pressure head*, and *velocity head* respectively. The sum of the pressure head and the elevation head, which remains same at all points of the cross-section for hydrostatic pressure distribution, as shown in Eq. (3.17), is called the *piezometric head*. Note that we have assumed that the velocity profiles at various sections are similar and, therefore, α is a constant at all sections. Using the chain rule of differentiation and the relationships $dz_0/dX = -S_0$, $dV^2/dy = -2Q^2T/A^3$, and $h_L/\Delta X = S_e$, where S_e is the slope of the energy line (representing the loss of mechanical energy per unit weight of fluid per unit length of channel), Eq. (3.21) becomes

$$\frac{dy}{dX} = \frac{S_0 - S_e}{\cos\theta_0 - \dfrac{\alpha Q^2 T}{gA^3}} \tag{3.22}$$

As we did earlier [Eq. (3.7)], we will assume that the energy slope is equal to that under uniform flow conditions at a flow depth of y and is given by the same expression.

Comparison of Eqs (3.15) and (3.22) brings out the differences in the GVF equations obtained by the applications of the momentum and energy equations: (i) the use of shear slope and energy slope in the numerator, and (ii) the use of momentum and energy correction factors in the denominator. At this point, however, we will make some simplifying assumptions so that both these equations become identical and then proceed towards a solution of this differential equation. The assumptions are as follows:

 (i) The velocity distribution over a cross section is fairly uniform and both the correction factors can be taken as unity.
 (ii) The shear slope and energy slope are equal (called the *friction slope* and denoted by S_f) and are obtained by using the uniform flow resistance formula (Manning's equation).
 (iii) The bed slope of the channel is very small[11] so that cosine of the angle is nearly equal to unity.

---

[11]Even without this assumption, Eqs (3.15) and (3.22) become identical. However, since most cases of open channel flow have small bed slope, we will make this simplifying assumption also. It should also be noted that the figures are drawn to an exaggerated vertical scale creating an impression of large slope. If we use the same scale in horizontal and vertical directions, most channels will appear to be horizontal (since the slope is 1 in 1000 or even smaller than that).

The governing equation for GVF is then

$$\frac{dy}{dX} = \frac{S_0 - S_f}{1 - \dfrac{Q^2 T}{gA^3}} \tag{3.23}$$

This equation forms the basis of analysis of flow profiles under various gradually varied conditions as described in the next section. We will use the interpretation of Eq. (3.23) based on the mechanical energy in the rest of this chapter but the reader should keep in mind the alternative interpretation based on the momentum equation.

3.3   Water Depth Variation, Critical Flow, and Specific Energy

Before we analyse the quantitative aspect of Eq. (3.23) to predict the flow depth at various locations, we should look at a qualitative analysis that gives a rough idea about the nature of variation of depth. One of the first questions that may be asked is whether the flow depth increases (called a *backwater* profile) or decreases (called *drawdown* profile) in the direction of flow. (Some authors use the term *backwater* to represent *all* profiles.) In other words, the question is whether dy/dX would be positive or negative for some given flow condition. In some cases, the answer would be quite obvious. For example, if we build a large dam on a river, we would expect the flow depth upstream of the dam to increase in the direction of flow. However, in a number of cases, the behaviour is not so obvious. We, therefore, look at Eq. (3.23) and see how it can help us in finding the qualitative behaviour of the flow depth.

The numerator $(S_0 - S_f)$ will be positive, zero, or negative, depending on whether the bed slope is larger than, equal to, or smaller than the friction slope. From Eq. (3.7), it is clear that the friction slope, S_f, decreases with increase in flow depth since the section factor[12] increases with increase in flow depth. Some readers may realize that it is not always true. (See Fig. 2.7 for circular channels. This special case will be discussed later.) It is also apparent that when the flow depth is equal to the normal depth, the friction slope will become equal to the bed slope. We may thus conclude that for flow depths greater than normal depth (for example, upstream of a dam) the term $(S_0 - S_f)$ would be positive and for flow depths less than normal depth (for example, flow upstream of a brink), it would be negative. Whether dy/dX

---

[12]The denominator of Eq. (3.7) may be recognized as the square of the section factor.

is positive or negative will also depend on the sign of the denominator, $1 - (Q^2 T/gA^3)$.

The denominator would be positive when $Q^2 T/gA^3 < 1$ and negative when $Q^2 T/gA^3 > 1$. A singularity in the equation of motion is caused when the denominator becomes zero, i.e., when

$$\frac{Q^2 T}{gA^3} = 1 \implies \frac{V}{\sqrt{g\dfrac{A}{T}}} = 1 \tag{3.24}$$

This condition is known as the critical condition[13] and the depth at which it occurs is known as the *critical depth* (denoted by y_c). The term A/T has dimensions of depth and is called the *hydraulic depth*, D. It will be shown in Chapter 7 that $\sqrt{gD}$ is the speed of a small wave in shallow water. The ratio of the flow velocity and the wave speed is called the *Froude number*, F_r, given by

$$F_r = \frac{V}{\sqrt{gD}} = \sqrt{\frac{Q^2 T}{gA^3}} \tag{3.25}$$

The denominator in Eq. (3.23) may, therefore, be written as $1 - F_r^2$ and the critical condition will occur at $F_r = 1$. From Eq. (3.25), one can see that with increase in flow depth the Froude number will increase since (for a given discharge) an increase in flow depth would lead to a decrease in velocity and increase in the hydraulic depth. Therefore, for flow depths greater than critical depth (for example, upstream of a dam), Froude number would be less than one and the denominator would be positive, and for flow depths less than critical depth (for example, flow just downstream of a sluice gate), Froude number would be more than one and the denominator would be negative.

The above discussion leads us to the conclusion that the variation of flow depth in a GVF will primarily depend upon the relationship of the flow depth with the normal depth and the critical depth. The normal depth has been discussed in detail in the previous chapter and the critical depth would be discussed later in this section. At this stage, however, we would introduce an alternative

---

[13]If the bed slope is large and the velocity distribution is nonuniform, the critical condition will be given by $V/\sqrt{gA \cos \theta_0 / \beta T} = 1$ from Eq. (3.15), and a similar expression with α replacing β, from Eq. (3.22). Liggett (1993) argued that the critical condition should be based on the singularity in equation of motion and not on the mechanical energy equation. In this book, we assume that the correction factors are equal to one and the distinction between the two conditions does not arise.

interpretation of the numerator in Eq. (3.23) which will lead to the important concept of specific energy and an alternative definition of the critical depth.

The term $(S_0 - S_f)$ represents the difference in the rate of fall of the bed and that of the energy line (see Fig. 3.7, H representing the total head at a section). We may also conceptualize this term as the rate of increase of the distance (shown as E) between the channel bed and the total energy line. The energy head (i.e., energy per unit weight of fluid), E, occurs frequently in computations of nonuniform flows and is called the *specific energy*. It is defined as the energy head measured with respect to the channel bed and is given by

$$E = H - z_0 = y + \frac{V^2}{2g} = y + \frac{Q^2}{2gA^2} \qquad (3.26)$$

from which,

$$S_0 - S_f = -\frac{dz_0}{dX} + \frac{dH}{dX} = \frac{dE}{dX} \qquad (3.27)$$

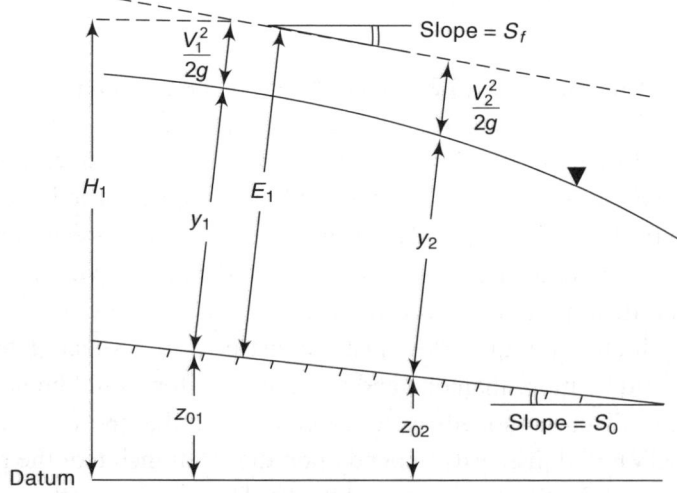

Fig. 3.7 Specific energy in an open channel

Clearly, a positive numerator of the right hand side of Eq. (3.23) implies that the specific energy is increasing with X and vice versa. From Eq. (3.26), it is obvious that for a given discharge, the variation of E with y is not monotonic. Considering the specific energy as the sum of the *pressure head* (y) and the *velocity head* ($V^2/2g$), E becomes very large at very large depths (large pressure head) and also at very small depths (large velocity head). There would be some flow depth at which the specific energy would be a minimum. This depth is obtained by the stationary point theorem[14] using Eq. (3.26) as[15]

$$\frac{dE}{dy} = 0 \;\Rightarrow\; 1 - \frac{Q^2 T}{gA^3} = 0 \;\Rightarrow\; F_r = 1 \tag{3.28}$$

implying that the specific energy for a given discharge would be minimum when the flow occurs at the critical depth. Figure 3.8 illustrates the variation of specific energy with flow depth for a trapezoidal channel (with bed width of 1 m and side slope 1H:1V) for two different discharges. The figure shows the critical point, regions of subcritical flow (Froude number less than 1, i.e., depth more than the critical depth) and supercritical flow (Froude number greater than 1, i.e., depth less than the critical depth), decomposition of specific energy into pressure and velocity heads, and the asymptotic approach of E to y as y becomes very large. Further analysis of the specific energy diagram and its applications are discussed in Chapter 5.

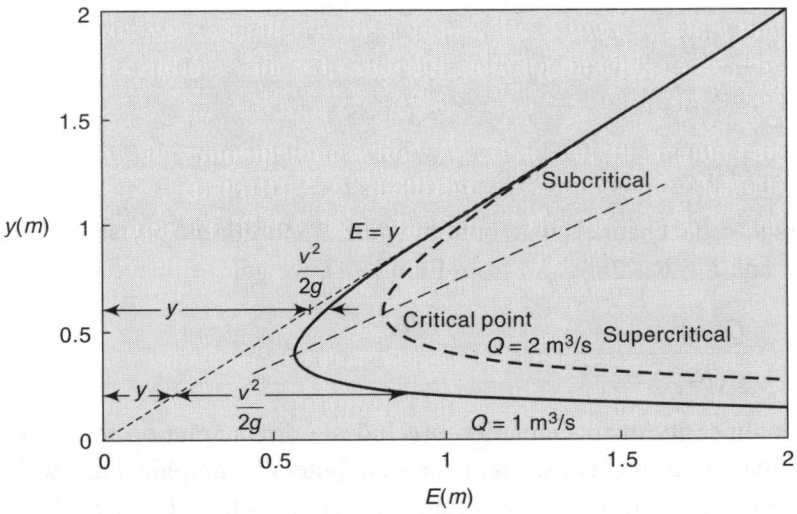

Fig. 3.8 Specific energy at different flow depths for a trapezoidal channel

Since the critical depth is an important parameter in the GVF profile computations, we now look at how to obtain the critical depth in channels of various cross-sectional shapes. The basic condition of criticality remains the

---

[14] If $x_0 \in (a, b)$ is a local extremum of $f(x)$ in (a, b) and f is differentiable at x_0 then $\left. \dfrac{df}{dX} \right|_{x=x_0}$

$= 0$.

[15] We have used the relation $dA/dy = T$ [see Eq. (3.14a)]. Also note that the derivative of E with respect to y could be obtained from Eq. (3.23) by noting that the numerator is dE/dX and, therefore, the denominator must be dE/dy.

same [Eqs (3.24) and (3.28)] but the computation of the critical depth becomes more involved as the shape changes from triangular or rectangular to trapezoidal, circular, or irregular.

Triangular channels

For triangular channels with side slopes of mH:1V, $A = my_c^2$ and $T = 2my_c$. So from Eq. (3.24), we get

$$\frac{Q^2 \, 2my_c}{gm^3 \, y_c^6} = 1 \Rightarrow y_c = \left(\frac{2Q^2}{gm^2}\right)^{1/5} \tag{3.29}$$

Rectangular channels

For rectangular channels with bed width B, $A = By_c$ and $T = B$. So from Eq. (3.24), we get

$$\frac{Q^2 \, B}{gB^3 \, y_c^3} = 1 \Rightarrow y_c = \left(\frac{Q^2}{gB^2}\right)^{1/3} \tag{3.30}$$

Trapezoidal channels

For trapezoidal channels with bottom width B and side slopes mH:1V, $A = By_c + my_c^2$ and $T = B + 2my_c$. So from Eq. (3.24), we get

$$\frac{Q^2 \, (B + 2my_c)}{g(By_c + my_c^2)^3} = 1 \tag{3.31}$$

The nonlinearity of this equation precludes a direct solution for the critical depth for a given discharge. In the pre-computer era, graphical methods were used to solve this problem by defining a section factor[16] for critical flow, Z_c, as

$$Z_c = \sqrt{\frac{A^3}{T}} \left(= \frac{Q}{\sqrt{g}} \right) \tag{3.32}$$

Since the section factor for a given section depends only on the depth, its variation with depth can be plotted. For the given discharge, the required value of Z_c is computed ($= Q/\sqrt{g}$) and the corresponding critical depth is

---

[16]Along the similar lines as the section factor for uniform flow, Z_u [see Eq. (2.15)].

obtained from the plot. As we did in Eq. (2.17), we nondimensionalize the section factor by writing Eq. (3.31) as

$$\frac{m^3 Q^2 (1+2\eta)}{gB^5 (\eta+\eta^2)^3} = 1$$

$$\Rightarrow \qquad \frac{m^{3/2} Z_c}{B^{5/2}} = \sqrt{\frac{(\eta+\eta^2)^3}{1+2\eta}} \left(= \frac{m^{3/2} Q}{g^{1/2} B^{5/2}} \right) \qquad (3.33)$$

where $\eta(= my_c/B)$ is the nondimensional critical depth. A plot of this equation is shown in Fig. 3.9 and its use is illustrated by an example.

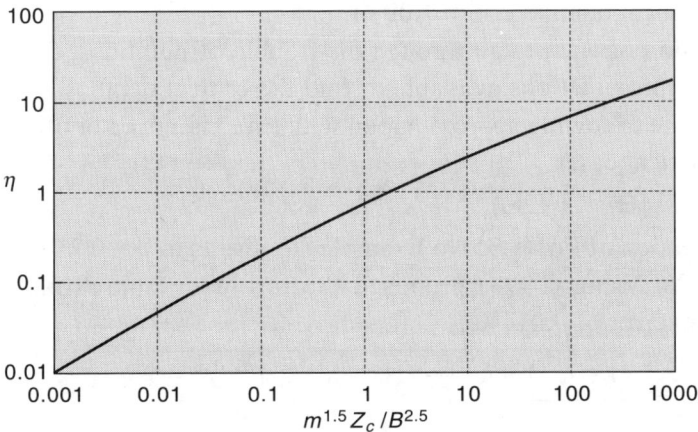

Fig. 3.9   Variation of nondimensional section factor with nondimensional depth for trapezoidal channels

Example 3.3   A trapezoidal channel with bottom width of 5 m and side slopes 1H:1V is laid at a longitudinal slope[17] of 1 in 1000. The channel is lined with brickwork and carries a discharge of 20 m³/s. What would be the critical depth of flow?

Solution

The nondimensional section factor $= \dfrac{m^{3/2} Q}{g^{1/2} B^{5/2}} = \dfrac{1 \times 20}{9.81^{1/2} \, 5^{5/2}} = 0.1142$

From Fig. 3.9, the nondimensional critical depth $\dfrac{my_c}{B} = 0.2$ and the critical depth = **1 m**.

---

[17]The bed slope and Manning's roughness are not needed for computation of the critical depth.

Easy availability of computers has made the graphical methods obsolete. Built-in solvers for nonlinear equations are available in several spreadsheets software. If one wants to incorporate the computation of critical depth in a computer program, the fixed-point or Newton's iterative method can be used. For example, writing a nondimensional discharge as $\Theta = m^{3/2}Q/g^{1/2}B^{5/2}$, Eq. (3.33) may be written as a fixed-point iterative scheme as follows:

$$\eta^{(k+1)} = \frac{\Theta^{2/3}\,[1 + 2\,\eta^{(k)}\,]^{1/3}}{1 + \eta^{(k)}} \tag{3.34}$$

For the problem of Example 3.3, i.e., $\Theta = 0.1142$, starting with $\eta^{(0)} = 0$, we obtain $\eta^{(1)} = 0.2354$, $\eta^{(2)} = 0.2167$, $\eta^{(3)} = 0.2181$, $\eta^{(4)} = \eta^{(5)} = 0.2180$. This gives a critical depth equal to **1.09 m**.

If we do not want to go through the trouble of iterations, there are some explicit approximations available to find the critical depth. Swamee (1993) suggests the following approximation with a maximum error of about 2% for most practical cases:

$$\eta = [\Theta^{-1.4} + (2\Theta)^{-0.84}]^{-0.476} \tag{3.35}$$

Application of Eq. (3.35) to Example 3.3 results in $\eta = 0.2189$, an error of only 0.4%. Wang (1998) proposed the following approximation with a maximum error of 0.014%:

$$\eta = \frac{\sqrt{1 + K\,\{1 + K\,[1 + K\,(1 + K)^{1/5}\,]^{1/6}\,\}^{1/6}} - 1}{2} \tag{3.36}$$

where $K = \dfrac{4\,mQ^{2/3}}{g^{1/3}\,B^{5/3}}$. For Example 3.3, we get $K = 0.9417$ and $\eta = 0.2181$.

Circular channels

For circular channels with diameter D_0,

$$A = \frac{D_0^2\,(\theta_c - \sin\theta_c\,\cos\theta_c)}{4} \quad \text{and} \quad T = D_0\sin\theta$$

where $\theta_c = \arccos\,[1 - (2y_c/D_0)]$ [see Fig. 2.6 and Eq. (2.25)]. From Eq. (3.24), we get

$$\frac{64Q^2\sin\theta_c}{gD_0^5\,(\theta_c - \sin\theta_c\,\cos\theta_c)^3} = 1 \Rightarrow \frac{Z_c}{D_0^{5/2}} = \frac{(\theta_c - \sin\theta_c\,\cos\theta_c)^{3/2}}{8\sin^{1/2}\theta_c}$$

$$\left(= \frac{Q}{g^{1/2}\,D^{5/2}} \right)$$

A plot of the dimensionless section factor with dimensionless depth is shown in Fig. 3.10 and its use is illustrated in Example 3.4. Note that unlike the section factor for uniform flow, Z_c is a monotonic function of the flow depth.

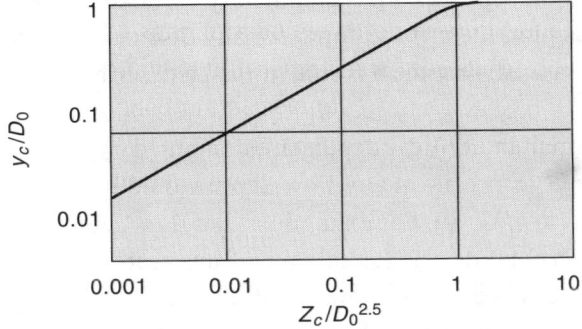

Fig. 3.10 Variation of nondimensional section factor with nondimensional depth for circular channels

---

Example 3.4 A circular channel of diameter 1 m carries a discharge of 0.5 m³/s. What would be the critical depth?

Solution
The nondimensional section factor is obtained as

$$\frac{Z_c}{D_0^{5/2}} = \frac{Q}{g^{1/2} D_0^{5/2}} = \frac{0.5}{9.81^{1/2} \, 1^{5/2}}$$

$$= 0.1596$$

From Fig. 3.10, corresponding $y_c/D_0 = 0.4$ giving a critical depth of **0.4 m** (the exact value is 0.3988 m).

To avoid graphical methods, one could use iterative techniques. However, an explicit approximation proposed by Swamee (1993) provides the critical depth with a maximum error of only about 1.25% and may be written as

$$\frac{y_c}{D_0} = \left[1 + 0.77 \left(\frac{Q}{g^{1/2} D_0^{5/2}} \right)^{-6} \right]^{-0.085} \tag{3.37}$$

For the channel in Example 3.4, we get $y_c = $ **0.4011 m**.[18]

---

[18]An approximation by Braine (1947) is simpler but less accurate: $y_c/D_0 = 0.083 + 1.032$ $(Q/g^{1/2}D_0^{5/2}/2)^{2/3}$ and results in a value of 0.3773 m for the critical depth. We prefer to use Eq. (3.37).

Irregular channels

For channels with irregular cross sections, the graphical method is likely to be the best (and probably the only) method. Variation of the section factor, Z_c, with the flow depth can be plotted graphically by obtaining the area and top width. From this plot, the critical depth for any discharge can be read. We do not describe the procedure since it is similar to that described for the trapezoidal channels.

Once the normal and critical depths are computed, we are in a position to say whether the GVF profile at any flow depth would be backwater ($dy/dX > 0$) or drawdown ($dy/dX < 0$). For example, if the flow depth is more than both the normal and critical depths (as in case of upstream of a dam), we would expect the profile to be a backwater profile since both the numerator and denominator in Eq. (3.23) would be positive ($y > y_n$ implies $S_f < S_0$ and $y > y_c$ implies $F_r < 1$). Obviously there are a number of possible ways in which the actual flow depth is related to the normal and critical depths. To study these in a systematic manner, in the next section we look at a method of classifying the GVF profiles on the basis of the interrelationship between the flow depth, normal depth, and critical depth.

3.4   Classification of GVF Profiles

The critical depth depends on the cross-sectional shape and discharge while the normal depth depends on, in addition to these two, the boundary roughness and bed slope. We consider the case of a laboratory flume with adjustable bed slope to look at the relationship between the normal and critical depths. Assuming that a constant discharge flows through such a channel with a given cross section (say, rectangular) and Manning's n, if we start from a very small bed slope (almost horizontal), the normal depth would be very large. The critical depth, however, will not depend on the bed slope (since we have assumed that discharge is maintained constant) and may be computed using the relevant equation [Eq. (3.30) for rectangular channels]. Since the normal depth is very large, it is reasonable to say that $y_n > y_c$. Now, if we go on increasing the bed slope, the normal depth will keep on decreasing and a stage would be reached when the normal depth becomes equal to the critical depth. This bed slope is known as the *critical slope, S_c*. Further increase in bed slope will cause the normal depth to become smaller than the critical depth. Thus we may define the critical slope as the bed slope for which the combination of discharge, channel shape, and roughness coefficient, will result in the normal depth being equal to the critical depth ($y_n = y_c$). A bed slope smaller than the critical slope ($S_0 < S_c$) is called *mild slope ($y_n > y_c$)* and that larger than the critical slope ($S_0 > S_c$) is called a *steep slope* ($y_n < y_c$). Obviously critical slope condition is very rare (out of the infinitely many values of bed slope, there is just one which is the critical slope). Most natural channels will have mild slope but steep slopes may be encountered in short reaches. There are two other classifications of the bed slope, which are not very common but are included here for the sake of completeness. These are the *Horizontal slope* ($S_0 = 0$) and *adverse slope* ($S_0 < 0$), which do not have a normal depth or we may say that the normal depth is *infinite* for horizontal slopes and a *complex number* for adverse slopes (see Exercise 2.3). These slopes are called *nonsustaining* slopes since the absence of a driving force (the component of weight in the flow direction being zero for horizontal channels and negative for adverse slope channels) implies that uniform flow cannot be sustained. The commonly used two-character classification of the GVF profiles (proposed by a Belgian engineer M. Boudin in 1861) uses the first letter of the slope (*A, H, M, C, S* for adverse, horizontal, mild, critical, and steep slopes) as its first character. The second character indicates the relationship of the flow depth in comparison with the normal and critical depths as described in the next paragraph.

If uniform flow occurs in a channel, the water surface will be parallel to the bed at a distance (measured perpendicular to the bed) of y_n. This line is called the *normal depth line* and is abbreviated as NDL. Similarly, if we draw a line parallel to the bed at a distance of y_c, it can be called the *critical depth line*, CDL. Irrespective of whether the NDL is above CDL or below it, these two lines divide the space above the channel bed into three zones (sometimes, only two zones, as will be discussed later), which are categorized as *zone 1* (the space above the topmost line), *zone 2* (the space between the two lines), and *zone 3* (the space between the channel bed and the lower line). The zone number is used as the second character[19] in the classification of GVF profiles, leading to 15 possible combinations ($A_1, A_2, A_3, H_1, \ldots, S_2, S_3$). Out of these, A_1 and H_1 do not exist because there is no zone 1 on adverse and horizontal slopes, which have no normal depth. The entire space above the CDL in these cases is, therefore, classified as zone 2. Similarly, the C_2 profile does not exist (or, even if it exists, it is not GVF but uniform flow) because for channels with bed slope equal to the critical slope, the NDL and CDL coincide and zone 2 collapses into a line. The space above this line is zone 1 and below it, zone 3. Figure 3.11 shows various zones for different slopes of the channel.

As stated earlier, the most common channel slopes are mild and steep. We would describe the profiles on these two slopes (*M, S*) and mention the others

---

[19]We will write the second character as a subscript, e.g., M_1. Some authors write the second character in full size, e.g., *M*1, as originally proposed by Boudin.

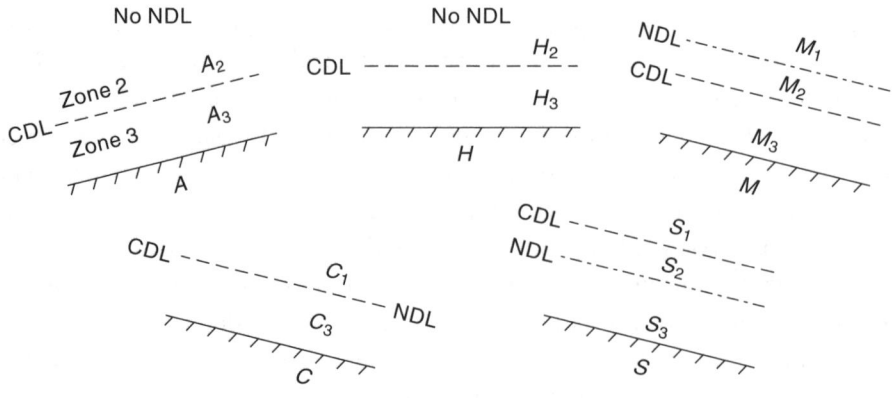

Fig. 3.11   Gradually varied flow zones

(A, H, C) very briefly.[20] The classification of profile gives us a fairly quick idea of the variation of the flow depth. For example, if a profile is classified as M_1, it indicates that the channel slope is mild (the first character being M), i.e., $y_n > y_c$, and the flow depth is more than both the normal and critical depths (because it is in zone 1), i.e., $y > y_n > y_c$. As discussed earlier, the flow depth would be increasing in the direction of flow ($dy/dX > 0$) since both $S_0 - S_f$ and $1 - F_r^2$ are positive. However, this information is not sufficiently detailed to enable us to sketch a qualitative water surface profile. For example, we do not know whether the profile is straight, concave, or convex; or how does it behave as it nears the normal depth or critical depth. Information about these aspects may be obtained from the differential equation of GVF as discussed next.

3.4.1   Salient Features of the Surface Profile

The gradually varied flow equation [Eq. (3.23)] is reproduced below in a slightly modified form

$$\frac{dy}{dX} = \frac{S_0 - S_f}{1 - F_r^2} \tag{3.38}$$

From this, the following inferences may be drawn about the GVF profile:
(i) *Near normal depth:* As $y \to y_n$, $S_f \to S_0$ and $dy/dX \to 0$. This implies that the water surface tends to become parallel to the bed. In other words, the flow depth approaches the normal depth asymptotically. Thus theoretically, the flow depth will never reach the normal depth. Therefore, the

---

[20]The profile for a horizontal slope may be inferred as a limiting case of a very mild slope and that for the critical slope would correspond to the upper limit of the mild slope and lower limit of the steep slope. The adverse slope also has the horizontal slope as its limiting case.

GVF computations are terminated when the flow depth is within 1% of the normal depth, i.e., 0.99 y_n when approaching from below (e.g., drawdown at a free overfall) and 1.01 y_n when approaching from above (e.g., backwater behind a dam). Beyond this depth, the flow is assumed to be uniform.

(ii) *Near critical depth:* As $y \rightarrow y_c$, $F_r \rightarrow 1$ and $dy/dX \rightarrow \infty$. This implies that the water surface tends to become vertical. However, the high curvature of streamlines as the flow depth approaches the critical depth renders Eq. (3.38) inapplicable since it is based on a hydrostatic distribution of pressure.[21]

(iii) *At large depths:* As $y \rightarrow \infty$, both S_f and $F_r \rightarrow 0$, and $dy/dX \rightarrow S_0 (=-dz_0/dX)$. This implies that $(y + z_0)$ approaches a constant value. In other words, the water surface tends to become horizontal as the GVF profile approaches a large depth (e.g., the depth upstream of a dam).

(iv) *At small depths:* As $y \rightarrow 0$, both S_f and $F_r \rightarrow \infty$, and $dy/dX \rightarrow S_f/F_r^2$. Using the Manning's equation to compute S_f, we may write

$$\frac{S_f}{F_r^2} = \frac{n^2 Q^2}{A^2 R^{4/3}} \frac{gA^3}{Q^2 T} = \frac{gn^2 P^{4/3}}{A^{1/3} T} \tag{3.39}$$

Thus for a rectangular channel, dy/dX approaches infinity at very small depths.[22] That is, the water surface becomes vertical[23] near the channel bed.[24]

(v) *Slope of the water surface:* A positive slope of the water surface (backwater curve) is obtained when both the numerator and denominator in

---

[21]Sometimes, it is argued that Eq. (3.38) should be written with the denominator moved from the right hand side to the left hand side to avoid the singularity at the critical depth. We continue to use the conventional form of the GVF equation keeping in mind that it is not valid near the critical depth.

[22]It is clearly true for any channel in which the wetted perimeter does not become zero as depth approaches zero. Even for a triangular channel, where P approaches zero, it may be shown that this statement is true.

[23]Since the water surface is vertical near the bed as well as at the critical depth, there would be a point of inflection near the channel bottom when $y < y_c < y_n$. Some studies have also suggested the presence of another point of inflection for $y > y_n > y_c$. However, since it is not very crucial in the computation of GVF profiles, we will not discuss this point further.

[24]If we use Chezy's equation, $dy/dX \rightarrow gP/C^2T$. Note that P/T approaches a constant value at small depths (e.g., for rectangular channels, the constant value is 1, and for triangular channels, it is $\sqrt{1 + m^2}/m$). However, since we prefer to use the Manning's equation and flow at very small depths is not very common, we do not discuss this aspect further.

Eq. (3.38) are of the same sign and a drawdown curve (negative dy/dX) is obtained when these are of opposite signs. We have already discussed that the numerator is positive when $y > y_n$ and the denominator is positive when $y > y_c$. We have also seen that zone 2 is the space between the two lines, the NDL and the CDL. Therefore, no matter which depth, normal or critical, is larger, profiles in zone 2 would have opposite signs for the numerator and denominator,[25] i.e., a drawdown curve. On the other hand, the profiles in zone 1 (y greater than both y_n and y_c, i.e., both the numerator and denominator positive) and zone 3 (y less than both y_n and y_c, i.e., both the numerator and denominator negative) will always be backwater.

(vi) *Curvature of the water surface:* The curvature of the water surface is decided by the value of the second derivative, d^2y/dX^2. A positive value will indicate increasing slope in the direction of flow, i.e., a concave (looking from above) profile. Similarly, a negative value for the second derivative implies a convex profile and a value of zero indicates a straight surface. Since the analysis for a general case becomes quite complicated, we analyse a simplified case here. (The results are valid for a general case also since we are only looking at the qualitative behaviour of the second derivative and not its actual value.)

We assume that both S_f and F_r^2 are proportional to y^{-3} (the constant of proportionality is obtained from the uniform flow and critical flow conditions, respectively).[26] We also assume that the channel slope is positive (i.e., horizontal and adverse slopes are excluded) and flow depth is non-critical, although extension to other cases is straightforward (see Exercise 3.6). Then,

$$\frac{d^2y}{dX^2} = \frac{d}{dy}\left[\frac{S_0 - (S_0 y_n^3) y^{-3}}{1 - (y_c^3) y^{-3}}\right] \times \frac{dy}{dX} = S_0 \frac{dy}{dX}\frac{d}{dy}\left[\frac{y^3 - y_n^3}{y^3 - y_c^3}\right]$$

$$= S_0 \frac{dy}{dX}\frac{3y^2 (y_n^3 - y_c^3)}{(y^3 - y_c^3)^2} \tag{3.40}$$

---

[25]For $y_n > y_c$ (i.e., a mild slope channel), zone 2 will have $y < y_n$ (numerator negative) and $y > y_c$ (denominator positive). For a steep channel, zone 2 will have $y > y_n$ (numerator positive) and $y < y_c$ (denominator negative).

[26]For a very wide rectangular channel, the friction slope is proportional to $y^{-10/3}$ and the square of the Froude number is proportional to y^{-3}. The assumption is, therefore, reasonably good.

The sign of the second derivative would, therefore, be the same as that of dy/dX if the normal depth is more than the critical depth (mild slope) and opposite to it when the normal depth is less than the critical depth (steep slope). For critical slopes ($y_n = y_c$), the curvature approaches zero indicating a fairly straight water surface.

To summarize these features of the flow profile, it would be instructive at this stage to plot the variation of the rate of change of flow depth along the longitudinal direction, dy/dX, with respect to the flow depth. Figure 3.12 shows a qualitative behaviour for a mild and a steep channel and Example 3.5 illustrates the quantitative aspects.

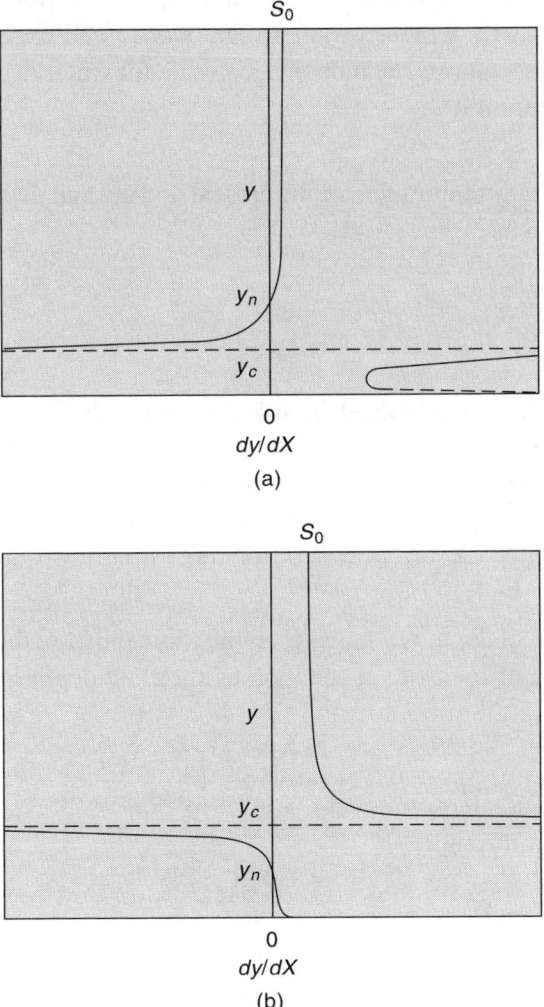

Fig. 3.12 Variation of the water depth gradient for (a) mild channel and (b) steep channel

From Fig. 3.12, it is clearly seen that
   (i) $dy/dX = 0$ at $y = y_n$;
   (ii) dy/dX is very large ($\pm\infty$) at $y = y_c$;
   (iii) dy/dX is equal to S_0 at large flow depths;
   (iv) dy/dX may become very large at small depths;
   (v) zones 1 and 3 have positive dy/dX and zone 2 has negative dy/dX.
Example 3.5 describes some quantitative aspects and also introduces the concept of *transitional depth*.

Example 3.5   A 2 m wide rectangular channel carries a discharge of 4 m³/s and is laid at a bed slope of 0.02. The Manning's roughness for the channel may be taken as 0.015. Plot the variation of the slope dy/dX expected at different water depths. Also enlarge the plot on a log scale for water depth range of 1 to 100 and comment on it.

Solution
The first step is the computation of the critical and normal depths. The critical depth is obtained as

$$y_c = \left(\frac{Q^2}{gB^2}\right)^{1/3} = 0.74 \text{ m}$$

and the normal depth is obtained from Eq. (2.24) with $\Theta = \left(\frac{nQ}{B^{8/3} S_0^{1/2}}\right)^{0.6} =$

0.1972 and $\eta_0 = \Theta(1 + 1.2\Theta)^{0.7826} = 0.2329$ as

$$y_n = B \times \frac{(1.2\eta_0 + 1)\Theta}{(2\eta_0 + 1)^{0.6} - 0.8\Theta} = 0.4587 \text{ m}$$

   Since the normal depth is smaller than the critical depth, the channel bed is steep. The value of dy/dX as a function of the flow depth is obtained from Eq. (3.23) as

$$\frac{dy}{dX} = \frac{S_0 - S_f}{1 - \dfrac{Q^2 T}{gA^3}} = \frac{0.02 - \dfrac{0.015^2 4^2 (2 + 2y)^{4/3}}{(2y)^{10/3}}}{1 - \dfrac{4^2 2}{9.81(2y)^3}}$$

$$= \frac{0.02 - 3.572 \times 10^{-4} \dfrac{(2 + 2y)^{4/3}}{y^{10/3}}}{1 - \dfrac{0.4077}{y^3}} \tag{3.41}$$

Figure 3.13 shows a plot of this equation and the normal depth (around 0.5 m), the critical depth (around 0.7 m) and the bed slope (about 0.02) are readily seen.

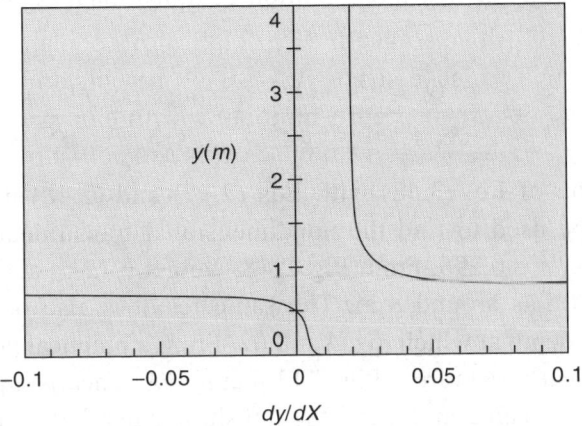

Fig. 3.13   *dy/dX* at various flow depths for the given rectangular channel

Figure 3.14 shows a log-scale plot of *dy/dX* and clearly shows that while *dy/dX* is equal to the bed slope, S_0, and the water surface is horizontal at very large depths, there is another depth at which the water surface becomes horizontal. This depth is called the *transitional depth* and has been used to argue for the existence of a point of inflection on the water surface. However, one should note the scale used for *dy/dX* and the extremely small differences in its values. Therefore, as far as computations of GVF profiles are concerned,

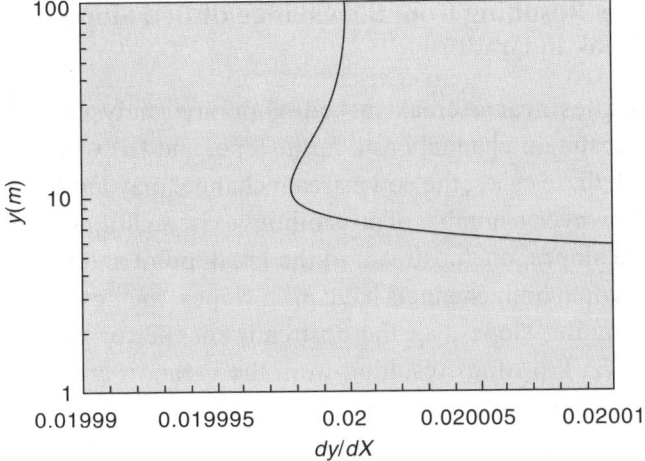

Fig. 3.14   Variation of *dy/dX* showing a point of inflection

the transitional depth is more of theoretical interest rather than of practical utility.[27] Mathematically, at the transitional depth, $dy/dX = S_0$, implying that $S_f/S_0 = F_r^2$. Using an analysis similar to the generalized Froude number [see Eq. (2.48) and Fig. (2.18)], the condition at transitional depth, y_t, can be written as

$$\frac{S_f}{S_0} = F_r^2 \implies \frac{n^2 Q^2 P^{4/3}}{S_0 A^{10/3}} = \frac{Q^2 T}{gA^3} \implies \frac{ng^{1/2}}{S_0^{1/2}} = \frac{T^{1/2} A^{1/6}}{P^{2/3}} \tag{3.42}$$

A comparison of Eq. (3.42) with Eqs (2.47) and (2.48) indicates that Fig. 2.18 can be used to find the nondimensional transitional depth with $F_* = ng^{1/2}/B^{1/6}S_0^{1/2} = 0.296$. From Fig. 2.18, we get y_t/B as about 4 and the transitional depth as around 8 m. The same result is also obtained from Fig. 3.14 as the depth at which $dy/dX = 0.02$. From a practical point of view, however, it is clearly seen from Fig. 3.13 that dy/dX is nearly equal to S_0 for flow depths larger than 2 m. Hence, the transitional depth does not play any significant role in GVF computations.

Using the characteristics of the flow profile discussed in this sub-section, we now illustrate various possible profiles on mild- and steep-slope channels with the help of a long prismatic channel followed by another long prismatic channel, but with a different bed slope. We believe this case, though not very common in practice, would be a good way of introducing the profiles. We will then move on to discuss more practical cases where different GVF profiles are observed.

3.4.2 Profiles Resulting from the Change of Bed Slope (or Break in Grade)

The GVF profiles near a break in bed slope are analysed for two cases: (i) when the upstream channel has a mild slope, and (ii) when it has a steep slope. For both these cases, the downstream channel may have a mild slope or steep slope. However, a number of possibilities exist regarding the interrelation of the two bed slopes, one upstream of the break point and the other after it. For example, when both channels have mild slopes, the downstream channel may have a smaller slope than the upstream channel or vice versa. These cases and the GVF profiles resulting from the break in grade are described further.

---

[27]In Chapter 6 on Spatially Varied Flow, we will re-visit the issue of transitional depth where it does find some application.

Mild slope followed by another mild slope

The three possibilities in this case include the downstream slope being (i) same as, (ii) smaller than, or (iii) larger than (but still mild, i.e., normal depth remains above the critical depth.) the upstream slope. In the first case, the flow would be uniform and there would not be any GVF profile. The other two cases are analysed below.

Figure 3.15 shows a mild slope channel followed by another mild slope channel, which is flatter than the first channel. We will use the terms *mild* and *milder* to denote two channels which are both mild but the milder channel has a smaller slope than the mild channel. Similarly, we use the terms *steep* and *steeper* to denote two channels which are both steep but the steeper channel has a larger slope than the steep channel. For a mild channel followed by a milder channel, the normal depth becomes larger after the junction or the break point (note that the critical depth remains the same since discharge is the same.) At a large distance, for both upstream and downstream of the break, the flow depth will be uniform (since the channels are assumed to be very long).

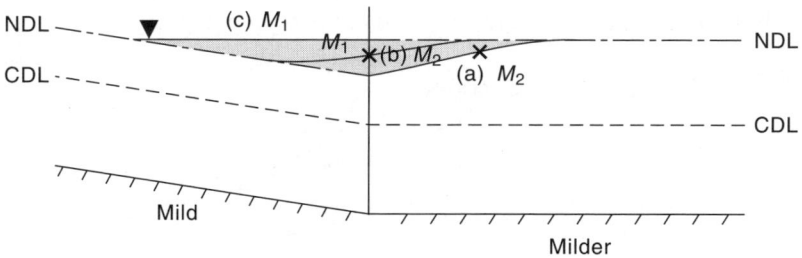

Fig. 3.15   GVF profiles for a mild slope followed by a milder slope

As shown in Fig. 3.15, there can be a number of possible profiles that may take the water surface from a depth equal to the normal depth in the first channel to the higher normal depth in the second channel. The three zones of flow are also shown on the figure and the possible profiles are (a) an M_2 profile on the second channel, (b) an M_1 profile on the first channel followed by an M_2 profile on the second, and (c) an M_1 profile on the first channel. However, the profile (a) is ruled out because profiles in zone 2 are always drawdown. Thus water depth cannot rise in the flow direction in zone 2. We may also interpret the infeasibility of the profile (a) on the basis of specific energy as follows:

Since flow depth is less than the normal depth, the friction slope would be larger than the bed slope, and the specific energy will decrease in the direction of flow. Moreover, since the flow depth is more than the critical depth, the flow is subcritical and the flow depth must decrease if the specific energy decreases (see Fig. 3.8). Therefore, the profile (a) is not possible.

We may similarly show that the profile (b) is also not feasible and the only possible GVF profile is (c), i.e., an M_1 profile on the first channel. From Eq. (3.40), since the normal depth is more than the critical depth, d^2y/dX^2 will also be positive indicating a concave upward profile. Also, the GVF profile will approach the normal depth on the first channel asymptotically. However, note that it will not approach the water depth in the second channel asymptotically since the profile is limited to the first channel and $S_0 \neq S_f$ as the flow approaches the junction. If the profile occurs on the second slope, as described a little later, it will approach its normal depth asymptotically. Also, if the second channel has a very small slope (i.e., a large normal depth), the water surface will approach it horizontally ($dy/dX = S_0$).

Figure 3.16 shows a mild slope channel followed by another mild slope channel, which has a larger slope than the first channel. Following our naming convention, we call it a milder slope followed by a mild slope. Arguments similar to those in the previous case show that the only possibility in this case is an M_2 profile on the first channel, which will be convex upwards and will approach the normal depth on the first channel asymptotically.

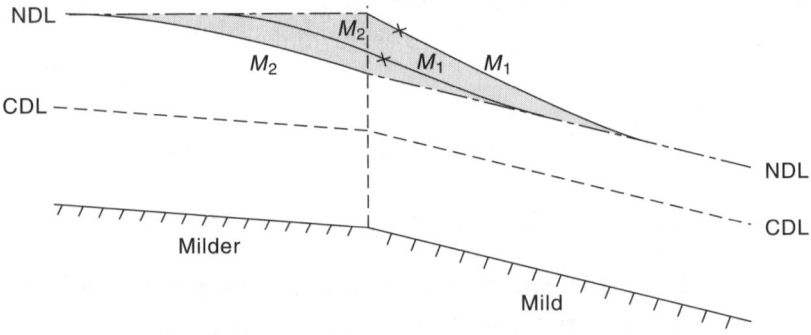

Fig. 3.16 GVF profiles for a milder slope followed by a mild slope

Mild slope followed by a steep slope

Figure 3.17 shows a mild slope channel followed by a steep slope channel. The three zones of flow are also shown in the figure and the possible profiles are (a) M_2 and M_3 profiles on the mild slope, (b) an M_2 profile on the mild slope followed by an S_2 profile on the steep slope, and (c) S_1 and S_2 profiles on the steep slope. Again, the profile (a) is ruled out because M_3 cannot be a drawdown curve and the profile (c) is also ruled out because S_1 cannot be drawdown. The only possibility is the profile (b), i.e., an M_2 profile on the mild slope followed by an S_2 profile on the steep slope, with the critical depth

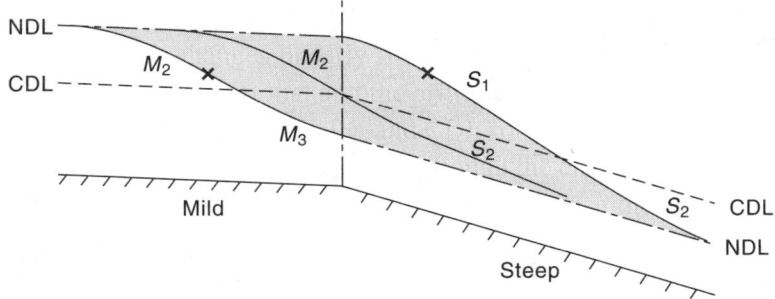

Fig. 3.17 GVF profiles for a mild slope followed by a steep slope

occurring at the junction. The profile will asymptotically approach the normal depths on both the channels and will be convex upwards on the mild slope and concave upwards on the steep slope.[28] It should be noted that the depth at the junction will not be equal to the *theoretical* critical depth (as obtained in Section 3.3) for the given channel shape because of the large curvature of streamlines when the flow depth is close to the critical depth. This depth would still correspond to the minimum specific energy but the pressure head in the specific energy will not be equal to the flow depth. More discussion on this topic is given in Chapter 4 (Section 4.4), where the flow in a free overfall[29] is described.

Steep slope followed by another steep slope

The three possibilities in this case include the downstream slope being (i) same as, (ii) larger than, (iii) smaller than the upstream slope. In the first case, the flow would be uniform and there would not be any GVF profile. The other two cases are analysed below.

Figure 3.18 shows a steep channel followed by a steeper channel. The three possible profiles are (a) an S_2 profile on the second channel, (b) an S_3 profile on the first channel followed by an S_2 profile on the second, and (c) an S_3 profile on the first channel. In this case, both the profiles (b) and (c) are ruled out because S_3 cannot be drawdown and the profile (a) is the only possibility. The profile will approach the normal depth on the steeper channel asymptotically and will be concave upwards.

Figure 3.19 shows a steeper channel followed by a steep channel. The three possible profiles are (a) an S_2 profile on the first channel, (b) an S_2 profile on

---

[28]From Eq. (3.49), since dy/dX is negative, d^2y/dX^2 will be negative for mild channel ($y_n > y_c$) and positive for steep channel ($y_n < y_c$).

[29]The free overfall in a mild slope channel may be thought of as a limiting case of a mild channel followed by a *very steep* channel.

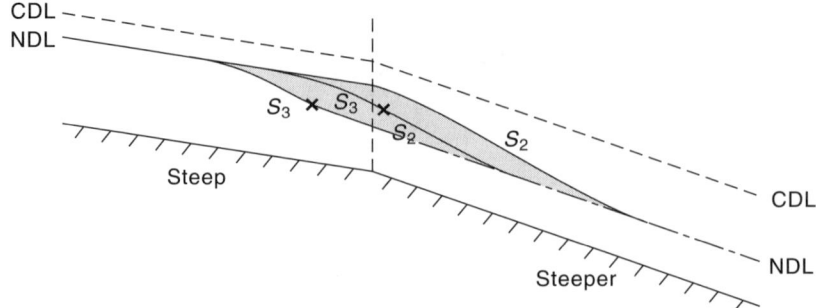

Fig. 3.18 GVF profiles for a steep slope followed by a steeper slope

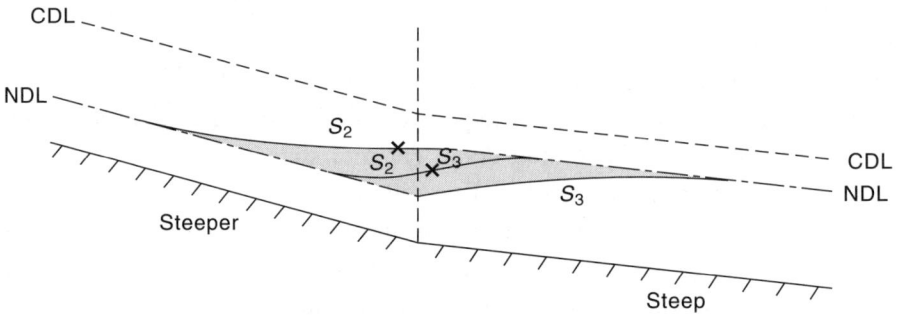

Fig. 3.19 GVF profiles for a steeper slope followed by a steep slope

the first channel followed by an S_3 profile on the second, and (c) an S_3 profile on the second channel. In this case, both the profiles (a) and (b) are ruled out because S_2 cannot be a rising profile and the profile (c) is the only possibility. The profile will approach the normal depth on the steep channel asymptotically and will be convex upwards.

Steep slope followed by mild slope

Figure 3.20 shows a steep channel followed by a mild channel. We may think of three possible profiles as shown in the figure as (a) S_2 and S_1 profiles on the steep channel, (b) an S_2 profile on the steep channel followed by an M_2 profile on the mild, and (c) M_3 and M_2 profiles on the mild slope. It turns out that none of these profiles is feasible because neither S_2 nor M_2 can be a rising curve. In this case, a rapidly varied flow phenomenon of hydraulic jump occurs and we will deal with this particular case of GVF in the next chapter after discussing the hydraulic jump in a little more detail.

The profiles for horizontal channels, or channels with adverse or critical slopes are similar to those for the mild and steep channels. Figure 3.21 shows a few of these profiles, but without detailed explanation.

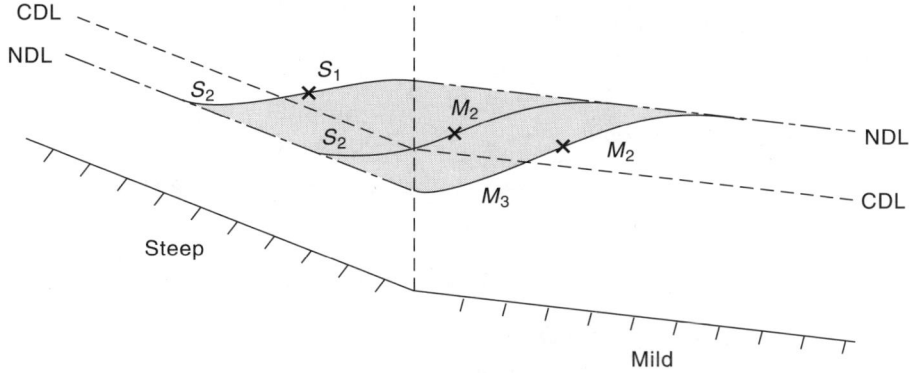

Fig. 3.20 GVF profiles for a steep slope followed by a mild slope

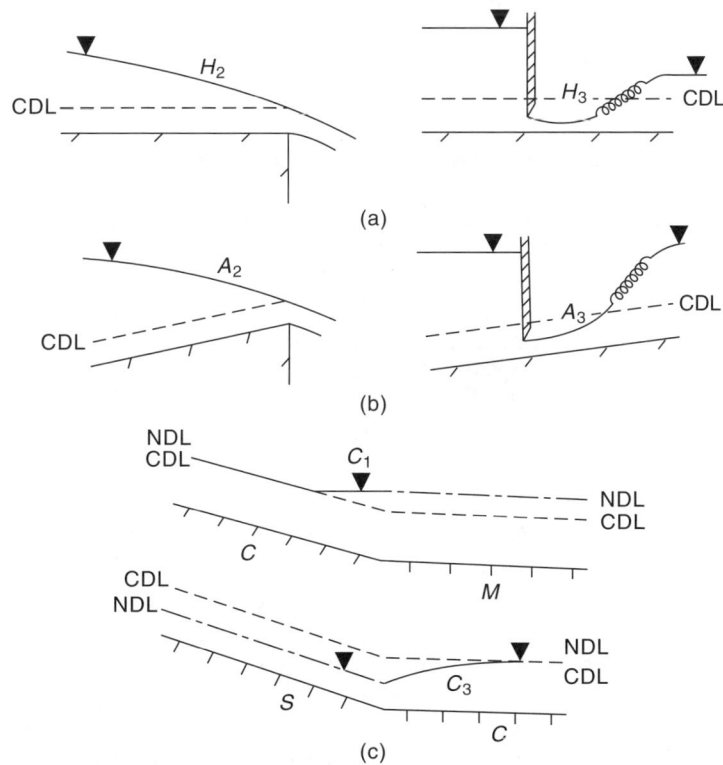

Fig. 3.21 GVF profiles in horizontal, adverse, and critical slope channels

files occurring under a variety of flow conditions. Figure 3.22 shows some situations frequently encountered in laboratory or field conditions and the GVF profiles obtained under these conditions and Table 3.1 summarizes the characteristics of various GVF profiles. Our goal, however, is to perform a quantitative analysis of the flow profiles, which is described in the next section.

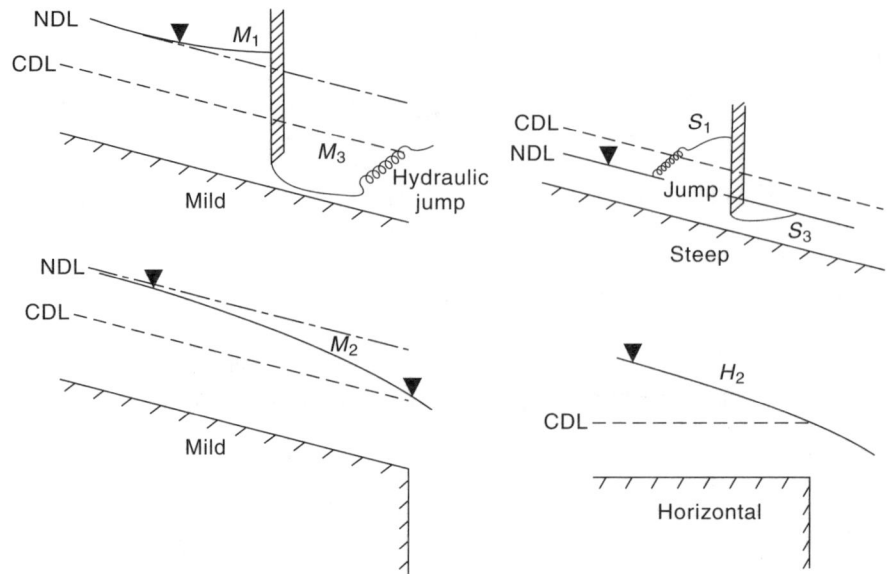

Fig. 3.22 Practical examples of gradually varied flow profiles

Table 3.1 Nature of water surface under gradually varied flow conditions

Bed slope	Zone	Classification	Flow depth relation	Nature of curve	Curvature (looking from the top)
Adverse	2	A_2	$y > y_c$	Drawdown	Convex
	3	A_3	$y < y_c$	Backwater	Concave
Horizontal	2	H_2	$y > y_c$	Drawdown	Convex
	3	H_3	$y < y_c$	Backwater	Concave
Mild	1	M_1	$y > y_n > y_c$	Backwater	Concave
	2	M_2	$y_n > y > y_c$	Drawdown	Convex
	3	M_3	$y_n > y_c > y$	Backwater	Concave
Critical	1	C_1	$y > y_c (= y_n)$	Backwater	Fairly straight
	3	C_3	$y < y_c (= y_n)$	Backwater	Fairly straight
Steep	1	S_1	$y > y_c > y_n$	Backwater	Convex
	2	S_2	$y_c > y > y_n$	Drawdown	Concave
	3	S_3	$y_c > y_n > y$	Backwater	Convex

3.5 Computations for Prismatic Channels

Computation of the water surface profile under gradually varied flow conditions involves the solution of Eq. (3.38), or any of its alternative forms, using analytical, graphical, semi-analytical, or numerical techniques. Since this is a first-order ordinary differential equation, its solution requires that one boundary condition be specified. This condition is generally in the form of a *known* depth (normal depth, critical depth, or any other given depth)[30] at a section. This section is called the *control section* and the computations normally proceed from this section towards the other section for which either the depth of flow is known (and one is asked to find the distance) or the distance from the control section is given (and one is asked to find the flow depth). Since the water surface approaches the normal depth asymptotically, typically we will not use a *uniform flow section* as a control section although there is a known relationship existing between the depth and the discharge. For supercritical flows, since a disturbance cannot move upstream, the control section for a given channel reach has to be an *upstream control*. Subcritical flows generally have a downstream control. In this section, we consider only prismatic channels, i.e., the bed slope and the cross section remains same and dy/dX is only a function of y and not X [see Eq. (3.38)].

The easiest method to solve Eq. (3.38) is, naturally, an analytical solution of the differential equation. However, it may not always be possible to do so. In fact, the nonlinear relationship between the friction slope and flow depth and between the Froude number and flow depth is such that the analytical solution is not possible except under very restrictive simplifying assumptions. The advantage of analytical solution, if it can be obtained, is that it is very quick and is very useful in analysing the sensitivity of the flow profiles to various parameters such as the bed slope, discharge, etc. Easy availability of computing facilities has made numerical solutions very useful. Even then, the analytical solutions are useful as a quick validation tool for these numerical models.

For most practical cases, it is difficult, or even impossible, to obtain analytical solutions. Semi-analytical methods, which perform analytical integration up to a certain stage and then use numerical integration or tables of functions, have been commonly used in such situations. The graphical methods can also be used to integrate the differential equation by plotting the derivative and measuring its area graphically. These involve tedious plotting as well as computation of the area and are not very suitable for the current day scenario where numerical methods have become very quick and reliable. We

---

[30]For example, a sluice gate opening may be given, which will fix the flow depth. Sometimes, we may be asked to find out the distance between two sections where the flow depths are specified with values other than the normal and the critical depth. In this case, either of these sections may be taken as the starting section for the computations. It is generally recommended that, to reduce the round-off errors in the solution, the computations should proceed upstream for subcritical flow and downstream for supercritical flow.

still describe all these methods but concentrate mostly on the numerical methods of solving Eq. (3.38).

The analytical and semi-analytical methods are based on the so-called hydraulic exponents which relate the section factors for uniform and critical flow with the flow depth. Since it has been shown that the definition of these exponents has some inconsistencies (Subramanya and Ramamurthy 1974), we will present a very brief description here and recommend against the use of the exponents. However, we would use them at a few places for simplifying the analysis of the GVF profiles. For very wide rectangular channels,[31] the section factors are written as

$$Z_u^2 = A^2 R^{4/3} = B^2 y^{10/3} \ (= C_u y^N)$$
$$Z_c^2 = A^3/T = B^2 y^3 \ (= C_c y^M) \tag{3.43}$$

where N is the hydraulic exponent for uniform flow, M is the hydraulic exponent for critical flow, and C_u and C_c are constants (e.g., for wide rectangular channels, from the expressions above, $C_u = C_c = B^2$, $N = 10/3$ and $M = 3$). It has been assumed that these constants are not functions of flow depth. From Eq. (3.43), it is seen that for wide rectangular channel, this assumption is true. However, for most practical cases, the coefficients are dependent on the flow depth and Eq. (3.43) is not consistent (see Patil, et al. 2001). The values of M and N are obtained by using Eq. (3.43). For example, to obtain the exponent M, we write

$$C_c y^M = Z_c^2 = A^3/T$$
$$\ln C_c + M \ln y = 3 \ln A - \ln T \tag{3.44}$$

and then differentiate this expression with respect to y to get an approximate (because we assume that C_c and M are not functions of y) relation

$$M = \frac{3Ty}{A} - \frac{y}{T}\frac{dT}{dy} \tag{3.45}$$

For a trapezoidal channel, a plot of M versus nondimensional flow depth, my/B, is shown in Fig. 3.23. Note that for rectangular channels, we get $M = 3$ and for triangular channels, $M = 5$.

Similarly, it can be shown that the hydraulic exponent for uniform flow, N, is given by

$$N = \frac{10}{3}\frac{Ty}{A} - \frac{4}{3}\frac{y}{P}\frac{dP}{dy} \tag{3.46}$$

---

[31]For other shapes also we can perform this analysis but the exponent will be function of flow depth.

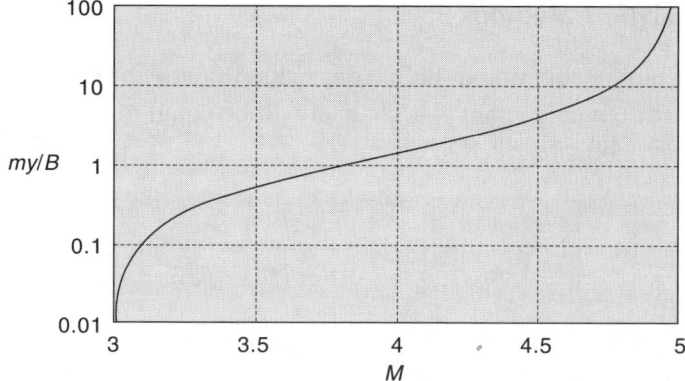

Fig. 3.23 Variation of the hydraulic exponent, *M*, for a trapezoidal channel

A plot of *N* versus *y/B* for a trapezoidal channel is shown in Fig. 3.24.

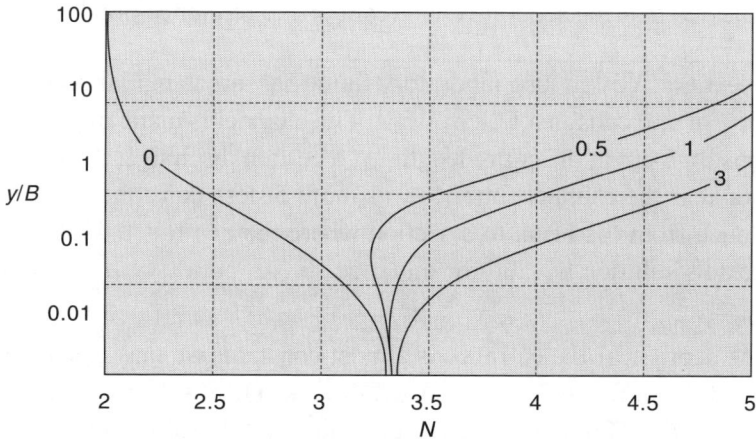

Fig. 3.24 Variation of the hydraulic exponent, *N*, for a trapezoidal channel. The numbers on the curve represent the side slope, *m* (0 indicating a rectangular channel)

By using the conditions for section factor at the normal depth [Eq. (2.15)] and critical depth [Eq. (3.43)], Eq. (3.38) may be written as

$$\frac{dy}{dX} = S_0 \frac{1 - \left(\dfrac{y_n}{y}\right)^N}{1 - \left(\dfrac{y_c}{y}\right)^M} \tag{3.47}$$

This simplification allows us to solve Eq. (3.38) using analytical or semi-analytical methods.

3.5.1 Analytical Methods

If we use Chezy's equation to define the section factor for uniform flow, the value of N for a wide rectangular channel will be equal to 3. For $N = M = 3$, Eq. (3.47) may be integrated to obtain

$$X = \frac{y_n}{S_0} \left\{ u - \left[1 - \left(\frac{y_c}{y_n} \right)^3 \right] \left[\frac{1}{6} \ln \frac{u^2 + u + 1}{(u-1)^2} - \frac{1}{\sqrt{3}} \arctan \frac{\sqrt{3}}{2u+1} \right] \right\} + C_I$$

(3.48)

in which u is the normalized depth, y/y_n, C_I is a constant of integration, and X represents the distance[32] to a section where the flow depth is y. This is known as the Bresse's solution after the French engineer, J.A.C. Bresse, who reported it in 1860. Note that X-axis is taken in the flow direction and a negative value of X indicates that the section is upstream of the reference point.

Example 3.6   A 50 m long laboratory flume has a rectangular section with a width of 2 m and ends in a free overfall. The channel is made of glass and the bed drops by 5 cm in the entire length. At a certain discharge, it was seen that the depth near the channel entrance was more or less constant at 0.5 m. Find the distance from the brink to a section where the depth is 0.45 m assuming that Bresse's solution is valid for this case.

Solution
Since the depth near the entrance is almost constant, we may assume it to be the normal depth for the given flow conditions. The discharge in the channel is obtained from Manning's equation (using $n = 0.01$ for glass) as

$$Q = \frac{1}{0.01}(2 \times 0.5) \left(\frac{2 \times 0.5}{2 + 2 \times 0.5} \right)^{2/3} \left(\frac{0.05}{50} \right)^{1/2} \text{ m}^3/\text{s} = 1.520 \text{ m}^3/\text{s}$$

and the critical depth is [Eq. (3.30)], $y_c = \left(\frac{1.520^2}{2^2 \times 9.81} \right)^{1/3}$ m $= 0.3891$ m

We assume that the end depth, i.e., the depth at the brink, is equal to the critical depth and compute the normalized flow depths at the two sections as
   $u_1 = 0.3891/0.5 = 0.7782$ and $u_2 = 0.45/0.5 = 0.9$
Using Eq. (3.48), we get
   $X_1 = 309.0 + C_I$ and $X_2 = 287.7 + C_I$

---

[32]Distance is measured from an arbitrary reference point (C_I is arbitrary). Since we are interested in the distance between the two sections where the depths are known, the difference in the values of X will give the distance (C_I will cancel out).

and the distance between the two sections as

$$X_2 - X_1 = -21.3 \text{ m,}$$

the negative sign indicating that the section 2 (depth equal to 0.45 m) is *upstream* of section 1 (brink). The constant of integration, C_I, is not required for the computations. However, if we assign it a value of -309.0 m, it is equivalent to fixing the origin of X at the brink. Similarly, if we fix the origin at the entrance of the channel, C_I will take a value of -259.0 m (since the length of the channel is 50 m).

Another solution for arbitrary values of M and N could be obtained for horizontal channels for which Eq. (3.38) may be written as

$$\frac{dy}{dX} = \frac{-S_f}{1 - F_r^2} = -\frac{\dfrac{n^2 Q^2}{C_u \, y^N}}{1 - \left(\dfrac{y_c}{y}\right)^M} \tag{3.49}$$

which results in

$$X = \frac{C_u}{n^2 Q^2}\left(\frac{y^{N-M+1} y_c^M}{N - M + 1} - \frac{y^{N+1}}{N + 1}\right) + C_I \tag{3.50}$$

For a wide rectangular channel, the value of C_u may be taken as B^2. For other cases, C_u may be obtained by computing the section factor at a representative depth, as illustrated in the example below.

Example 3.7 Since the bed slope of the channel in Example 3.6 is very small, it may be approximated as a horizontal channel. Find the distance from the brink to a section where the depth becomes 0.45 m.

Solution

Since the depth variation is from 0.3891 m to 0.45 m, we may compute C_u based on an average depth of 0.42 m. The value of the hydraulic exponents are taken as $M = 3$ (for rectangular channels) and $N = 10/3$ (for wide rectangular channels).[33] Thus

$$C_u = \left.\frac{A^2 R^{4/3}}{y^{10/3}}\right|_{y = 0.42} = \frac{0.84^2 \times 0.2958^{4/3}}{0.42^{10/3}} = 2.506$$

---

[33]Although the channel in this example is not very wide, we use an approximate value of N to illustrate the procedure. Since we believe this method should not be used, we do not go into details of finding exact value of the exponents.

From Eq. (3.50), at section 1 (brink depth of 0.3891 m), $X_1 = 94.23$ m and for section 2 ($y = 0.45$ m), $X_2 = 86.60$ m. Thus a distance of only **7.63 m** is predicted, which is quite different from that obtained in Example 3.6. We will see later that the result obtained by using the Bresse's method is close to the actual solution. For this case, therefore, assumption of a horizontal channel is not very good. For the channels that have a very flat slope, say 1 in 10,000, this assumption is likely to work better.

Both the analytical solutions discussed here are valid only under very restrictive assumptions and are not very useful for practical applications. Also note that these solutions compute the distance for a given flow depth. Since the solutions are nonlinear in y, obtaining the depth at a given distance requires iterations. Other analytical solutions (e.g., Gill 1976) may be derived for general values of the exponent but become quite elaborate. However, these are of utility in validation of numerical methods or comparison of errors of different methods. Some of the restrictions of these analytical solutions have been relaxed in the semi-analytical methods as described next.

3.5.2   Semi-analytical Methods

The semi-analytical methods are typically based on a *varied flow function* defined and tabulated by Russian academician Bakhmeteff in 1910s. Here we will list one of these methods, proposed by Chow (1955), which solves (Eq. 3.47) to obtain the distance to a section as

$$X = \frac{y_n}{S_0}\left[u - F(u, N) + \frac{J}{N}\left(\frac{y_c}{y_n}\right)^M F(v, J)\right] + C_I \tag{3.51}$$

where u is the normalized depth (y/y_n), $J = N/(N - M + 1)$, $v = u^{N/J}$, and $F(a, b)$ is the varied flow function defined by

$$F(a, b) = \int_0^a \frac{dx}{1 - x^b} \tag{3.52}$$

Extensive tables of this function are available (Chow 1955, HDC 2006) and the use of the method is illustrated in the following example.

Example 3.8   For the channel in Example 3.6, find the distance of the section with $y = 0.45$ m using the Chow's method.

Solution
For a rectangular channel, $M = 3$. N can be obtained as 10/3 as in Example 3.7. We can also obtain its value from Eq. (3.46) or read its value from

Fig. 3.24 as about 2.9. However, to illustrate another possibility, here we obtain the value of N corresponding to a flow depth of 0.42 m by using a small perturbation in Eq. (3.43) to write

$$\left(\frac{0.42+\delta}{0.42}\right)^N = \frac{Z_u^2\big|_{0.42+\delta}}{Z_u^2\big|_{0.42}} = \frac{(A^2 R^{4/3})_{0.42+\delta}}{(A^2 R^{4/3})_{0.42}}$$

With $\delta = 0.01$ m (we may use a smaller value but the result is almost same), we get $N = 2.94$ and $J = 3.13$. The following table shows the steps of computations [linear interpolation has been used to obtain the varied flow function, $F(a, b)$, from the tables given in Chow (1955)]:

Table 3.2   Computation of GVF profile using the Chow's method

Section	Flow depth (m)	$u = y/y_n$	$v = u^{N/J}$	$F(u, N)$	$F(v, J)$	X (m)
Brink	0.3891	0.78	0.79	0.916	0.920	162.6
u/s	0.45	0.90	0.91	1.228	1.236	145.8

Thus the depth of 0.45 m occurs **16.8 m** upstream of the brink.

Another semi-analytical approach without using the hydraulic exponents is that proposed by Keifer and Chu in 1955 for circular channels, which also uses varied flow functions (called Keifer and Chu functions). This approach has also been extended to other shapes of channels but we do not discuss these in detail because of the need of using tables of these functions and availability of, what we feel, better methods. The use of the hydraulic exponents method is not recommended due to the inconsistencies in this approach and use of methods using varied flow functions is not recommended due to their need of tabular values.

3.5.3   Graphical Method

The graphical method is not affected by the limitations of the analytical or semi-analytical methods but is rather time consuming.
Equation (3.23) may be written as

$$\frac{dy}{dX} = \frac{S_0 - \dfrac{n^2 Q^2}{A^2 R^{4/3}}}{1 - \dfrac{Q^2 T}{gA^3}} \Rightarrow X_{1,2} = \int_{y_1}^{y_2} \frac{1 - \dfrac{Q^2 T}{gA^3}}{S_0 - \dfrac{n^2 Q^2}{A^2 R^{4/3}}} \, dy \qquad (3.53)$$

where $X_{1,2}$ represents the distance between two sections where the flow depths are y_1 and y_2. The integral can be obtained graphically by plotting the variation of the function within the integral, i.e., dX/dy, with y as illustrated in the following example.

Example 3.9   Obtain the length of the profile for Example 3.6 using graphical integration.

Solution

Figure 3.25 shows the plot of the integrand of Eq. (3.53) for the depth range of 0.39 m to 0.45 m. The area enclosed is *counted* as 217 rectangles, i.e., $-217 \times 50 \times 0.002$ m. Thus the length of the profile is **−21.7 m**.

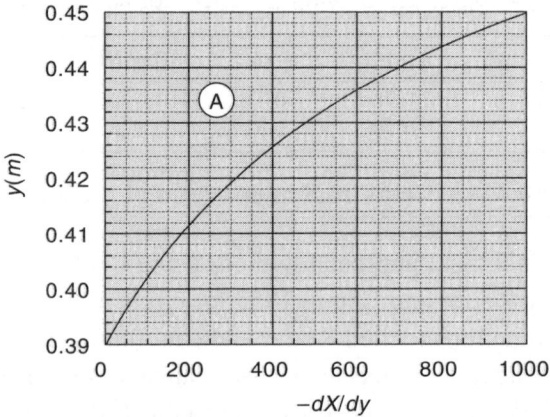

Fig. 3.25   Variation of the inverse slope with flow depth

3.5.4   Numerical Methods

With easy availability of computing facilities and the high speed of computers, the numerical methods have become the method of choice for almost all engineers and academicians dealing with profile computations for gradually varied flow. There are a number of numerical techniques available for solving a first order ordinary differential equation (Appendix C). We describe here two methods, one that computes the distance between the two sections where depths are specified, and the other that computes the depth at a specified distance from a control section.

An obvious numerical method would be based on Eq. (3.53) with the area obtained by using the trapezoidal rule rather than graphically (see Exercise 3.10). Since it is conceptually similar to the graphical method, we do not discuss it further. The numerical methods that we describe here aim at obtaining an algebraic equation in variables by replacing the derivative with finite differences using the Taylor's series expansion of the variable (y) in terms of its derivatives. The easiest method would be to replace dy/dX by $(y_2 - y_1)/(X_2 - X_1)$ in, say, Eq. (3.38). The right hand side, however, contains the friction slope and the Froude number, which are both functions of y. One can compute these parameters corresponding to some *average* depth over the channel reach (e.g., arithmetic, geometric, or harmonic mean of y_1 and y_2). Another alternative would be to compute the parameters at both ends of the reach (i.e., at depths y_1 and y_2) and then use a *mean* value. However, we can avoid this procedure for the Froude number by writing the differential equation in terms of specific energy [Eq. (3.27)] and using the method of finite difference. This method is called the direct step method and was one of the earliest numerical methods used for computing the GVF profiles.

Direct-step method

The distance between the two sections, 1 and 2, is obtained from Eq. (3.27) as

$$\Delta X = \frac{\Delta E}{S_0 - \bar{S}_f} \tag{3.54}$$

where ΔE is the difference in specific energy at the two sections $(E_2 - E_1)$ and $\bar{S}_f$ is the average friction slope over the channel reach. There have been some studies to see which method of averaging is the best and suggestions have been made about whether to use average depth to compute the friction slope or to use average of the two friction slopes. However, as long as the depths at the two sections are not significantly different, we believe both the methodologies would be equally good. Therefore, we use the average of the friction slopes, i.e., $\bar{S}_f = 0.5(S_{f1} + S_{f2})$ in all cases. The use of average friction slope introduces an approximation in the analysis that may cause the numerical result to be different from the exact solution if the depths at the two sections are widely different. Therefore, we divide the reach into smaller subsections[34] in such a way that the depth change over a subsection is not very large. The following example illustrates the use of the direct-step method.

Example 3.10   Use the direct-step method to obtain the length of profile in Example 3.9. Use two equal depth increments.

---

[34]Note that in the analytical or graphical method, we do not need to subdivide the reach. If we assume constant values of M and N, the results would be the same whether we use a single reach or divide it in subsections. However, if M and N are considered as varying with depth, we should use subdivisions with different values of the hydraulic exponents. Since we do not recommend the use of hydraulic exponents, we will not discuss it here.

Solution

The following table shows the computations, and the required length is obtained as **20.45 m**.

Table 3.3 Computation of GVF profile using the direct-step method

Section	y (m)	A (m²)	R (m)	V (m/s)	F (m)	S_f	ΔE (m)	$\bar{S}_f$	ΔX (m)
Brink	0.3891	0.7782	0.2801	1.954	0.5836	0.002082			
Middle	0.42	0.8400	0.2958	1.810	0.5869	0.001662	0.003331	0.001872	−3.819
u/s	0.45	0.9000	0.3103	1.689	0.5954	0.001358	0.008482	0.001510	−16.63

The direct-step method is based on truncation of the Taylor's series (of specific energy, E) after the linear term [see Eq. (C.9)] and hence is only first order accurate. In other words, the error in the numerical solution would vary linearly with the step size, ΔX and we may need to use a very small step size to obtain an acceptable accuracy. Higher order methods, therefore, have been proposed and the fourth order Runge–Kutta method, which is probably the most commonly used method for numerical solution of a first-order differential equation, is described next.

Fourth order Runge–Kutta method

We write Eq. (3.53) as

$$\frac{dy}{dX} = \frac{S_0 - \dfrac{n^2 Q^2}{A^2 R^{4/3}}}{1 - \dfrac{Q^2 T}{gA^3}} = f(X, y) \tag{3.55}$$

For prismatic channel, dy/dX is not a function of X. However, we write it as $f(X, y)$ to elaborate the difficulties associated with the solution for a nonprismatic channel. The fourth order Runge–Kutta (R–K) method, starting from a known depth, y_i, at the location X_i, computes the depth y_{i+1} at a distance ΔX from X_i, as

$$y_{i+1} = y_i + \frac{\Delta X}{6}(k_1 + 2k_2 + 2k_3 + k_4) \tag{3.56}$$

where the k's represent the slope, dy/dX, i.e., $f(X, y)$, at different locations as follows:

$$k_1 = f(X_i, y_i)$$

$$k_2 = f\left(X_i + \frac{\Delta X}{2}, y_i + k_1 \frac{\Delta X}{2}\right) \tag{3.57}$$

$$k_3 = f\left(X_i + \frac{\Delta X}{2}, y_i + k_2 \frac{\Delta X}{2} \right)$$

$$k_4 = f(X_i + \Delta X, y_i + k_3 \Delta X)$$

We reiterate that for prismatic channels, the k's are only functions of depth and not X. The following example shows the application of the R–K method.

Example 3.11   For Example 3.10, using the fourth order Runge–Kutta method obtain the water depth 20 m downstream of the section where the depth is 0.45 m.

Note: We have avoided the critical depth since $f(X, y)$ becomes infinite at critical flow condition. If we start from the brink, k_1 in Eq. (3.57) will become infinite. In the methods discussed earlier, dX/dy was used, which becomes zero at critical depth, (see Fig. 3.25) and does not result in a singularity. However, near the normal depth, dX/dy becomes infinite. If one were to use the fourth order R–K method near the critical depth, one can start at a depth of 1.01 y_c and take very small length step initially. However, since the GVF equation is not valid near the critical depth, nothing much is gained by this.

Solution

Table 3.4 shows the computations with $\Delta X = 20$ m, and the required depth is obtained as $0.45 + (-0.001012 - 2 \times 0.001466 - 2 \times 0.001733 - 0.004034)/6$, i.e., **0.4119 m** from Eq. (3.61). Note the large difference in the value of k_4 compared to the other slopes indicating that a smaller ΔX should have been used.

Table 3.4   Computation of GVF profile using the fourth order Runge–Kutta method

Section	y (m)	A (m²)	R (m)	V (m/s)	S_f	k [Eq. 3.57)]
u/s	0.45	0.9	0.3103	1.689	0.001358	−0.001012
Middle, 1	0.4399	0.8798	0.3055	1.728	0.001451	−0.001466
Middle, 2	0.4354	0.8707	0.3033	1.746	0.001496	−0.001733
d/s, 1	0.4154	0.8307	0.2935	1.830	0.001718	−0.004034

In the above example, the length step, ΔX, was specified. However, most of the times, we will have to decide its value based on the given data. For example, while computing the length of the backwater curve behind a dam or weir, we would know the flow depth at the obstruction and the normal depth but have no idea about the total profile length. If we choose too small a value for ΔX, it will need a large number of steps (the results will be very accurate, though). On the other hand, if we use a large value for ΔX, the results may not have the desired accuracy.[35] Some estimate of the total profile length will, therefore, be helpful in deciding the step size. Some empirical relationships have been proposed but for very special cases (e.g., for M_1 profile in a wide rectangular channel having small Froude numbers). A rough estimate for the length of the M_1 profile upstream of a large dam may be obtained by considering a level pool and is given by y_d/S_0, where y_d is the water depth at the dam. The length step may then be taken as 1/10th, 1/50th, ... of this length depending on the accuracy desired. Another option can be to choose a depth increment and then obtain ΔX using the value of dy/dX at the 'known' section. The following example clarifies these points.

Example 3.12   A trapezoidal channel, with bed width of 20 m and side slopes of 2H:1V, has a bed slope of 0.0005 and roughness coefficient of 0.015. Uniform flow occurs at a flow depth of 3 m. A small dam across the channel raises the flow depth to 5 m at that section. How much length upstream of the dam will be affected by the backwater?

Solution
The discharge in the channel is obtained from Manning's equation as

$$Q = \frac{1}{0.015} \frac{(20 \times 3 + 2 \times 3^2)^{5/3}}{(20 + 2\sqrt{1 + 2^2} \times 3)^{2/3}} \sqrt{0.0005} \text{ m}^3/\text{s} = 204.6 \text{ m}^3/\text{s}$$

The corresponding critical depth is obtained from Eq. (3.35), with $\Theta = 0.1033$, as 2.05 m. Since the normal depth is more than the critical depth, the channel has a mild slope and the profile would be M_1. A rough estimate of the backwater profile length is obtained by dividing the flow depth at the dam by the bed slope as 10,000 m. We decide to use 10 steps of 1000 m each. Table 3.5 shows the computations (some details, which are similar to Table 3.4, have not been shown), with $\Delta X = -1000$ m and the required length (where the depth is 1% more than the normal depth) is obtained as a little more than 8000 m.[36]

---

[35]We assume here that ΔX is kept constant throughout the computation. In practice, it may be a good idea to increase ΔX as we move farther upstream of the obstruction and the slope, dy/dX, becomes smaller (since the accuracy is governed by the difference in flow depths at the sections at either end of the length step, it would reduce the number of steps without significantly reducing the accuracy).

[36]One may do a linear interpolation to obtain the location where the depth is exactly 3.03 m (computed as 8167 m). However, since this limit of 1.01 y_n is itself an approximation, we can say that the profile length is 8 km (or 9 km, if we want to be on the safer side).

Table 3.5 Computation of backwater curve using the fourth order Runge–Kutta method

X (m)	y (m)	k_1	k_2	k_3	k_4	$\bar{k}$
dam	5.000	4.449E-04	4.338E-04	4.341E-04	4.204E-04	4.335E-04
-1000	4.566	4.204E-04	4.035E-04	4.043E-04	3.836E-04	4.033E-04
-2000	4.163	3.837E-04	3.585E-04	3.603E-04	3.300E-04	3.586E-04
-3000	3.805	3.304E-04	2.948E-04	2.991E-04	2.583E-04	2.961E-04
-4000	3.509	2.592E-04	2.152E-04	2.234E-04	1.762E-04	2.188E-04
-5000	3.290	1.783E-04	1.345E-04	1.460E-04	1.022E-04	1.403E-04
-6000	3.150	1.056E-04	7.228E-05	8.323E-05	5.125E-05	7.798E-05
-7000	3.072	5.498E-05	3.492E-05	4.238E-05	2.351E-05	3.885E-05
-8000	3.033	2.625E-05	1.595E-05	2.003E-05	1.040E-05	1.810E-05
-9000	3.015	1.196E-05	7.105E-06	9.086E-06	4.557E-06	8.150E-06

An alternative method uses depth increments of roughly 0.2 m, i.e., 10 steps between the normal depth of 3 m and the depth of 5 m at the dam. Using the dy/dX value at the dam (4.449E-04), we get a length step of about 450 m. Table 3.6 shows the steps of computations with the step length being increased at each step on the basis of the slope computed at the beginning of the step [note from Eq. (3.57) that the computation of k_1 does not require the value of ΔX]. The length of the profile is obtained as 7685 m.

Table 3.6 Fourth order Runge–Kutta method with variable step length

X (m)	y (m)	k_1	$\Delta X \, (=-0.2/k_1)$	k_2	k_3	k_4	$\bar{k}$
dam	5.000	4.449E-04	-449	4.403E-04	4.403E-04	4.352E-04	4.402E-04
-449	4.802	4.352E-04	-460	4.217E-04	4.222E-04	4.057E-04	4.214E-04
-909	4.608	4.234E-04	-472	4.071E-04	4.078E-04	3.879E-04	4.069E-04
-1382	4.416	4.088E-04	-489	3.892E-04	3.902E-04	3.663E-04	3.890E-04
-1871	4.226	3.907E-04	-512	3.670E-04	3.686E-04	3.400E-04	3.670E-04
-2383	4.038	3.680E-04	-543	3.394E-04	3.419E-04	3.080E-04	3.398E-04
-2926	3.853	3.393E-04	-589	3.052E-04	3.091E-04	2.697E-04	3.063E-04
-3516	3.673	3.027E-04	-661	2.631E-04	2.688E-04	2.245E-04	2.652E-04
-4176	3.498	2.559E-04	-782	2.117E-04	2.200E-04	1.728E-04	2.153E-04
-4958	3.329	1.954E-04	-1023	1.507E-04	1.617E-04	1.163E-04	1.561E-04
-5981	3.170	1.174E-04	-1704	8.166E-05	9.302E-05	5.856E-05	8.755E-05
-7685	3.020						

Figure 3.26 shows the backwater curve (note the exaggerated vertical scale) obtained through the numerical methods.

There are some other numerical methods that can be used to obtain the GVF profile. Some of these methods use an adaptive control of the step size, ΔX, to keep the numerical errors within a specified tolerance. Others use a predictor-corrector type of iterative refinement where the predicted value of depth is used to estimate a *better* value of slope to re-compute a more correct

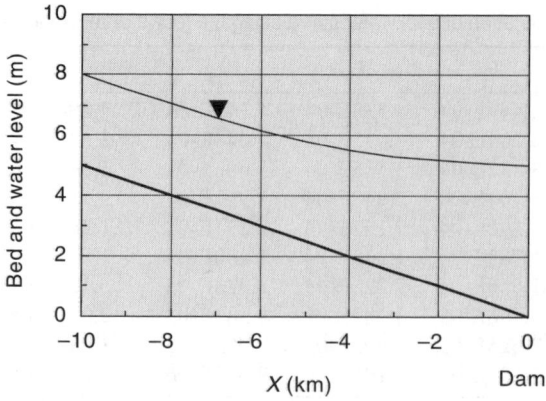

Fig. 3.26 Backwater curve behind a small dam

value of depth. However, we will not describe these methods here (since we believe that the fourth order R–K method is sufficiently accurate) and refer the reader to a good book on numerical methods (see Appendix C).

3.6 More Complex GVF Profiles

The discussion in this chapter so far will enable one to compute the GVF profile for most prismatic channels. However, there are some cases which require additional discussion. For example, a circular channel has multiple normal depths for some discharges which leads to a non-monotonic variation of friction slope and a compound channel may have multiple critical depths. For nonprismatic channels, not only do the section and the bed slope vary spatially, but also the information about this variation may not be available at all locations. For example, information about the geometric elements of a river cross section is usually available at some selected locations only. This would mean that we can not use methods that compute the distance for a given flow depth, since the distance is fixed on the basis of availability of the cross-sectional information. Even for methods that compute the distance for a given ΔX, we may require the information at the middle of the reach [see Eq. (3.57), expressions for k_2 and k_3], which is not available. The other complications that may arise are the presence of an island in the channel causing bifurcation of the flow resulting in the discharge being unknown. Sometimes, channels form a network where not only the amount of flow but also the direction may be unknown. Another scenario that we have not considered is the effect of a change in discharge on the GVF profile. For example, an irrigation canal may be designed for some discharge but the actual discharge will vary over a wide range depending on the requirements. Since both the normal and critical depths depend on the discharge, the GVF profile may be different for different discharges. Yet another complexity arises in the case of a channel joining two water bodies (lakes, reservoirs, or rivers) in that the discharge (also called the *delivery* of the channel) and the GVF profile will vary depending on the water surface elevation at either end of the channel. Although these topics may be a little advanced for an introductory course, a brief discussion is provided in this section.

3.6.1   Circular Channels

The discussion on the salient features of GVF profiles (see section 3.4.1) was based on the fact that both the friction slope and Froude number are monotonic functions of flow depth, i.e., both decrease with increase in flow depth for a given discharge, roughness, and bed slope. For a circular channel, or other channels with a closing top, the Froude number variation is monotonic (as may be inferred from Fig. 3.10) but the friction slope variation is not (Fig. 2.7). As long as the normal depth is less than about $0.82D_0$, the previous analysis of profiles would be valid since for depths greater than y_n, friction slope (though not monotonic) would be smaller than the bed slope. However, for cases where there are two normal depths (one larger than about $0.94D_0$ and the other between $0.82D_0$ and $0.94D_0$), we have to look at the differential equation again to determine the nature of profile.

Denoting the lower normal depth by y_n^l and the upper normal depth by y_n^u, there are three possible cases: (a) $y_n^u > y_n^l > y_c$, (b) $y_n^u > y_c > y_n^l$, and (c) $y_c > y_n^u > y_n^l$. Clearly, the friction slope is equal to the bed slope when the flow depth, y, is equal to either of the two normal depths. When $y < y_n^l$ or $y > y_n^u$, the friction slope would be larger than the bed slope (since the section factor is smaller than the value at the normal depths) and when $y_n^u > y > y_n^l$, the friction slope would be smaller than the bed slope. Using this information, Fig. 3.27 shows the expected GVF profiles for circular channels (since the arguments are similar to those for noncircular channels, they are not repeated here).

3.6.2   Compound Channels

It has been shown (Blalock and Sturm 1981, Chaudhry and Bhallamudi 1988) that a compound channel may have minimum specific energy at two different depths—one when the flow is confined to the main channel and the other when it spills onto the floodplains (see Fig. 3.28). Using the definition of subcritical and supercritical flows based on the variation of specific energy with flow depth (i.e., a subcritical flow has increasing specific energy with

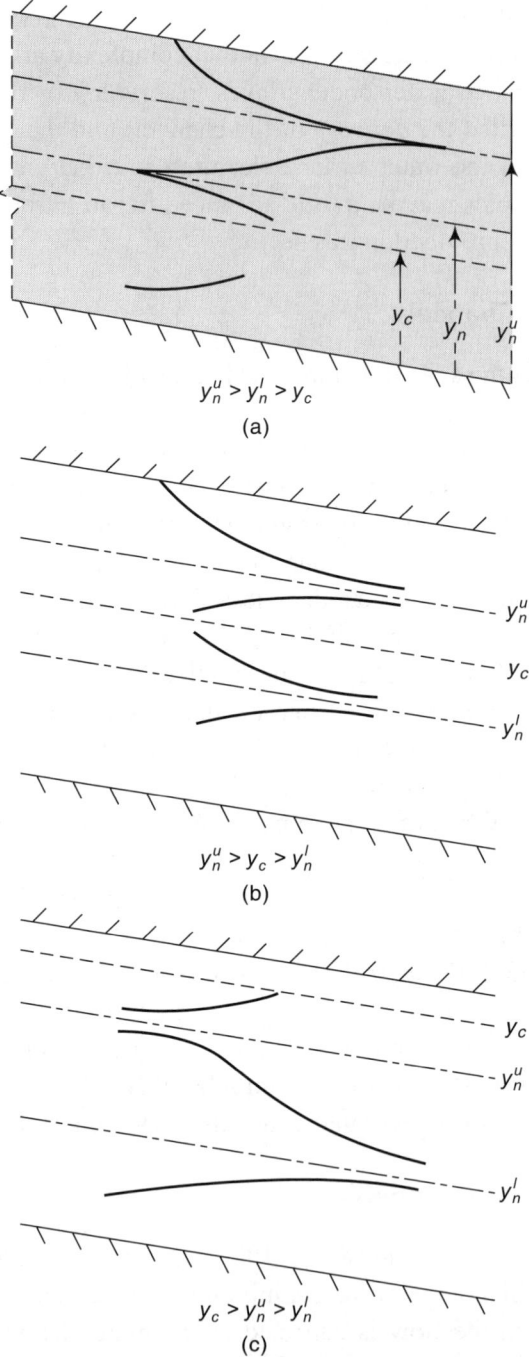

Fig. 3.27 Flow profiles in a circular channel

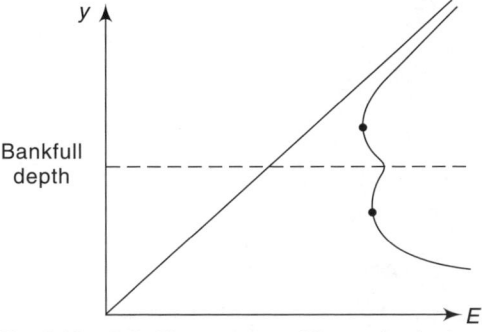

Fig. 3.28 Specific energy at different depths in a compound channel

increasing flow depth and a subcritical flow has an opposite nature), it is seen that the flow at depths larger than the *higher critical depth* would be subcritical and the flow at depths smaller than the *lower critical depth* would be supercritical. However, between these two critical depths. there would be a change from subcritical flow to supercritical flow as the depth increases to the bankfull depth and then above it. This implies that the GVF profiles at depths between these two depths would be different depending on whether it is above the point of local maximum specific energy or below it. The analysis of these profiles, though a little more complicated, may be performed using the methods discussed in this chapter. We do not discuss the details here. Chaudhry (1993) provides a more detailed discussion of this flow situation.

3.6.3   Nonprismatic Channels

For nonprismatic channels, the cross section survey is done only at a few locations along the length of the channel. Thus the GVF profile computation aims at finding the water depth at these sections starting from a known depth at the control section. None of the methods discussed in Section 3.5 is suitable for nonprismatic channels because they compute the distance to a section of specified depth (except for the fourth order R–K method, but that also requires information at the mid-section, which is not available). Another factor that enters into the computations for nonprismatic channels is the energy loss due to the channel expansion, contraction, or change in alignment. We will now describe the standard-step method, which is designed to take care of these factors.

Standard-step method

In this method, we write the energy equation between sections 1 and 2 (Fig. 3.29) as

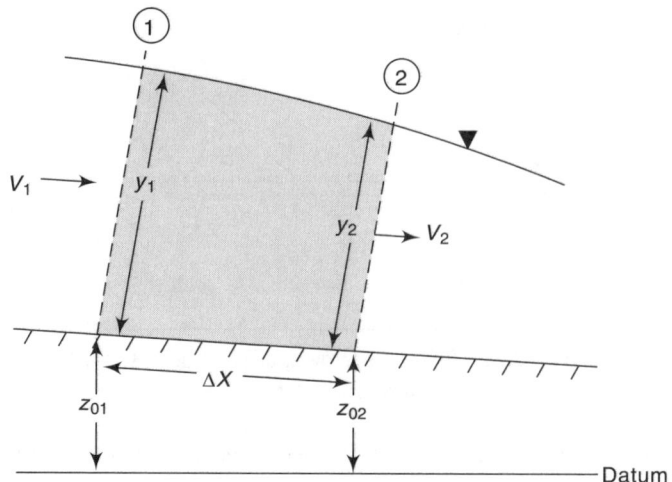

Fig. 3.29 Application of energy equation in the standard step method

$$y_1 + z_{01} + \frac{V_1^2}{2g} = y_2 + z_{02} + \frac{V_2^2}{2g} + \overline{S}_f \Delta X \pm K_l \left(\frac{V_2^2 - V_1^2}{2g} \right) \qquad (3.58)$$

where K_l is a loss coefficient, which depends on the change in the cross section and/or alignment of the channel between the sections 1 and 2 (a value of 0.1 is commonly used). This head loss is assumed to be proportional to the difference in the velocity heads at the two sections.[37] The ± sign on the last term in Eq. (3.63) is used to ensure that the loss is always positive. If V_2 is less than V_1 (generally for backwater profiles), we use the negative sign.

Assuming that the conditions at section 1 are known (either it is a control section or the computations have already been carried out up to this section), and using the average friction slope, we may write

$$F(y_2) = y_2 + (1 \pm K_l) \frac{V_2^2}{2g} + 0.5 S_{f2} \Delta X$$

$$- \left[y_1 + (1 \pm K_l) \frac{V_1^2}{2g} + z_{01} - z_{02} - 0.5 S_{f1} \Delta X \right] = 0 \qquad (3.59)$$

---

[37]This would mean that the head loss will be zero if the velocities at the two sections are same. Variations of cross section and alignment through the channel reach are, therefore, not accounted for. It may be better to use the velocity head, e.g., $V_1^2/2g$, rather than the difference with a smaller value of the loss coefficient. We follow the standard practice here.

which is a nonlinear equation in y_2 and may be solved using an iterative technique.[38] For example, the Newton method can be used to write the following iterative scheme:

$$y_2^{(k+1)} = y_2^{(k)} - \frac{F[y_2^{(k)}]}{(dF/dy_2)_{y=y_2^{(k)}}} \tag{3.60}$$

with a reasonable guess for $y_2^{(0)}$ (in absence of other information, it may be taken equal to y_1). The evaluation of the derivative requires differentiating the velocity head and the friction slope with respect to the flow depth, which may be quite involved for general channel shapes. However, a simplification may be made by treating the channel as a wide rectangular channel. [One should note that the derivative in Eq. (3.60) does not have to be *exact*. If we approximate it by using the assumption of wide rectangular channel, it may require a few more iterations to converge but the converged solution would still be correct]. The example below uses this simplification to obtain the derivatives.

Example 3.13   For the channel in Example 3.6, find the flow depth at a section 20 m upstream of the brink[39] using the standard-step method. Neglect the loss coefficient (since it is a prismatic channel).

Solution

Taking the brink as section 1 and the upstream section as section 2, we have $\Delta X = -20$ m, and Eq. (3.59) may be written as

$$F(y_2) = y_2 + \frac{0.02944}{y_2^2} - \frac{2.293 \times 10^{-4} \, (2 + 2y_2)^{4/3}}{y_2^{10/3}} - 0.5844 = 0$$

The derivative may be obtained approximately[40] by replacing $2 + 2y_2$ in the third term with 2 (i.e., assuming it to be a wide channel) to get

$$F'(y_2) = 1 - \frac{0.05888}{y_2^3} + \frac{1.926 \times 10^{-3}}{y_2^{13/3}}$$

The iterations are shown in Table 3.7, starting with the guess value $y_2^{(0)} = y_1 = 0.3891$ m. Convergence up to the fourth significant figure is achieved

---

[38]We can, of course, use this method for prismatic channels also. However, the fourth order R–K method is preferred because it is accurate and non-iterative.

[39]In this case, we can start from the critical depth at the brink because we are writing the energy equation directly and not in terms of the slope, dy/dX.

[40]For this case, the exact derivative is not very difficult to obtain. However, for trapezoidal channels, it would be a little more cumbersome.

in five iterations and the flow depth 20 m u/s of the brink is obtained as
0.4557 m.

Table 3.7   Computation of drawdown curve using the standard-step method

Iteration no., $k \rightarrow$	0	1	2	3	4	5
$y_2^{(k)}$	0.3891	0.5769	0.4728	0.4559	0.4557	0.4557
$F(y_2)$	−0.02165	0.07431	0.008363	5.472E-05	−3.197E-06	−
$F'(y_2)$	0.1153	0.7141	0.4923	0.4362	0.4357	−

3.6.4   Branching and Networks

Figure 3.30 shows a channel divided into two branches by an island. The
discharge in the reaches *AB* and *EF* will be known (and is same) but that in
reaches *BCE* and *BDE* is unknown (if we add another channel *CD*, to make it
a channel network, the amount as well as the direction of flow in *CD* would
be unknown). Hence, we cannot use the methods described so far which had
considered the discharge to be known. Under these conditions, the *simultaneous
solution approach* is utilized, which treats both discharge and flow depth as
unknown at each section.

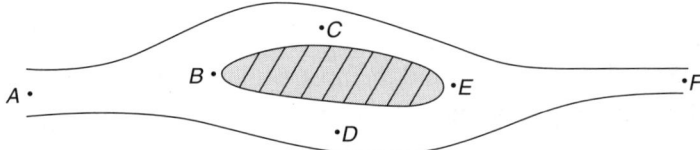

Fig. 3.30   Branching of an open channel

In the simultaneous solution approach, we divide each reach into subsections
and use the subscript (i, j) to represent the *j*th section (or node) of the *i*th
reach. For example, *AB* can be called reach 1, *BCE* reach 2, *BDE* reach 3, and
EF reach 4. The reach *AB* could comprise N_1 sections, i.e. $N_1 - 1$ sub-reaches
(or elements), *BCE*, ..., and *EF* $N_4 - 1$ sub-reaches. The energy equation for
the *j*th sub-reach [i.e., enclosed by the *j*th and $(j + 1)$th sections] of the *i*th
reach can then be written as [see Eq. (3.59), assuming $K_l = 0$ and ΔX to be
constant]

$$F_{i,j} = y_{i,\,j+1} - y_{i,j} - S_0 \Delta X + \frac{1}{2g}\left(\frac{Q_{i,\,j+1}^2}{A_{i,\,j+1}^2} - \frac{Q_{i,j}^2}{A_{i,j}^2}\right)$$

$$+ \frac{n_i^2 \Delta X}{2}\left(\frac{Q_{i,\,j+1}^2}{A_{i,\,j+1}^2 R_{i,\,j+1}^{4/3}} + \frac{Q_{i,j}^2}{A_{i,j}^2 R_{i,j}^{4/3}}\right) = 0 \qquad (3.61)$$

The continuity equation for the same reach can be written as

$$Fq_{i,j} = Q_{i,j+1} - Q_{i,j} = 0 \tag{3.62}$$

Since the ith reach has $N_i - 1$ sub-reaches, the total number of these equations for the ith reach will be $2(N_i - 1)$, half for energy equation (3.61) and the other half for the continuity equation (3.62). The number of unknowns for the ith reach will be $2N_i$, i.e., the depth and discharge at each section (or node). Therefore, for the branching channel shown in Fig. 3.30, the number of unknowns would be $2(N_1 + N_2 + N_3 + N_4)$ and the number of equations is $2(N_1 + N_2 + N_3 + N_4 - 4)$. In order to solve for the unknowns, we would require eight additional equations. The boundary conditions at the control point supply two of these conditions (e.g., depth and discharge may be given at the downstream end). The other six conditions are obtained by applying the energy equation and continuity equation at the junctions, B and E. For example, at B,

$$y_{1,N_1} + z_{0_{1,N_1}} + \frac{Q_{1,N_1}^2}{2\,gA_{1,N_1}^2} = y_{2,1} + z_{0_{2,1}} + (1 + K_j)\frac{Q_{2,1}^2}{2\,gA_{2,1}^2} \tag{3.63a}$$

$$y_{1,N_1} + z_{0_{1,N_1}} + \frac{Q_{1,N_1}^2}{2\,gA_{1,N_1}^2} = y_{3,1} + z_{0_{3,1}} + (1 + K_j)\frac{Q_{3,1}^2}{2\,gA_{3,1}^2} \tag{3.63b}$$

$$Q_{1,N_1} = Q_{2,1} + Q_{3,1} \tag{3.63c}$$

where K_j is the junction loss coefficient. Thus we have a set of $2(N_1 + N_2 + N_3 + N_4)$ nonlinear equations, which can be solved to obtain the flow depth and discharge at all nodes. For network problems, where the direction of flow in some channels may not be known, care must be taken to ensure that the head loss occurs in the direction of flow. In Eq. (3.61), for example, the last term representing the friction slope must be modified to include the direction of flow, i.e., Q^2 should be replaced by $Q|Q|$. Chaudhry (1993) provides more details of the GVF profiles in channel networks.

3.6.5 Change in Discharge

The discharge in a channel may vary over a range from zero (when, say, the channel is closed for repairs) to the design value due to changes in demand and/or supply. A channel which has a mild slope at the design discharge may behave as a steep channel at a smaller discharge. To analyse this, the critical slope, defined earlier, is utilized. Using the Manning's equation and the condition for criticality [Eq. (3.24)], the critical slope, S_c, is written as

$$S_c = \frac{n^2 Q^2}{A_c^2 R_c^{4/3}} = \frac{n^2 P_c^{4/3} g A_c^3}{A_c^{10/3} T_c} = \frac{g n^2 P_c^{4/3}}{A_c^{1/3} T_c} \tag{3.64}$$

where the subscript c indicates that the quantities correspond to the critical depth, which, in turn, depends on the discharge and the channel geometry. [Compare with Eq. (2.47). Naturally, these equations are identical with the Froude number being 1 at critical flow and the critical slope, S_c, replacing the bed slope, S_0.] For example, for a rectangular channel, a nondimensional critical slope may be written as

$$S_c^* \left(= \frac{B^{1/3} S_c}{g n^2} \right) = \frac{(1 + 2\eta_c)^{4/3}}{\eta_c^{1/3}} \tag{3.65}$$

where η_c is the dimensionless critical depth, y_c/B, which is related to the discharge [see Eq. (3.30)] as $\eta_c^3 = Q^2/gB^5$. Figure 3.31 shows the variation of the critical slope with critical depth for a rectangular channel using *Manning's equation*.[41] For a rectangular channel, it can be easily seen that the critical depth is proportional to the two-thirds power of discharge, Q; while the normal depth is proportional to Q for very large Q and to three-fifths power of Q for very small Q. Therefore, at both limits of discharge (very large or very small), the normal depth will be more than the critical depth and the channel will act as a mild slope channel. Consequently, both the regions, above the upper part of the curve and below the lower part in Fig. 3.31, represent mild slope, the points on the curve indicate critical slope, and the region enclosed by the

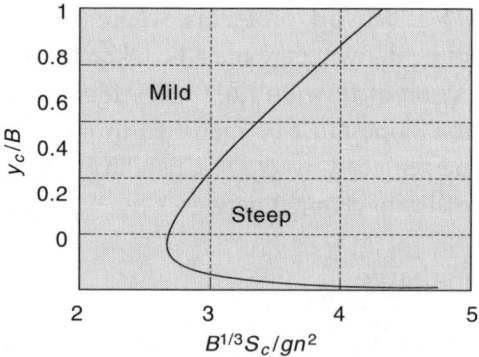

Fig. 3.31 Variation of critical slope with critical depth for rectangular channels

---

[41]If we use Chezy's equation, and assume Chezy's constant to be independent of depth, the nondimensional critical slope, $C^2 S_c/g$, would be equal to P_c/T_c (i.e., $1 + 2\eta_c$ for a rectangular channel).

curve represents steep slope. For any given bed slope, e.g., $B^{1/3}S_0/gn^2 = 3$, say, for y_c/B less than about 0.06, the slope would be mild, from 0.06 to 0.4, it would act as steep, and beyond 0.4, it would again be mild. The GVF profiles would accordingly be classified as *M* or *S*. Also note that there is a minimum value of critical slope [from Eq. (3.65), it can be shown that the minimum critical slope is obtained at $y_c/B = 1/6$ and is equal to $8gn^2/3B^{1/3}$] beyond which the slope would always be mild irrespective of the discharge. This slope may be called the *limit slope* and may be used to design a channel in such a way that the flow remains subcritical (i.e., the slope remains mild) for all discharges. However, this concept is not valid for trapezoidal channels (even for sides which are *almost* vertical, see Exercise 3.16) and, we believe, is not very useful for practical applications.

3.6.6 Delivery of Channels

Figure 3.32 shows a channel connecting two water bodies that have water levels at heights of y_u and y_d measured from the corresponding channel bed level. If we assume that the channel slope remains mild for all discharges, and neglect the minor losses at the entrance and exit, the GVF profile in the channel would be M_1 when $y_d > y_u$ and M_2 when $y_d < y_u$. At some combination of the water levels, it may happen that critical flow occurs at the downstream end. Further reduction in the downstream water level (keeping the same upstream level) will not affect either the discharge or the water surface profile in the channel. On the other hand, if the channel is long enough such that uniform flow occurs in the upstream portion even when critical condition occurs at the downstream section, the discharge will not be affected (but the surface profile will change) as the downstream depth increases from the critical depth to the normal depth. Since this case is not frequently encountered in practice, we do not describe the details here. Interested readers may refer to Chow (1981).

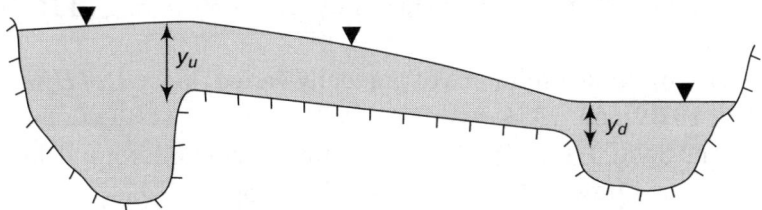

Fig. 3.32 Flow in a channel connecting two water bodies

SUMMARY

In this chapter, we have looked at the variation of water depth under different gradually varied flow conditions. The momentum and energy equations have been used to obtain the differential equation of GVF. A qualitative analysis has been performed to help in obtaining a quick estimate of the nature of the profile expected in a variety of situations. Quantitative estimate of the profile has then been discussed using various methods, viz., analytical, semi-analytical, graphical, and numerical. Critical depth has been defined and its computation for different sections has been discussed. Some more complicated cases of gradually varied flow have been briefly presented but have not been discussed thoroughly as they are thought to be somewhat advanced for an introductory course. In all this discussion, we have assumed that the spatial variation of flow depth is gradual and the curvature of the streamlines is so small that pressure distribution may be assumed to be hydrostatic. There are numerous situations encountered in open channel flow where these assumptions would not be valid. For example, the flow close to the brink, flow near a weir, flow at a hydraulic jump, all exhibit a rapid variation of flow depth. These cases of rapidly varied flow are discussed in the next chapter.

REFERENCES

Braine, C.D.C. (1947): 'Draw-down and other factors relating to the design of storm-water outflows on sewers', *J. of Inst. Civ. Eng.*, **28**(6), pp. 136–63.

Chaudhry, M.H. (1993): *Open-Channel Flow*, Prentice-Hall, Englewood Cliffs, New Jersey, USA.

Chow, V.T. (1955): 'Integrating the equation of gradually varied flow', *Proc. of ASCE*, **81**, pp. 1-32.

Chow, V.T. (1981): *Open-Channel Hydraulics*, McGraw-Hill International, Tokyo, Japan.

Dey, S. and Lambert, M.F. (2005): 'Reynolds stress and bed shear in nonuniform unsteady open-channel flow', *J. of Hydr. Eng.*, **131**(7), pp. 610–14.

Gill, M.A. (1976): 'Exact solutions of gradually varied flow', *J. of Hydr. Div.*, ASCE, **102**(9), pp. 1353–64.

Graf, W. H. and Song, T. (1995): 'Bed-shear stress in nonuniform and unsteady open-channel flows', *J. of Hydr. Res.*, **35**(5), pp. 699–704.

HDC (2006): 'Hydraulic Design Criteria', US Army Corps of Engineers, *http://chl.erdc.usace.army.mil/%5CMedia/2/7/6%5C000.pdf* (last accessed on Oct. 24, 2006).

Liggett, J.A. (1993): 'Critical depth, velocity profiles, and averaging', *J. Irrig. and Dra. Eng.*, **119**(2), pp. 416–22.

Patil, R.G., Diwanji, V.N., and Khatsuria, R.M. (2001): 'Integrating Equation of Gradually Varied Flow', *J. of Hydr. Eng.*, **127**(7), pp. 624–25.

Subramanya, K. and Ramamurthy, A.S. (1974): 'Flow profile computation by direct integration', *J. of Hydr. Div.*, ASCE, 100(9), 1306–11.

Swamee, P.K. (1993): 'Critical depth equations for irrigation canals', *J. of Irrig. and Dra. Eng.*, **119**(2), pp. 400–09.

EXERCISES

3.1 In Example 3.2, the velocity profile was influenced by the distance from the sidewalls. However, for wide rectangular channels, we may assume the velocity profile to be independent of Z and use a power law form as $v = v_{max}(Y/y)^{1/7}$. Find the momentum and energy correction factors for this distribution and compare with the previously obtained values. [1.016, 1.045]

3.2 A trapezoidal channel, with bed width of 1 m and side slopes 1H: 1V carries a discharge of 1 m³/s. What would be the specific energy when the depth of flow is 0.6 m? Is the flow supercritical or subcritical at this depth? What would be another flow depth which will have the same specific energy (see Fig. 3.8)? [0.655 m, subcritical, 0.29 m]

3.3 From Fig. 3.9, it can be seen that for a trapezoidal channel, at very low and very high depths, the plot of section factor versus critical depth is a straight line. This implies that the section factor is proportional to some power of critical depth. Find this power at both the limits and comment on these values.

$$\left[\text{Low depth}: \frac{3}{2}; \text{Large depth}: \frac{5}{2} \right]$$

3.4 For the modified triangular section shown in Fig. 2.16, would you expect the critical depth to be higher or lower than that for the unmodified triangular section? Find the expression relating the critical depth to the discharge and side slope and verify.

$$\left[\text{Lower}, \ y_c = \left(\frac{2Q^2}{gm^2} \right)^{1/5} \left[\frac{m^2\sqrt{1+m^2}}{\left(m + \tan^{-1}\frac{1}{m} \right)^3} \right]^{1/5} \right]$$

3.5 A 1.25 m diameter circular concrete pipe is laid at a longitudinal slope of 1 in 2000. The normal depth of flow for a certain discharge is equal to 0.8 times the diameter. If the flow depth at a section under gradually varied conditions is 0.75 m, what would be the nature of the flow profile?

$$[M_2]$$

3.6 In Eq. (3.40), the curvature of the water surface was obtained for the special case of a wide rectangular channel. For a triangular channel, show that the friction slope is inversely proportional to the 16/3 power of the flow depth and the Froude number is inversely proportional to the 5/2 power of the depth. Using this result, obtain an equation similar to Eq. (3.40) and comment on the relationship between curvature and slope.

3.7 A brick-lined trapezoidal channel with bed width 2.5 m and side slope 3H:1V carries a discharge of 3 m^3/s. At a section, the bed slope changes from 0.001 to 0.002. What GVF profile will occur near this section?

[M_2 before the junction]

Note: The problem can be solved by finding the normal depth of flow for both the slopes and comparing these with the critical depth. However, a quicker way is to find the critical slope and compare with the bed slope.

3.8 Sketch the possible GVF profiles for the following situations:
(a) Mild slope followed by a milder slope and a steep slope
(b) Mild slope followed by a short horizontal channel and a steep slope
(c) Steep slope followed by a critical slope and a mild slope

[(a): M_1, M_2, S_2; (b): M_1, H_2, S_2; (c): C_3, C_1]

3.9 A concrete-lined trapezoidal channel ($B = 2$ m, $m = 3$H:1V) has a bed slope of 1 in 1000 and carries a discharge of 3 m^3/s. Due to an obstruction in the channel, it was found that the water depths at two sections, $S1$ and $S2$, were 1.2 m and 1.35 m, respectively. It was also observed that the hydraulic exponents, M and N, were both close to 4. Assuming that $M = N = 4$, obtain an expression for the length of profile similar to Eq. (3.48) and estimate the distance between the sections $S1$ and $S2$.   [153.6 m]

3.10 The graphical integration method was used in Example 3.9 to obtain the distance between two sections where the flow depths were known. Generate the values of the integrand in Eq. (3.53) between the flow depths of 0.39 m and 0.45 m, at an interval of 0.01 m, and use the trapezoidal rule to estimate the length of the profile. Repeat the computations with an interval of 0.005 m and compare the results.   [22.347 m, 22.166 m]

3.11 In Example 3.10, the computations were started from the brink and proceeded upstream. Recompute the profile length starting from the upstream section at which the flow depth is 0.45 m and proceeding downstream and compare the results. What problems would be encountered if computations start further upstream where the depth is close to normal depth?

[22.347 m]

3.12 If the channel in Example 3.6 is not prismatic but is converging such that the width changes linearly from 2 m at the upstream end to 1 m at the brink, both the normal and the critical depths will change along the channel length. Plot the NDL[42] and CDL for this case assuming that the discharge is same (1.52 m³/s). Even though the channel is nonprismatic, the fourth order R–K method can be used because the cross section is known at all sections. What complications may arise due to the varying normal and critical depths? Since the channel width is not constant, the flow area (and the top width and wetted perimeter) is a function of both y and X. Obtain the GVF equation for such cases (which includes a term involving dB/dX).

$$\left[\frac{dy}{dx} = \frac{S_0 - S_f + \dfrac{Q^2 y}{gA^3}\dfrac{dB}{dx}}{1 - \dfrac{Q^2 B}{gA^3}} \right]$$

3.13 In Example 3.11, the depth at 20 m d/s of the specified location was computed using a single step of 20 m. Recompute this depth and also find the depth 30 m d/s using steps of 10 m each. Which result should be more reliable? Also compute the depth at 30 m using a single step of 30 m and argue that the results are unreasonable.
Hint: look at the value of k_4.

$$[0.4118 \text{ m}, 0.7022 \text{ m}, 0.4387 \text{ m}]$$

3.14 Fig. 3.26 shows the water surface profile (note that Table 3.3 shows the water depth measured from the channel bed and not from a horizontal datum). It is obvious that even though the water depth is increasing in the flow direction, the water surface is falling. Using the hydraulic exponents with the assumption $M = N$, obtain an expression for $d(y + z)/dX$ and show that the water surface will fall in the direction of flow for M_1 and S_3 profiles.

$$\left[\frac{d(y+z)}{dx} = S_0 \frac{y_c^N - y_n^N}{y^N - y_c^N} \right]$$

3.15 In the circular channel of Exercise 3.5, the discharge increases in such a way that the normal depth becomes 88% of the diameter. What would be the other normal depth? What GVF profile is expected at a section where the flow depth is 1.2 m? $\qquad$ [1.228 m, rising profile]

---

[42]Note that uniform flow does not occur because the channel is nonprismatic. However, we would assume that the Manning's equation can be used to obtain a *normal* depth.

3.16 For a trapezoidal channel with bed width B and side slope $m\text{H}:1\text{V}$, using Eq. (3.64), obtain an expression relating the side slope and the nondimensional critical depth, my_c/B, for which the critical slope would be a minimum.[43] Prepare a plot similar to Fig. 3.31 for a trapezoidal channel with side slope $m = 0.5$ (use y_c/B values from 0.1 to 30) and show that a limit slope does not exist.

$$\left[\eta_c^3 + (10M - 1)\eta_c^2 + (10M - 1.5)\eta_c + M = 0, \text{ where } M = \frac{m}{4\sqrt{1 + m^2}} \right]$$

---

[43]Note that it is a cubic equation and may result in more than one stationary points for small side slopes. Although it involves tedious mathematical manipulations, it can be shown that for side slopes less than about 0.466H:1V, there would be both a local minimum and a local maximum in the critical slope versus critical depth curve (also see Exercise 2.14). For flatter sides, there would be no stationary point.

4 Nonuniform Flow: Rapidly Varied

4.1 Introduction: Occurrence and Importance

As stated earlier, uniform flow in a channel is quite uncommon. The nonuniform flow may show a gradual variation in flow depth (and velocity) in the longitudinal direction, as described in the previous chapter, which allows us to use a hydrostatic pressure distribution. However, there are a number of flow situations where the assumption of a gradual change in flow depth, i.e., a small curvature of streamlines, would not be valid. Some examples are flow over a weir, under a sluice gate, near a canal fall, or in a hydraulic jump (Fig. 4.1). A rapid variation in depth complicates the analysis because the pressure distribution becomes nonhydrostatic. On the other hand, it sometimes simplifies the analysis because the variation in flow depth occurs over a

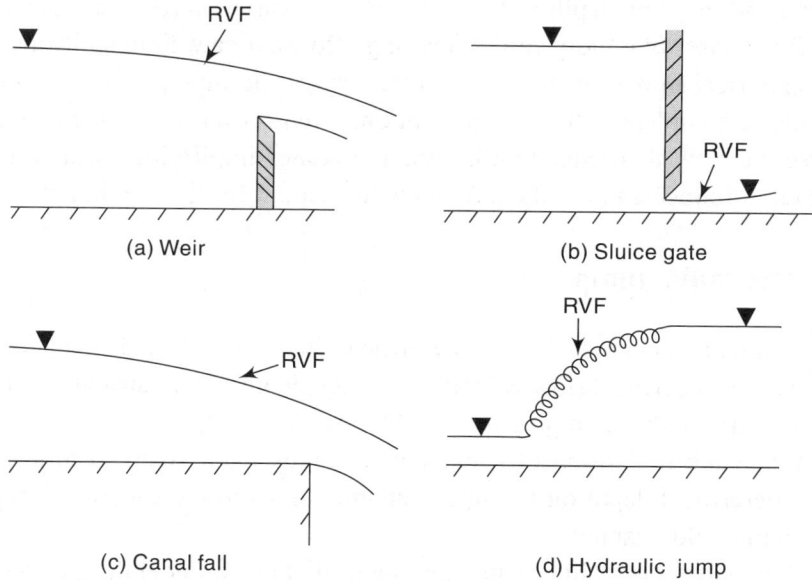

(a) Weir

(b) Sluice gate

(c) Canal fall

(d) Hydraulic jump

Fig. 4.1   Examples of rapidly varied flow

short reach for which frictional forces and the gravitational force component in the flow direction may be so small as to be negligible.

Depending on the parameters that we want to find out, the effect of nonhydrostatic pressure distribution may or may not affect the analysis. For example, in a hydraulic jump, where we are interested in finding out (for known pre-jump conditions) the flow depth after the jump and the energy loss in the jump, we may choose a control volume enclosing but not intersecting the jump. Although the pressure is nonhydrostatic in the jump, it is reasonable to assume a hydrostatic pressure distribution a little before and a little after the jump. Similarly, for flow through a sluice gate, where we need to find the discharge for a given gate opening, a proper selection of the control volume will do away with the need of estimating the pressure distribution in the rapidly varied portion of the flow. However, for some flow situations, e.g., flow immediately upstream of a drop, the nonhydrostatic pressure distribution must be known (through, say, experimental observation) in order to estimate the flow depth by applying the momentum or energy equation. We would first discuss the hydraulic jump for which the commonly required parameters may be estimated theoretically without resorting to experimentally or numerically generated data. As we will see later, some parameters, e.g., the length of the jump, do require the use of experimental data. Flow through a sluice gate, which involves experimental data to correlate the gate opening and the flow depth just after the rapidly varied portion, is discussed next. Other cases of rapidly varied flow, e.g., flow over a weir or spillway, are more empirical in nature since the complexity of the flow pattern generally precludes a completely theoretical solution. Even for these cases, we would analyse the flow theoretically after making some simplifying assumptions, and then introduce empirical coefficients to account for these assumptions.

4.2   Hydraulic Jump

A hydraulic jump occurs when a supercritical (high velocity and small depth) flow meets a subcritical (low velocity and large depth) flow causing a 'jump' in flow depth. Some examples are given below (Fig. 4.2):

(a) When the bed slope of a channel changes from steep to mild, the supercritical depth on the steep channel jumps to the subcritical depth on the mild channel.

(b) Downstream of a sluice gate, the supercritical flow under the gate meets the uniform flow depth (typically subcritical) in the downstream channel.

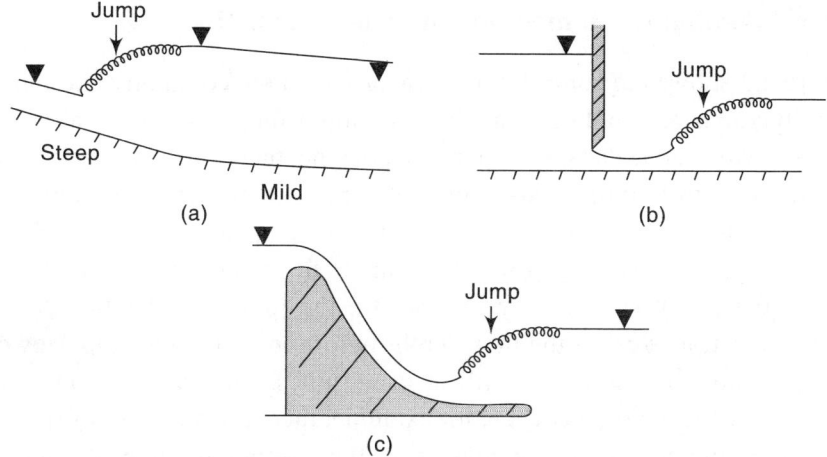

Fig. 4.2 Occurrence of hydraulic jump

(c) At the toe of a spillway, where the high velocity flow over the spillway meets the larger flow depth downstream.

Generally the formation of the jump involves considerable air entrainment, mixing, and energy loss, which leads to various practical benefits such as

(a) dissipation of the energy of water falling over a weir or spillway (sometimes in combination with other devices);

(b) thorough mixing of a chemical (e.g., a purifying agent) in water by introducing it upstream of the jump;

(c) improvement of the water quality by increasing the dissolved oxygen content through air-entrainment;

(d) holding back of high tailwater downstream of a sluice gate in order to keep it from getting submerged and causing a reduction in discharge;

(e) reducing the uplift pressure on an apron on the channel bed by forming a jump on or before the apron thereby increasing the flow depth on the apron.

The important parameters of a hydraulic jump, therefore, are the energy loss, flow depths, length of the jump, location of jump, surface profile of the jump (as it determines the downward force counterbalancing the uplift on an apron), and the amount of air entrained by the jump. While some of these characteristics can be obtained theoretically using the basic equations of continuity, momentum, and energy, some other are obtained based on the empirical relations developed on the basis of extensive experiments (mostly in rectangular channels). We first discuss the application of the basic equations and then describe the empirical relationships.

4.2.1 Continuity, Momentum and Energy Equations

Figure 4.3 shows the control volume over which the continuity and momentum equations are applied to analyse the jump for a given discharge in the channel. We assume that the pre-jump depth (and, therefore, velocity) is known and we have to find the post-jump depth and velocity. The continuity equation provides us a relation between the post-jump depth and velocity. We need one more equation in order to estimate the post-jump depth. Note that due to the (unknown) energy loss within the jump, we cannot apply the energy equation to relate the flow depths before and after the jump. However, the momentum equation may still be used (unless there are unknown forces within the control volume when, for example, there are blocks constructed on the bed as energy dissipating devices). After computing the post-jump flow depth, one can use the energy equation to estimate the energy loss within the jump. We describe the analysis for a general cross section in a sloping channel including the frictional forces at the boundary. However, since most of the experimental work pertains to horizontal rectangular channels and ignoring friction, we would later concentrate on the simplified version of the momentum equation.

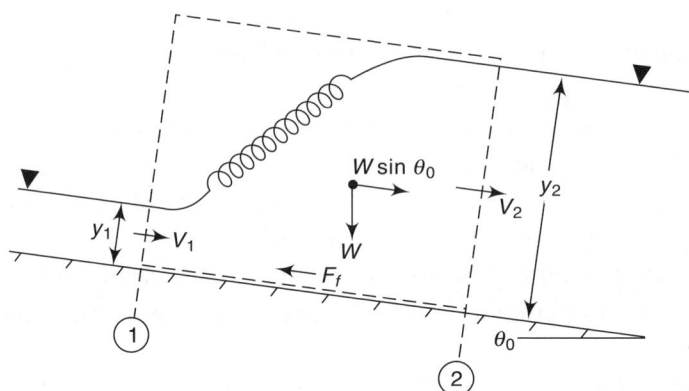

Fig. 4.3 Control volume for the analysis of hydraulic jump

Application of the continuity equation over the control volume results in

$$Q = V_1 A_1 = V_2 A_2 \tag{4.1}$$

and the momentum equation provides

$$P_1 + W \sin \theta_0 - P_2 - F_f = \rho Q(V_2 - V_1) \tag{4.2}$$

in which P denotes the pressure force on the control surface, W is the weight of water in the control volume, and F_f is the friction force acting on the boundary of the control volume (we neglect the air resistance on the air water interface, which is generally very small). Since the rapidly varied flow typically occurs over a short length and the slope of the channel is generally small, it would not be unreasonable to assume that $W \sin \theta_0 - F_f \approx 0$.[1] Moreover, if sections 1 and 2 are sufficiently distant from the rapidly varying zone of the jump, we may assume the pressure distribution at these sections to be *hydrostatic*. With these assumptions, and using the continuity equation, Eq. (4.2) is written as

$$\rho g A_1 \bar{y}_1 - \rho g A_2 \bar{y}_2 = \rho Q^2 \left(\frac{1}{A_2} - \frac{1}{A_1} \right)$$
(4.3a)

i.e.,
$$\left(A\bar{y} + \frac{Q^2}{gA} \right)_1 = \left(A\bar{y} + \frac{Q^2}{gA} \right)_2$$
(4.3b)

We now define a *specific force* (Sometimes called the momentum function or force plus momentum) F_s, as

$$F_s = A\bar{y} + \frac{Q^2}{gA}$$
(4.4)

and conclude that the pre-jump and post-jump specific force values should be the same (provided the assumptions made in the analysis are satisfied). The specific force, as seen from Eq. (4.4), has dimension L^3 and may be thought of as the sum of pressure force and momentum flux per unit specific weight of the fluid.[2] The problem of finding the post-jump depth for known pre-jump depth, therefore, amounts to finding out the depth that would have the same specific force as the pre-jump depth. A plot of specific force versus flow depth is very useful for this purpose. We also observe that, similar to the specific energy, the specific force contains one term which increases with increase in the flow depth and another which decreases. For very large as well as very small flow depths, the specific force becomes infinite and there would be an intermediate depth at which the specific force will attain a minimum. We look at the characteristics of the specific force in the next subsection and then discuss the computation of the post-jump depth.

---

[1]The two terms being neglected, the weight component and the friction force act in the opposite direction; the assumption would be good even when the two terms are not very small but are of the same order.

[2]We could have defined the specific force as the sum of the pressure force and the momentum force or the sum of these two forces per unit mass density. However, it is more convenient to define it in such a way that it is independent of fluid properties and only involves the length dimensions.

4.2.2   Characteristics of Specific Force

A plot of Eq. (4.4) for a trapezoidal channel (bottom width of 5 m, side slopes of 1H:1V, and carrying a discharge of 20 m³/s) is shown in Fig. 4.4. As postulated in the previous subsection, the specific force attains a minimum value (of about 9.5 m³) at a flow depth of about 1.1 m. The specific energy variation with depth is also plotted in this figure and clearly shows that the minimum specific energy (of about 1.5 m) also occurs at the same depth (which we have already defined as the critical depth). This fact may be mathematically established by minimizing F_s through Eq. (4.4), and using the relations $d(A\bar{y})/dy = A$ and $dA/dy = T$ [see Eq. (3.14)], to obtain the stationarity condition

$$\frac{Q^2 T}{gA^3} = 1 \tag{4.5}$$

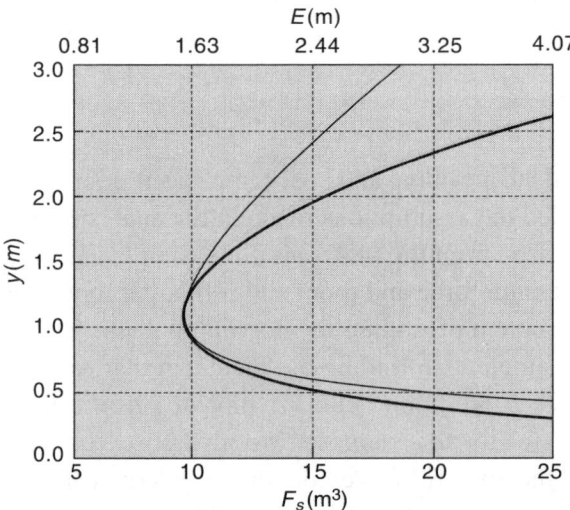

Fig. 4.4   Specific force (thick line) and specific energy (thin line) in a trapezoidal channel

same as Eq. (3.24)[3]. For a specific force greater than the minimum value, there would be two flow depths, one supercritical and the other subcritical,[4] which will result in the same specific force. These depths are called the

---

[3]If we include the energy and momentum correction factors in these equations, there would be slight difference between the two, and in that case Eq. (3.22) would be used to define the critical condition.

[4]Earlier we had loosely called the hydraulic jump as a jump from supercritical depth to subcritical depth. Now we have shown through Eq. (4.5) that the pre-jump depth is supercritical and the post-jump depth is subcritical.

conjugate depths or the *sequent depths*. From Fig. 4.4, we see that for a given combination of conjugate depths, there would be an associated energy loss. For example, for the upstream supercritical depth of 0.5 m, the specific force is about 15 m³. The sequent depth may be read from the figure as about 1.95 m. (Note that the actual value of the specific force is not used in finding the sequent depth, what is needed is the equality of specific force at two depths.) The specific energies corresponding to these flow depths are about 3.25 m and 2 m, respectively, indicating an energy loss of about 1.25 m per unit weight of water.[5] Note that as the pre-jump depth approaches the critical depth, the post-jump depth also approaches the critical depth and the energy loss becomes smaller. When pre-jump depth is equal to the critical depth, there is no jump and the flow depth is critical throughout.

The specific force plot comes in handy for obtaining the post-jump depth for given pre-jump conditions. However, the way we have plotted Fig. 4.4, it would require a new plot if channel dimensions or discharge changes. It would be desirable to have a generally applicable specific force diagram. For a trapezoidal channel of bed width B and side slopes mH:1V, Eq. (4.4) may be manipulated to obtain a nondimensional specific force (F_s^*) as follows:

$$F_s = \frac{Q^2}{g(By + my^2)} + By\frac{y}{2} + my^2\frac{y}{3} \tag{4.6}$$

from which, we get

$$F_s^* \left(= \frac{m^2 F_s}{B^3}\right) = \frac{\Theta^2}{\eta(1+\eta)} + \frac{\eta^2}{2} + \frac{\eta^3}{3} \tag{4.7}$$

where the nondimensional discharge $\Theta = m^{3/2}Q/g^{1/2} B^{5/2}$ and the nondimensional depth $\eta = my/B$. Figure 4.5 shows a plot of this equation. For the channel used for generating Fig. 4.4, we have $\Theta = 0.1142$ and for a pre-jump depth of 0.5 m, $\eta = 0.1$. From Fig. 4.5, we get $F_s^* = 0.12$ and the nondimensional sequent depth as 0.4, i.e., a post-jump depth of 2 m.

It is obvious that Fig. 4.5 is not applicable to rectangular or triangular channels for which B and m respectively become zero. (For rectangular channels, as we will see in subsection 4.2.4 the post-jump depth can be analytically computed.) Another option for preparing a comprehensive plot, which is valid for these limiting cases also, is the so-called SDR (sequent depth ratio) plot,

---

[5]With a discharge of 20 m³/s and specific weight of 9810 N/m³, it amounts to about 245 kW. However, assuming that the entire energy loss is utilized in raising the temperature of water, it causes a rise of only 0.003 °C (specific heat capacity of water = 4.18 kJ/kg/K).

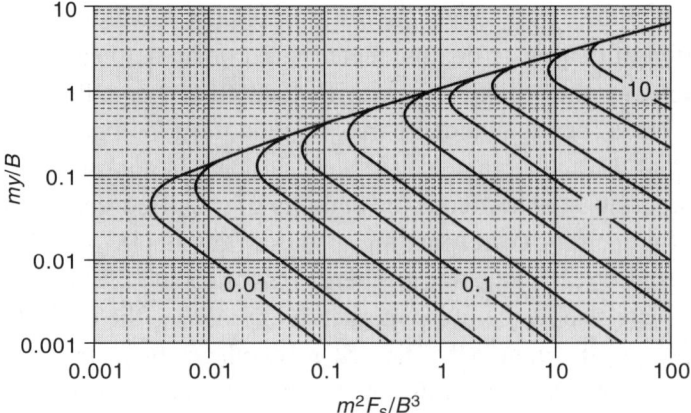

Fig. 4.5 Nondimensional specific force variation in a trapezoidal channel [numbers on the curves represent the values of nondimensional discharge (0.01, 0.02, 0.05, 0.1,...,10)]

which relates the ratio of the post-jump and pre-jump flow depths to the pre-jump values of Froude number and dimensionless depth. It can be seen from Eq. (4.6) that the SDR, r ($= y_2/y_1$), is related to upstream Froude number, F_1, $[= Q\sqrt{B + 2my_1}\, / \sqrt{g\,(By_1 + my_1^2)^{3/2}}\,]$ and nondimensional depth, η_1 ($= my_1/B$), through

$$\frac{r + \eta_1 r^2}{1 + \eta_1\,(1+r)}\left(\frac{1+r}{2} + \eta_1\,\frac{1+r+r^2}{3}\right) = \frac{(1+\eta_1)^2\,F_1^2}{1 + 2\eta_1} \tag{4.8}$$

which is shown in Fig. 4.6 (note that $\eta_1 = 0$ corresponds to rectangular channel, and ∞, to triangular channels). For the channel described earlier, we have $F_1 = 3.4$ and, for $\eta_1 = 0.1$, get $y_2/y_1 = 4$, i.e., a post-jump depth of 2 m.

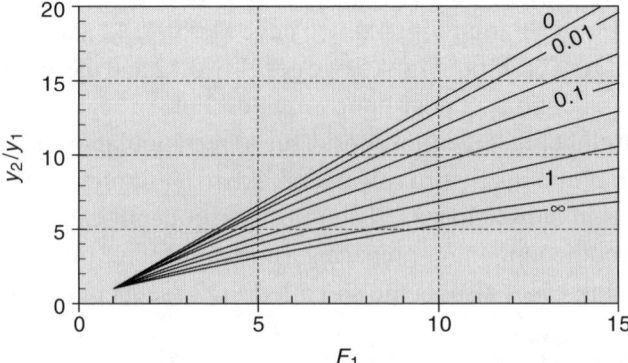

Fig. 4.6 Sequent depth ratio in a trapezoidal channel [Numbers on the curves represent the values of nondimensional pre-jump depth (0, 0.01, 0.05, 0.1, 0.2, 0.5, 1, 5, ∞).]

Since the graphical determination of sequent depth involves (sometimes significant) errors in reading the values, it would be desirable to have techniques or formulae to compute it. In the next subsection, we discuss some aspects of computing the sequent depth.

4.2.3 Computation of Sequent Depth

Since Eq. (4.8) is a quartic equation, its analytical solution is quite involved (Abramowitz and Stegun, 1958). Iterative techniques based on fixed point iteration or Newton's method can be employed to solve for the sequent depth ratio. Approximate expressions have been suggested (Hager 1992) for the limiting cases of rectangular and triangular channels, *for pre-jump Froude numbers greater than 2.5*, as follows:

$$\text{Rectangular: } \frac{y_2}{y_1} = \sqrt{2}\, F_1 - 0.5$$

$$\text{Triangular: } \frac{y_2}{y_1} = \left(\frac{3F_1^2}{2} - 1 \right)^{1/3}$$

(4.9)

For trapezoidal channels, an approximation, with an error bound of less than 2% within the *Froude number range of 2.5 to 15*, is obtained as

$$\frac{y_2}{y_1} = C_1 (F_1^{2/3} - 0.5)^{C_2}$$

(4.10)

in which the coefficients C_1 and C_2 depend on the nondimensional depth, η_1, and are given by

$$C_1 = \frac{2.058 + 1.53\, \eta_1}{1 + \eta_1}$$

(4.11a)

$$C_2 = \frac{1.333 + 4.585\, \eta_1}{1 + 5.285\, \eta_1}$$

(4.11b)

For a circular channel of diameter D_0, a plot of the sequent depth ratio is shown in Fig. 4.7. The numbers near the curve represent the values of y_1/D_0. Note that the curves stop at $y_2 = D_0$, i.e., $y_2/y_1 = 1/(y_1/D_0)$, since the post-jump depth cannot exceed the pipe diameter. It is also seen that there is a very slight dependence of the SDR on y_1/D_0.

Hager (1992) provides an approximation, valid for $y_1/D_0 < 0.7$, to compute the SDR for a circular channel. Similar expressions for other shape of channels are available in literature. However, most practical cases of hydraulic jump occur in channels which are (or can be approximated as) rectangular.

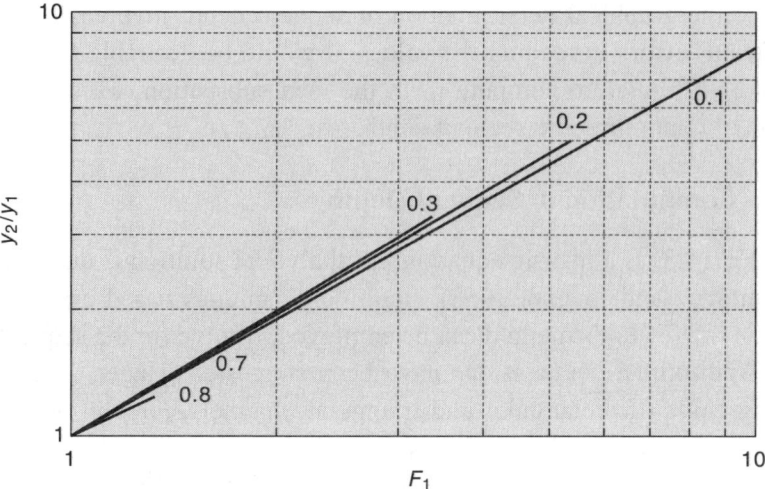

Fig. 4.7 Sequent depth ratio in a circular channel [Numbers on the curves represent the values of nondimensional pre-jump depth, y_1/D_0; the lines corresponding to $y_1/D_0 = 0.4, 0.5$, and 0.6 are virtually identical.]

(Another advantage of considering rectangular section is that the analytical solution for the sequent depth ratio is rather straightforward.) Therefore, after an example demonstrating the computation of sequent depth in non-rectangular channels, we would concentrate on the characteristics of the jump in rectangular channels only.

Example 4.1 A discharge of 25 m³/s passes through channels of various shapes such that a hydraulic jump is formed. Find the post-jump depth for (a) a trapezoidal channel with bottom width of 5 m, side slopes of 1H:1V, and pre-jump depth of 0.5 m, (b) a 5 m wide rectangular channel with pre-jump depth of 0.5 m, (c) a triangular channel with side slopes of 3H:1V, and pre-jump depth of 1.0 m, and (d) circular channel of diameter 4 m, and pre-jump depth of 1 m.

Solution

The pre-jump Froude number is given by $F_1 = \dfrac{V_1}{\sqrt{gD_1}} = \left(\dfrac{Q^2 T_1}{gA_1^3}\right)^{1/2}$

(a) *For the trapezoidal channel:*
Using Fig. 4.5: The dimensionless pre-jump depth, $\eta_1 = my_1/B = 0.1$ and the nondimensional discharge, $\Theta = m^{3/2}Q/g^{1/2}B^{5/2} = 0.1428$. From the figure, we also get my_2/B equal to about 0.5, thus resulting in a post-jump depth of **2.5 m**.

Using Fig. 4.6: The pre-jump Froude number, $F_1 = 4.29$ and the sequent depth ratio is about 5, indicating a post-jump depth of **2.5 m**.

Using Eqs (4.10) and (4.11): For $\eta_1 = 0.1$, the coefficients C_1 and C_2 are obtained as 2.01 and 1.172 respectively. The sequent depth ratio is then obtained from Eq. (4.10) as 4.9 and the post-jump depth as **2.45 m**.

All the methods give nearly identical results, which are close to the exact solution of 2.476 m.

(b) *For the rectangular channel:*
Using Fig. 4.6: The pre-jump Froude number is $F_1 = 4.515$ and the sequent depth ratio is about 6, indicating a post-jump depth of **3 m**.

Using Eq. (4.9): For $F_1 = 4.515$, the sequent depth ratio is 5.89 and the post-jump depth is obtained as **2.94 m**.

Using Eqs (4.10) and (4.11): For $\eta_1 = 0$, the coefficients C_1 and C_2 are obtained as 2.058 and 1.333 respectively. The sequent depth ratio is then obtained from Eq. (4.10) as 6.00 and the post-jump depth as **3.00 m**.

All the methods give nearly identical results, which are close to the exact solution of 2.953 m. As we will see shortly, this value may be directly obtained from the Froude number.

(c) *For the triangular channel:*
Using Fig. 4.6, with a flow depth of 1.0 m, the pre-jump Froude number is $F_1 = 3.763$ and the sequent depth ratio is about 2.7, indicating a post-jump depth of **2.7 m**.

Using Eq. (4.9), for $F_1 = 3.763$, the sequent depth ratio is 2.73 and the post-jump depth is obtained as **2.73 m**.

Using Eqs (4.10) and (4.11), for $\eta_1 = \infty$, the coefficients C_1 and C_2 are obtained as 1.53 and 0.868 respectively. The sequent depth ratio is then obtained from Eq. (4.10) as 2.69 and the post-jump depth as **2.69 m**.

All the methods give nearly identical results, which are close to the exact solution of 2.682 m.

(d) *For the circular channel:*
Using Fig. 4.7, with a flow depth of 1 m and diameter of 4 m, the pre-jump Froude number is $F_1 = 3.858$, $y_1/D_0 = 0.25$, and the sequent depth ratio is about 3.7, indicating a post-jump depth of **3.7 m**. This value is close to the exact solution of 3.725 m obtained by equating the specific force values at the two depths.

4.2.4　Hydraulic Jump in Rectangular Channels

For a rectangular channel, Eq. (4.8) reduces to a quadratic equation

$$r(1 + r) = 2F_1^2 \tag{4.12}$$

with $F_1 = V_1/\sqrt{gy_1}$, from which the sequent depth ratio, $r(= y_2/y_1)$, is obtained as

$$\frac{y_2}{y_1} = \frac{\sqrt{1 + 8F_1^2} - 1}{2} \tag{4.13}$$

This equation is called the *Belanger equation*, since it was first obtained by the Frenchman, J.B. Belanger, in early 1800s. A similar equation can be derived to compute the pre-jump depth when the post-jump conditions are given as

$$\frac{y_1}{y_2} = \frac{\sqrt{1 + 8F_2^2} - 1}{2} \tag{4.14}$$

Eq. (4.13) or (4.14) also provides us with a means of estimating the discharge in a channel by measuring the sequent depths. Using any of these equations, it can be shown that the discharge per unit width, $q(= V_1y_1)$, is given by

$$q = \sqrt{gy_1 y_2 \frac{(y_1 + y_2)}{2}} \tag{4.15}$$

(The same expression can be obtained directly from application of Eq. (4.3) to a rectangular channel.) Once the sequent depth is computed, the energy loss is obtained by computing the energy head before and after the jump. However, a direct relation between the energy loss, E_L, and the sequent depths can be obtained by writing

$$E_L = y_1 + \frac{V_1^2}{2g} - y_2 - \frac{V_2^2}{2g}$$

$$= \frac{q^2}{2g}\left(\frac{1}{y_1^2} - \frac{1}{y_2^2}\right) - (y_2 - y_1)$$

and then using Eq. (4.15) to obtain

$$E_L = \frac{(y_2 - y_1)^3}{4 y_1 y_2} \tag{4.16}$$

[If we want to express the energy loss in terms of pre-jump parameters only, we can similarly obtain $E_L = y_1(\sqrt{1 + 8F_1^2} - 3)^3/16(\sqrt{1 + 8F_1^2} - 1)$.]

In most cases, the actual energy loss may not be of interest and we may require dissipation of a certain percentage of the incoming (pre-jump) energy. The relative energy loss[6] can be expressed in terms of the pre-jump Froude number as

$$\frac{E_L}{E_1} = \frac{(y_2 - y_1)^3}{4\,y_1 y_2\,[\,y_1 + (q^2/2\,gy_1^2)]} = \frac{(r-1)^3}{2r(2 + F_1^2)}$$

$$= \frac{(\sqrt{1 + 8F_1^2} - 3)^3}{8(2 + F_1^2)(\sqrt{1 + 8F_1^2} - 1)} \quad (4.17)$$

Figure 4.8 shows the variation of the relative energy loss with the upstream Froude number. Obviously since we aim at dissipating a large amount of energy, the Froude number should be as high as possible. However, based on extensive experiments (USBR 1955), it has been shown that for $F_1 > 9$, the downstream surface may become unacceptably rough and wavy (classified as a *strong jump*). Moreover, the rate of increase in energy dissipation also becomes quite small beyond this Froude number. Therefore, it is preferable to keep the pre-jump Froude number below 9. On the other hand, it was found that for Froude number between 2.5 and 4.5, the incoming high-velocity jet oscillates irregularly between the channel bed and water surface (called *oscillating jump*) and may generate waves on the surface. Also, the energy loss for Froude numbers less than 4.5 is less than 45%. Therefore, the Froude number should be kept above 4.5. The desired type of jump will have a Froude number

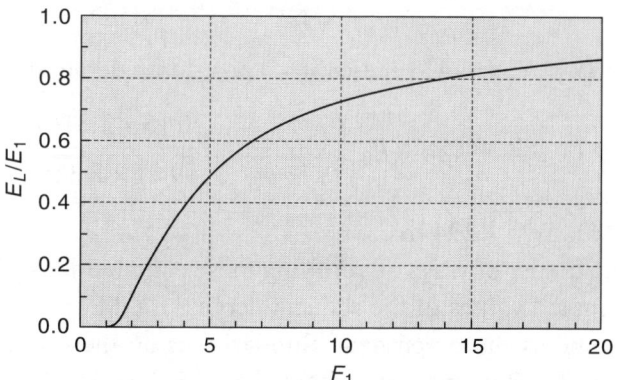

Fig. 4.8 Relative energy loss in a rectangular channel

---

[6]Sometimes, the term *efficiency*, defined as the ratio of post-jump and pre-jump specific energies, is used to provide an idea of the relative energy loss $(E_2/E_1 = 1 - E_L/E_1)$.

ranging from 4.5 to 9, and is called a *steady jump*. As seen from Fig. 4.8, the relative energy loss in this type of jump is from 45 to 70%.

From Eqs (4.15) and (4.16), it should be noted that both the discharge and the energy loss are nonlinear functions of the pre-jump and post-jump depths. Therefore, for a given discharge, q, and a specified energy loss, E_L, one would need to perform iterations to obtain the flow depths. Swamee and Prasad (1977) and Swamee, et al. (1996) provide approximate equations for determination of the flow depths in such cases. The following equation has been found to work well within the range of upstream Froude numbers from 1 to 20:

$$\frac{y_2}{y_c} = 1 + \frac{1.4\,(E_L/y_c)^{0.415}}{1 + 0.476(E_L/y_c)^{0.2}} \tag{4.18}$$

The critical depth is obtained from the given discharge $[y_c = (q^2/g)^{1/3}]$ and once y_2 is computed from Eq. (4.18), it is a simple matter to obtain other jump characteristics [Eq. (4.14) is used to obtain y_1 and the Froude number and specific energy at both sections are then easily computed]. The following example illustrates the computation of the jump characteristics.

Example 4.2   A hydraulic jump is formed in a 5 m wide rectangular channel carrying a discharge of 20 m^3/s. The pre-jump depth is 0.5 m. Find the post-jump depth, post-jump Froude number, and energy loss in the jump.

Solution

The pre-jump Froude number, $F_1 = \dfrac{V_1}{\sqrt{gy_1}} = \dfrac{20/(5 \times 0.5)}{\sqrt{9.81 \times 0.5}} = 3.612.$

From Eq. (4.13), $y_2/y_1 = 4.633$ indicating a post-jump depth of **2.316 m**.

The post-jump Froude number, $F_2 = \dfrac{V_2}{\sqrt{gy_2}} = \dfrac{20/(5 \times 2.316)}{\sqrt{9.81 \times 2.316}} = 0.362.$

From Eq. (4.16), $E_L = $ **1.294 m**.

There are a number of ways of verifying these computations. For example, using the computed values of the sequent depths, Eq. (4.15) results in $q = 4$ m^3/s/m, the same as the given data. Similarly, using the pre-jump specific energy value of 3.762 m and the computed energy loss of 1.294 m, we have the relative energy loss of 0.344, the same as obtained from Eq. (4.17)[7] with $F_1 = 3.612$. Equation (4.18) can also be checked from these values. With the

---

[7]Figure 4.8 can also be used to obtain (approximately) the relative energy loss.

critical depth $y_c = (q^2/g)^{1/3} = 1.177$ m, we have $E_L/y_c = 1.099$. Equation (4.18) results in $y_2/y_c = 1.98$ and $y_2 = 2.331$ m, very close to the exact value of 2.316 m.

In addition to the sequent depth and the energy loss, the other important characteristics of the hydraulic jump are its length, location, profile, and amount of air entrained. These are typically based on empirical equations as discussed below.

Length of the jump

The channel bed within the jump has to be protected against the erosive action of the high velocity jet, and the computation of the length of the jump becomes important.[8] Although the beginning of the jump is well-defined by the start of the roller, it is rather difficult to pinpoint the end since the water surface is almost flat near the end of the roller. Sometimes, the end of the roller is taken as the end of the jump and sometimes, it is the point where the high velocity jet at the bottom separates from the channel bed (for the *steady jump*, these two points are virtually in the same vertical plane). Based on the experimental data of USBR (1955), the jump length, L_j, may be given in terms of the pre-jump parameters by

$$\frac{L_j}{y_1} = 10(F_1 - 1) - 0.02(F_1 - 1)^{2.53} \qquad (4.19)$$

However, it is more convenient to express the length of the jump in terms of the post-jump depth, y_2, since experimental data suggest that the ratio, L_j/y_2, is fairly constant over a wide range of F_1 (4.5 to 15). It varies from about 5.9 at Froude no. of 4.5 to about 6.2 at 9 and then 6.0 at 15. Therefore, for most practical cases, the jump length is taken as about 6.2 times the post-jump depth. Yet another alternative is to use the expression $L_j = 6.9\ (y_2 - y_1)$, indicating that the length of the jump is almost 7 times its height.

Profile of the jump

The pressure distribution at the channel bottom is needed for an efficient design of the floor[9] and the water surface profile is needed to decide the height

---

[8]If the length turns out to be unacceptably large, we would have to use other energy dissipating devices, e.g., steps, baffle blocks, sills, etc. Since our main concern here is the hydraulic jump and not energy dissipating devices, we refer the reader to USBR (1955) and Hager (1992) for further reading.

[9]For a jump-formed downstream of a spillway on permeable foundation, an uplift pressure acts on the floor due to the seepage of water from the reservoir. This head has to be balanced by the weight of the apron and the water pressure acting on it.

of the training walls. Although the pressure distribution within the jump is not hydrostatic, the pressure at the bottom of the channel has been observed to correspond closely to the water surface profile in the jump. Rajaratnam and Subramanya (1968) analysed a large number of experimental data and proposed a dimensionless jump profile, which may be closely approximated as (see Fig. 4.9 for the coordinate system used)

$$\frac{y - y_1}{y_2 - y_1} = 1.03 \tanh\left(\frac{2.15\,x}{L_j}\right) \tag{4.20}$$

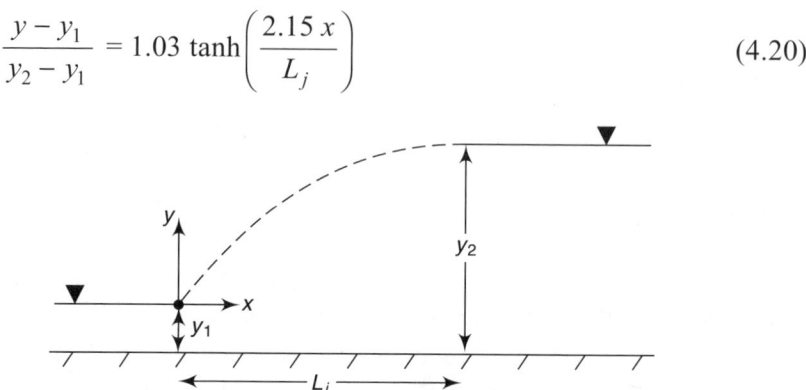

Fig. 4.9 Elements of a hydraulic jump

Location of the Jump

Going back to the problem of gradually varied flow profile computation in case of a steep slope followed by a mild slope (see Section 3.4.2, Fig. 3.20), we now know that a hydraulic jump would be formed since the uniform flow on the steep channel is supercritical and that on the mild slope is subcritical. The location of the jump, whether it is formed on the steep channel or the mild channel, would depend on the normal depths in the two channels. Assuming that the channel slopes are so small that they may be treated as horizontal,[10] the sequent depths would correspond to the same specific force. The specific-force diagram is shown in Fig. 4.10, in which the two normal depths, y_{n1} (on steep channel) and y_{n2} (on mild channel), are also shown. Clearly, if y_{n1} and y_{n2} are sequent depths [Fig. 4.10(a)], the jump would form right at the junction of the two slopes. If, however, y_{n2} is larger than the sequent depth corresponding to y_{n1} [Fig. 4.10(b)], the specific force on the downstream side is larger than that required for the formation of the jump. Hence, the jump would be pushed upstream and would form on the steep channel. The jump will be followed by an S_1 profile, which would increase the flow

---

[10]The analysis is valid even without this assumption. However, the component of weight in the flow direction has to be accounted for. Jump on a sloping floor will be discussed a little later.

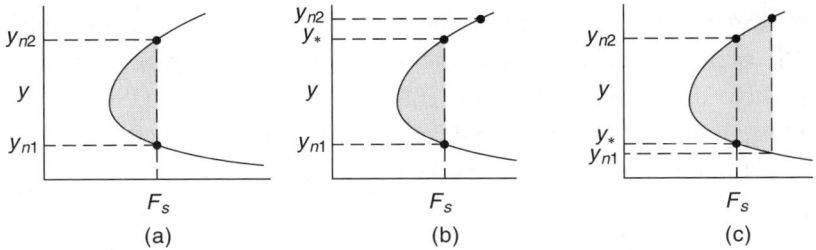

Fig. 4.10 Specific-force diagram to analyse the location of a jump

depth from the sequent depth, y_*, to y_{n2}. Similarly, if y_{n2} is smaller than the sequent depth corresponding to y_{n1} [Fig. 4.10(c)], the specific force on the downstream side is smaller than that required for the formation of the jump. Hence, the jump would be pushed downstream and would form on the mild channel. The jump will be preceded by an M_3 profile, which would increase the flow depth from y_{n1} to the sequent depth, y_*, corresponding to y_{n2}.

Computation of the profile length can be done using any of the methods discussed in the previous chapter. Similarly, the location of jump downstream of a spillway is also determined. Sometimes, however, there may be constraints on the movement of the jump. For example, considering the supercritical flow issuing from a sluice gate, which jumps to the subcritical tailwater depth, if we increase the tailwater depth, the jump would be pushed upstream. The presence of the gate, however, restricts the movement of the jump and a *submerged* or *drowned* jump is formed.

Air entrainment

The amount of air entrained by the jump is an important factor from the point of view of the water quality. Empirical equations, relating the volumetric flow rate of air to that of water, have been suggested based on the analysis of the data obtained from laboratory experiments and field observations. Since there is not much difference between these relations and there is no theoretical basis for preferring one to the others, we list here only one of these (Rajaratnam 1967):

$$\frac{Q_a}{Q_w} = 0.018(F_1 - 1)^{1.245}$$

in which the subscript a refers to air and w to water. Wisner (1965) and Renner and Naudascher (1975) proposed similar relationships—Wisner (based on the data covering a wider range of the Froude number) using different coefficients in the equation above and Renner and Naudascher relating the ratio of discharges to square of the Froude number rather than $(F_1 - 1)$.

Example 4.3   A 5 m wide rectangular channel with a roughness coefficient of 0.01 carries a discharge of 20 m³/s. A hydraulic jump is formed near a point where the channel bed slope changes from 0.002 to 0.001. Would the jump form before or after the break in slope? What would be the length of the jump? Estimate the water depth at the mid-length.

Solution

The normal depths for the two bed slope values can be computed using any of the methods described in Chapter 2. For the steeper channel, the normal depth is y_{n1} = 1.080 m and for the milder channel, y_{n2} = 1.372 m. Since the critical depth is 1.177 m (see Example 4.2), it is obvious that the flow is supercritical before the break and subcritical after it (we can also arrive at the same conclusion by finding the critical slope as S_c = 0.00155). A jump would, therefore, form near the break in slope, as specified in the problem. To determine whether the jump forms on the steep slope or the mild slope, we need to compute the sequent depth for the flow depth of y_{n1} (it would also work if we compute the sequent depth corresponding to y_{n2} and compare it with y_{n1}).

The Froude number is obtained as

$$F_1 = \frac{20/(5 \times 1.080)}{\sqrt{9.81 \times 1.080}} = 1.14$$

and the sequent depth, using Eq. (4.13), as 1.280 m. Since this is smaller than the downstream normal depth, the jump would form before the break on the steep slope from a **pre-jump depth of 1.08 m** to the **post-jump depth of 1.280 m**. It would be followed by an S_1 profile from the depth of 1.280 m to a depth of 1.372 m, the length of which could be computed using the methods described in Chapter 3.

The length of the jump is obtained from Eq. (4.19) as **1.5 m**, which is 7.5 times its height of 0.2 m, close to the correlation of the length being 6.9 times the height. However, it is very different from another correlation which states that the length is 6.2 times the post-jump depth. This is because the pre-jump Froude number is very small while the correlation is valid for a range of F_1 from about 4 to 16. From Eq. (4.20), the flow depth at the middle of the jump is obtained from

$$\frac{y - y_1}{y_2 - y_1} = 1.03 \tanh \left(\frac{2.15 \times 0.5 L_j}{L_j} \right) = 0.8151$$

resulting in y = **1.243 m**.

The discussion so far enables us to analyse hydraulic jumps in rectangular channels with *horizontal bed* with the assumption that the only forces acting on the control volume enclosing the jump are the *pressure forces*. While these would hold good in most cases, there are a number of situations where the effect of channel bed slope or *other* forces would be considerable. For example, a jump formed on the downstream face of a spillway must account for the steep slope of the face and a jump with additional energy dissipating devices, e.g., baffle blocks, end sill, etc., must account for the forces acting on these devices. Due to the difficulties in a theoretical formulation[11] empirical relations have to be utilized to arrive at a solution for various parameters (e.g., sequent depth ratio, energy loss, jump length). The next section describes some of these relations for the jump in sloping channels.

4.2.5   Hydraulic Jump in Sloping Channels

For a hydraulic jump to form in a steep channel, there must be some downstream control to make the depth subcritical. In most practical cases, this control is in the form of a horizontal (or nearly horizontal) bed that follows the steep sloped channel. Depending on the velocity and depth of flow in the sloping channel and the downstream flow depth, the jump may form completely in the steep channel, may start in the steep channel and end in the flat channel, or may be pushed down completely to the flat channel. The third case, of course, poses no problem since it may be treated as a jump in a horizontal channel. The second case is quite complex and has not been studied extensively. It is the first case, the jump being completely on the steep slope, which has been investigated in detail by several researchers and is described here.

Figure 4.11 shows the formation of hydraulic jump on a sloping floor. If we assume it to be a free jump, such that the tailwater depth does not influence the post-jump depth, we may write a momentum equation in the direction of flow as [see Eq. (4.2)]

$$P_1 + W \sin \theta_0 - P_2 - F_f = \rho Q (V_2 - V_1)$$

Neglecting friction forces, considering unit width of a *rectangular* channel, and using the hydrostatic pressure force in a steep channel, we get

$$\frac{1}{2} \gamma y_1^2 \cos \theta_0 + W \sin \theta_0 - \frac{1}{2} \gamma y_2^2 \cos \theta_0 = \rho q^2 \left(\frac{1}{y_2} - \frac{1}{y_1} \right) \qquad (4.21)$$

---

[11]For example, for a sloping channel, the weight component in the direction of flow has to be included in the momentum equation, but is not known since it will depend on the length of the jump as well as its profile. Similarly, the forces acting on baffle blocks would not be known.

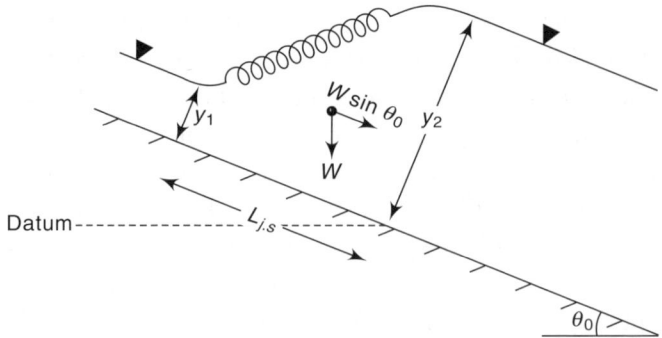

Fig. 4.11 Hydraulic jump on a sloping floor

with q being the discharge per unit width. The weight of water within the control volume may be approximated using the trapezoidal rule and applying a correction factor, κ, for the actual shape of the jump, as $W = \kappa \gamma L_{j,s}(y_1 + y_2)/2$ with $L_{j,s}$ being the length along the sloping floor. (Clearly, $\kappa > 1$. Experimental results suggest that its value is about 1.06 for most practical cases.). The next approximation is made to eliminate the (unknown) length of the jump from this expression by treating it as some multiple of the height of the jump, i.e., $L_{j,s} = \iota(y_2 - y_1)$ (comparing with the horizontal channel, the value of ι should be about 6.9 for $\theta_0 = 0$. The expression $\iota = 6.9/(1 + 5 \tan \theta_0)$ has been found to fit the data very well for bed slopes up to about 20°). Equation (4.21) is then written as

$$\frac{1}{2} \gamma y_1^2 \cos \theta_0 + \kappa \iota \frac{\gamma (y_2^2 - y_1^2)}{2} \sin \theta_0 - \frac{1}{2} \gamma y_2^2 \cos \theta_0 = \rho q^2 \left(\frac{1}{y_2} - \frac{1}{y_1} \right)$$

(4.22)

which results in a sequent depth ratio

$$r \left(= \frac{y_2}{y_1} \right) = \frac{\sqrt{1 + [8 F_1^2/(1 - \kappa \iota \tan \theta_0)]} - 1}{2}$$

(4.23)

in which the pre-jump Froude number for a large bed slope is defined as [see Eq. (3.24) and its footnote] $F_1 = V_1/\sqrt{g y_1 \cos \theta_0}$

Once the sequent depth is computed, the energy loss is readily obtained from

$$E_L = E_1 + L_{j,s} \sin \theta_0 - E_2$$

in which E_1 and E_2 are the *specific energies* at sections 1 and 2 respectively and E_L is the loss in *total energy* in the jump.[12] To compute the relative energy loss, we must choose a datum to define the energy at the pre-jump section. Choosing the datum as shown in Fig. 4.11, the relative energy loss is written as

$$\frac{E_L}{H_1} = \frac{E_1 + L_{j,s}\sin\theta_0 - E_2}{E_1 + L_{j,s}\sin\theta_0} \tag{4.24}$$

Using the fact that $E = y\cos\theta_0\left(1 + \dfrac{F^2}{2}\right)$ and $L_{j,s} = \iota(y_2 - y_1)$, we get[13]

$$\frac{E_L}{H_1} = \frac{(r-1)\left(\dfrac{r+1}{r^2}F_1^2 - 2 + 2\iota\tan\theta_0\right)}{2 + F_1^2 + 2\iota(r-1)\tan\theta_0} \tag{4.25}$$

Example 4.4 A 50 m wide spillway chute is designed to carry a maximum discharge of 100 m³/s and has a bed slope of 3H:1V. A hydraulic jump is formed on the chute at a point where the pre-jump flow depth is 0.3 m. What would be the post-jump depth? Estimate the length of jump and the energy loss.

Solution
We will use the value of κ as 1.06 and, since the slope is less than 20°,

$$\iota = \frac{6.9}{1 + 5\tan\theta_0} = \frac{6.9}{1 + 5\dfrac{1}{3}} = 2.59$$

The pre-jump Froude number is obtained as

$$F_1 = \frac{V_1}{\sqrt{gy_1\cos\theta_0}} = \frac{100/50/0.3}{\sqrt{9.81 \times 0.3 \times 0.95}} = 4.00$$

Eq. (4.23) results in a sequent depth ratio of 18.9, i.e., a post-jump flow depth of **5.7 m**.

The length of the jump, $L_{j,s} = \iota(y_2 - y_1) = 2.59(5.7 - 0.3)$ m = **14 m**

---

[12]For the horizontal channel, the channel bed was chosen as the datum and the specific energy and total energy were interchangeable. For sloping channels, the two most obvious choices of datum are the bed level at the pre-jump location or that at the post-jump section. Generally, the lower datum corresponding to the post-jump bed elevation is used.

[13]Clearly, Eq. (4.25) should reduce to Eq. (4.17) for horizontal channels.

The energy loss in the jump,

$$E_L = E_1 + L_{j,s} \sin \theta_0 - E_2 = \frac{V_1^2}{2g} + y_1 \cos \theta_0 + L_{j,s} \sin \theta_0 - \frac{V_2^2}{2g}$$
$$- y_2 \cos \theta_0$$
$$= \mathbf{1.56 \ m}$$

One should note that the specific energy at the pre-jump section (2.55 m) turns out to be smaller than that at the post-jump section (5.41 m).

4.2.6 Advanced topics related to the Hydraulic Jump

The assumptions made in the analysis of hydraulic jump may not hold good in various practical situations. For example,

(a) the boundary friction may be significant in some cases;

(b) the velocity distribution may be nonuniform necessitating the inclusion of momentum correction factors;

(c) baffle blocks, floor raising, floor lowering, or cross jets may be used to force the formation of the jump at a given location; or

(d) the channel may have an expansion or contraction at the jump location.

Most of these situations require some empirical parameters (e.g., drag force on the baffle blocks, pressure distribution at a sudden drop in bed level) and would not be discussed in this book. The interested reader is referred to Rajaratnam (1967), Chow (1981), or Ranga Raju (1993) for a more detailed description of the jump. For the use of jump in energy dissipation and design of different types of *stilling basins*, the US Bureau of Reclamation (USBR 1955) is a good resource, though various recent studies are also available. Although we feel that this topic belongs more in books related to hydraulic structures, we describe here a commonly used design of a stilling basin since it is the most common use of the jump.

Stilling basin

A stilling basin (also called a stilling box) is used to reduce the velocity of the flow. The simplest type of basin utilizes just the hydraulic jump to dissipate the energy and, obviously, it should have a minimum length equal to the length of the jump. However, it may become very expensive and various attempts have been made to reduce the basin length by providing additional devices such as chute blocks, baffle blocks, and end-sill (see Fig. 4.12). These accessories not only reduce the required basin length but also provide additional stability by enabling adequate dissipation over a wide range of flow conditions.

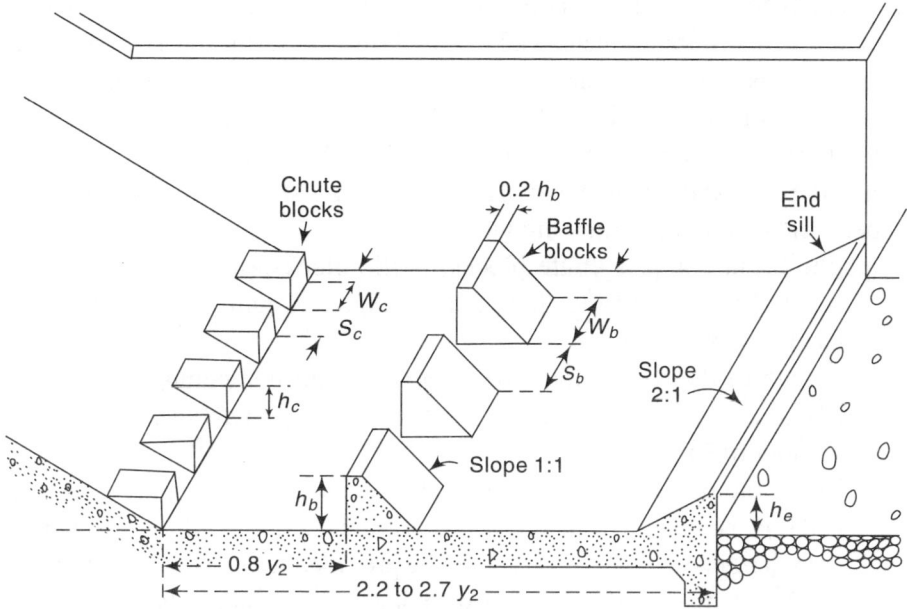

Fig. 4.12 Stilling basin with accessories

The chute blocks lift a portion of the incoming high velocity jet above the basin floor, increase the dissipating action of the jump, and result in a shorter jump. Baffle blocks provide most of the energy dissipation by the impact action and the end-sill prevents scouring near the basin end by lifting the outgoing stream. Their dimensions and locations should be chosen carefully since these affect the basin efficiency significantly. For example, if chute blocks have a large height, they would tend to throw the jet over the baffle blocks resulting in smaller impact loss and larger jump length. Similarly, if baffle blocks are located far away from the upstream end, of the basin, the dissipation would be smaller and if they are very close to the upstream end, it may cause an undesirable wave action. If these blocks are too high, it may produce a cascade and if it is too low, a rough water surface is obtained. Based on extensive experiments, USBR (1955) suggested designs of various types of basins. A Type I basin has no accessories and relies on the hydraulic jump to dissipate the energy. In a Type II basin, the chute blocks and a dentated (broken) end sill are used while a Type III basin uses chute blocks, baffle blocks, and a continuous end sill. The length of the basin required, naturally, is largest for Type I, smaller for Type II, and smallest for Type III. For example, at an incoming Froude number, F_1, of 10 (for higher Froude numbers, stilling basins are not generally used and a bucket-type dissipator may be more economical), the lengths of these basins are about 6.1, 4.3, and 2.7 times the *theoretical* post-jump depth, y_2 (theoretical here indicates the conjugate depth in absence of any appurtenances). Here we briefly describe the Type III basin. The US Department of Agriculture, Agricultural Research Service, developed the design of a very short stilling basin at the St. Anthony Falls (SAF) laboratory. This SAF basin was modified by USBR (1955) to propose the Type III basin and extensive analysis of existing structures and laboratory experiments led to the following guidelines:

1. The basin contains chute blocks, baffle blocks, and an end sill (Fig. 4.12). The slope of the chute does not have a significant effect on the basin performance.

2. Length of the basin should be 2.2 to 2.7 times the conjugate depth, y_2 (the smaller ratio of 2.2 at $F_1 = 4.5$, increasing to about 2.6 at Froude number of 7).

3. The chute blocks should have height, width, and clear spacing (h_c, w_c, and s_c) all equal to the incoming supercritical depth y_1. However, a minimum height of about 20 cm should be provided for construction convenience. A half-spacing is provided adjacent to the walls. Also, slight variation is allowed to avoid fractional blocks.

4. The baffle blocks are located at a distance of $0.8y_2$ downstream of the chute blocks and have a height (h_b) dependent on the Froude number, F_1. It varies linearly from about $1.3y_1$ at $F_1 = 4$ to $3y_1$ at $F_1 = 14.5$. The thickness at the top is 20% of the height, the width (w_b) is three-fourths of the height, and the spacing (s_b) is equal to the width. The downstream face has a 1:1 slope, however, cube-shaped baffle blocks have also been found to work equally well. The corners should not be rounded as sharp corners lead to higher energy dissipation. Sometimes, it is recommended that baffle blocks and chute blocks should follow a staggered arrangement. However, not much gain has been observed with this practice.

5. The end sill has a slope of 2:1 and its height (h_e) varies with the Froude number. Recommended heights follow a linear variation from about $1.2y_1$ at $F_1 = 4$ to about $1.8y_1$ at $F_1 = 14.5$.

6. The required tailwater depth has been found to be 15 to 18 percent smaller than the conjugate depth, y_2. If the tailwater depth is smaller than this, the front of the jump starts to move away from the chute blocks and the functioning of the basin is adversely affected. However, as an additional factor of safety, it is recommended that the tailwater depth be kept equal to y_2.

The analysis of hydraulic jump involved the use of the momentum equation because of the presence of an unknown energy loss and absence (or disregard) of unknown forces. There are some situations where the energy loss may be negligible but an unknown force may be present. In the next section, we discuss one such case—that of flow under a sluice gate.

4.3 Sluice Gate

Figure 4.13 shows the flow under a sluice gate, which is a common type of underflow gate when water needs to be carried over an structure (e.g., an spillway) or into a channel (e.g., from a river or reservoir into a canal). The streamlines immediately after the gate show a high curvature and the flow is rapidly varied. For free flow, at section 2, a nearly parallel flow is obtained (this section of minimum flow depth is called *vena contracta*) and beyond that the flow depth increases gradually till it attains a depth conjugate to the tailwater depth, y_t, and a jump is formed.[14] For submerged flow, the high velocity jet from the gate is drowned by the tailwater and a reverse flow occurs above this jet. The pressure head close to the gate under these conditions may be assumed to be equal to the water depth at the gate, y_0. Though other types of problems may be encountered, the most common issue is finding the discharge for a given upstream depth, y_1, a particular gate opening, a, and a tailwater depth, y_t. Obviously for free flow, the tailwater depth does not affect the discharge. We first analyse the free flow and then look at the effect of submergence.

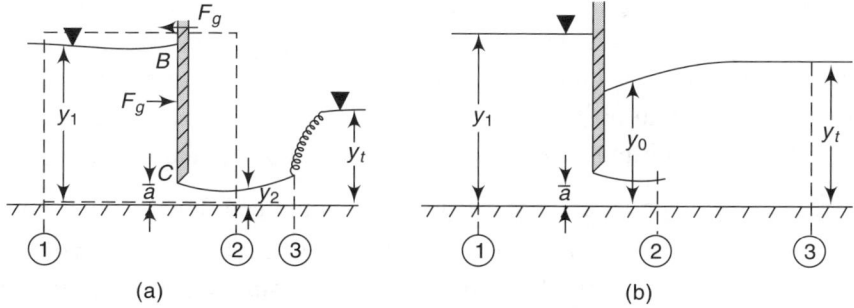

Fig. 4.13 Flow under a sluice gate: (a) free and (b) submerged

The gate experiences a force, F_g, in the direction of flow and applies an equal and opposite reaction on the control volume (chosen in such a way that the streamlines are parallel at section 2 and the pressure distribution there is

---

[14]We assume that the streamline curvature is small from section 2 to section 3 and Eq. (4.14) is applicable.

hydrostatic) as shown in Fig. 4.13a. Since the pressure distribution along line *BC* is not hydrostatic (the pressure will be atmospheric at both *B* and *C*), this force is not known. Therefore, we cannot apply the momentum equation to obtain the discharge. However, it is reasonable to assume that the energy loss between the sections 1 and 2 is negligible.

4.3.1 Free Flow

Assuming a rectangular, frictionless, horizontal channel, we may write (Fig. 4.13a)

$$y_1 + \frac{q^2}{2\,gy_1^2} = y_2 + \frac{q^2}{2\,gy_2^2} \tag{4.26}$$

Due to streamline curvature as water flows under the gate, the flow depth, y_2, will be smaller than the gate opening, a, and a contraction coefficient, C_c, is introduced such that $y_2 = C_c a$. This coefficient is a function of the relative gate opening, a/y_1, and has to be obtained empirically based on physical or numerical experiments. Values of C_c obtained over a wide range of parameters indicate that, for the relative gate openings, typically encountered in practical problems (a/y_1 less than about 0.5), it may be assumed to be constant at 0.61. Obviously for large values of a/y_1, the curvature of streamlines would be smaller and C_c would be higher. In the limiting case of the gate opening being just equal to the upstream depth, there would be no contraction and the downstream depth would be same as the upstream depth. Application of the sluice gate equation to find the discharge in such a condition would be meaningless. However, if we look at Eq. (4.26) as relating two alternate depths—the subcritical depth upstream and the supercritical depth downstream—we can think of the limiting case as a critical flow and obtain the discharge as $q = \sqrt{gy_1^3}$. On the other hand, if we consider it a submerged flow (to be discussed next), the degree of submergence would be 1 and the discharge would be zero. In any case, as we will see in the next paragraph, C_c is absorbed in the discharge coefficient, C_d, and its individual determination is not very critical. Using the contraction coefficient, Eq. (4.26) is written as

$$q = C_d a \sqrt{2\,gy_1} \tag{4.27}$$

where C_d is the discharge coefficient given by

$$C_d = \frac{C_c}{\sqrt{1 + \dfrac{C_c a}{y_1}}} \tag{4.28}$$

and is obtained through laboratory and field experiments involving measurement of upstream depth, gate opening, and corresponding discharge (theoretical and numerical studies have also been conducted to obtain C_d, but these are also generally compared with experiments to verify their results). The experimental value has been found to be a little smaller than that given by Eq. (4.28) with $C_c = 0.6$ for a/y_1 from 0.1 to 0.6, and a little larger for smaller a/y_1 values. The following equation is found to be accurate within 2% of the observed values for a/y_1 up to 0.7:

$$C_d = \frac{0.6}{\sqrt{1 + 0.8\dfrac{a}{y_1}}} \tag{4.29}$$

4.3.2 Submerged Flow

For free flow to occur, the flow depth at *vena contracta* should be smaller than the conjugate depth corresponding to the tailwater depth. In other words, the specific force at *vena contracta* should be greater than that at the tailwater depth. If the gate opening is kept same and the tailwater depth increases,[15] the specific force at section 3 [see Fig. 4.13(b)] becomes larger and the corresponding conjugate depth becomes smaller. Therefore, the jump moves upstream and will keep on moving further upstream till the flow depth at *vena contracta* is equal to the conjugate depth. Any further increase in the tailwater depth will lead to the drowning of the jet and will affect the discharge. This condition is the submergence threshold, i.e., it separates the free and submerged flows and is known as the *modular limit*,[16] *transitional submergence,* or *distinguishing condition.*

It may be assumed that under submerged conditions, the lower part of section 2 contains moving water (in the form of the jet issuing from the gate) and the upper part has turbulent but virtually stagnant water. If we further make the approximation that the energy loss between sections 1 and 2 is negligible, we would be able to apply the energy equation between sections 1 and 2 and the momentum equation between sections 2 and 3 (Fig. 4.13b). However, in

---

[15]Similar analysis is valid if the tailwater depth remains same but the gate opening is increased. In most cases, however, both the gate opening and the tailwater depth would change simultaneously because the tailwater depth is typically dependent on the discharge.

[16]The flow over or through a discharge measuring device is called *modular* if it is not affected by the downstream water level.

applying these equations, we must account for the fact that only part of section 2 has moving water. Therefore, for velocity computation, we should use the depth y_2, but for pressure computation, the entire depth, y_0, should be used. The energy and momentum equations are then written as

$$y_1 + \frac{q^2}{2\,gy_1^2} = y_0 + \frac{q^2}{2\,gy_2^2}$$

(4.30)

$$\frac{y_0^2}{2} + \frac{q^2}{gy_2} = \frac{y_t^2}{2} + \frac{q^2}{gy_t}$$

(4.31)

Elimination of y_0 from these equations leads to a quadratic equation in the variable q^2, from which q may be obtained and then the discharge coefficient, C_d, may be obtained using Eq. (4.27). The submergence limit (or the distinguishing condition), beyond which one would use the free discharge equation, is obtained by the condition that the depth at section 2 is the conjugate of the tailwater depth. Using this approach and defining a submergence ratio (s) as y_t/y_1, and a nondimensional depth at section 2 as $\eta = C_c a/y_1$, the theoretical values of submergence limit (s^*) and the discharge coefficient (C_d) are (Bos 1989, Lin, et al. 2002)

$$s^* = \frac{\eta}{2}\left(\sqrt{1 + \frac{16}{\eta(1+\eta)}} - 1\right)$$

(4.32)

$$C_d = C_c \frac{\sqrt{\xi - \sqrt{\xi^2 - (1 - s^2)\left(\frac{1}{\eta^2} - 1\right)^2}}}{\frac{1}{\eta} - \eta}$$

(4.33)

where $\xi = \left(\frac{1}{\eta} - 1\right)^2 + 2\left(\frac{1}{s} - 1\right)$.

A plot of the discharge coefficient (using $C_c = 0.6$) is shown in Fig. 4.14 with C_d obtained from Eq. (4.33) for $s > s^*$ and from Eq. (4.28) for lower submergence ratio. Note that the curves become almost vertical as the submergence limit is approached indicating that a slight increase in the tailwater depth causes a significant decrease in C_d for a fixed upstream depth and gate opening. This is a consequence of the assumption that the jump gets drowned as soon as the submergence limit is exceeded. In practice, the transition would not be so sudden and the jump will slowly move from the *vena contracta*

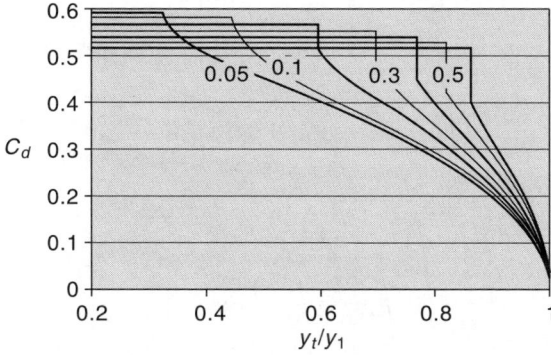

Fig. 4.14 Discharge coefficient for a sluice gate [Numbers on the curve show the values of a/y_1 (the values for unmarked curves are 0.2, 0.4, and 0.6).]

towards the gate. Therefore, the use of the theoretical curve when the tailwater depth is slightly greater than the submergence limit would not be proper.

As was the case for free flow, the experimental data deviate slightly from the theoretical curve. Swamee (1992) suggested a relation between the submergence limit and relative gate opening as $s^* = 1.13(a/y_1)^{0.42}$, which slightly overpredicts (Lin, et al. 2002) the value for large gate openings ($a/y_1 \geq 0.5$). The following relation is found to represent the experimentally observed submergence limit within a 2% error:

$$s^* = \frac{1.446}{\sqrt{1.091 + \dfrac{y_1}{a}}} \tag{4.34}$$

Swamee (1992) obtained an approximate expression for the discharge coefficient, which reproduces the experimental data very well. A slightly modified form is given below:

$$C_d = \frac{0.6}{\sqrt{1 + 0.8\dfrac{a}{y_1}}} \frac{1}{1 + 0.42\dfrac{(s - s^*)^{0.46}}{(1 - s)^{0.85}}} \tag{4.35}$$

Figure 4.15 shows a plot of Eq. (4.35) along with the experimental curves (Henry 1950). Note that the axis and the third parameter are different from those used in Fig. 4.14. The lines in this figure represent the variation of C_d, as the upstream depth is changed keeping the same gate opening and the tailwater level.

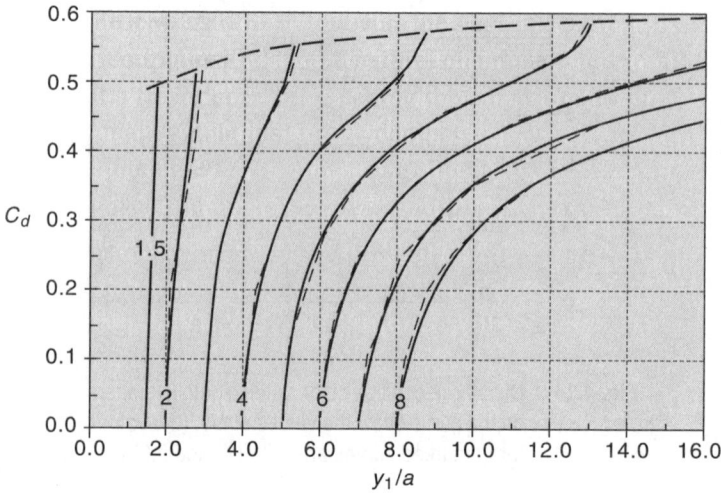

Fig. 4.15 Discharge coefficient for a sluice gate under submerged conditions [The curves correspond to the values of y_t/a equal to 1.5, 2, 3, 4, 5, 6, 7, and 8. Solid lines obtained from Eq. (4.35) and dashed lines represent the experimental data.]

From Eq. (4.30), it is easy to see that the effect of submergence may also be accounted for by modifying the discharge equation, Eq. (4.27), to express the submerged discharge, q_s, as

$$q_s = C_{ds} a \sqrt{2g(y_1 - y_0)} \qquad (4.36)$$

with C_{ds} given by

$$C_{ds} = \frac{C_c}{\sqrt{1 - \left(\dfrac{C_c a}{y_1}\right)^2}} \qquad (4.37)$$

However, it necessitates the measurement of y_0, which may be subjected to larger errors (due to turbulence) than those in measurement of y_t.

Once the discharge through the gate is obtained, the momentum equation can be applied to estimate the force acting on the gate. For example, under free flow conditions (Fig. 4.13a), application of the momentum equation over the control volume results in[17]

$$F_g = \frac{\rho g y_1^2}{2} - \frac{\rho g y_2^2}{2} - \rho q^2 \left(\frac{1}{y_2} - \frac{1}{y_1}\right) \qquad (4.38)$$

in which F_g represents the force acting on the gate per unit width.

---

[17]The usual assumptions of horizontal and frictionless channel and hydrostatic pressure distribution have been invoked.

Sometimes, the sluice gates are constructed over a raised platform and the discharge would be different from that obtained above under free and submerged conditions due to separation of flow at the upstream end of the crest. Ranga Raju and co-workers (1979, 1981) found that the discharge becomes 3 to 4 percent smaller under free flow conditions and also suggested a procedure for obtaining the discharge under submerged conditions. However, the results would be applicable only for the range of parameter values covered in the experiments. Similarly, if the gate lip is not sharp but bevelled, the discharge coefficient increases. However, model studies have to be performed to determine the variation of this coefficient.

Example 4.5   A sluice gate is located in a 2 m wide rectangular channel. When the gate opening is 0.10 m, the upstream depth is measured as 0.80 m. What is the discharge through the gate if the tailwater depth is 0.3 m? What force would act on the gate? Find the modular limit and the discharge when the tailwater depth is 0.7 m.

Solution

To find out whether the flow is occurring under free or submerged conditions, we would first find out the submergence limit, s^*. From Eq. (4.34), with $y_1 = 0.8$ m and $a = 0.1$ m, we get $s^* = \mathbf{0.48}$.[18] Therefore, when the tailwater depth is 0.3 m, i.e., a submergence ratio of 0.375, we will have free flow conditions.

Equation (4.29) results in a coefficient of discharge, C_d, equal to 0.572 and the discharge per unit width is obtained using Eq. (4.27) as $q = 0.2266$ m³/s per m. The total discharge is, therefore, $Q = \mathbf{0.4532}$ **m³/s**.

Using a coefficient of contraction of 0.6, the depth at *vena contracta*, y_2, is equal to 0.06 m. The force on the gate is obtained from Eq. (4.38) as $F_g = 6602$ N per m width, i.e., total force of **13.2 kN**.

Since the modular limit has been computed as 0.48, the tailwater depth of 0.7 m ($s = 0.875$) will cause submerged flow. Using Eq. (4.35), $C_d = 0.220$ and the discharge is obtained as **0.174 m³/s**, a reduction of more than 60% from the free discharge value. We can also use Fig. 4.15 (with $y_1/a = 8$ and $y_t/a = 7$) to obtain $C_d = 0.22$ or use Eq. (4.33) to obtain the *theoretical* coefficient of discharge (with $s = 0.875$, $\eta = 0.075$, $\xi = 152.4$, $C_c = 0.6$) as $C_d = 0.232$. The depth just downstream of the gate, y_0, may be obtained using Eqs (4.36) and (4.37). With $C_c = 0.6$, we get $C_{ds} = 0.602$, and from Eq. (4.36) with $q_s = 0.087$ m³/s per m, we get $y_0 = 0.69$ m, almost same as the tailwater depth.

We have assumed that under the submerged conditions, the discharge gets reduced. Another possibility is that the discharge remains same but the water

---

[18]Using Eq. (4.32) with $C_c = 0.6$, we get $s^* = 0.49$.

level upstream of the gate increases. An iterative solution would be needed for such cases to compute the value of y_1. Once we assume a value of y_1, the discharge can be obtained with the help of Fig. 4.14 or 4.15. Depending on the value of the discharge, y_1 may be increased or decreased and the procedure repeated till the discharge matches with the given value of 0.2266 m²/s. For example, starting with a guess of $y_1 = 1$ m, we get $s^* = 0.43$, $s = 0.7$, $C_d = 0.35$, and $q = 0.156$ m²/s. With $y_1 = 1.5$ m, we get $s^* = 0.36$, $s = 0.47$, $C_d = 0.47$, and $q = 0.253$ m²/s. Thus the correct value would be between 1 m and 1.5 m. After a few more iterations, we obtain $y_1 = 1.34$ m, for which $s^* = 0.38$, $s = 0.52$, $C_d = 0.44$, and $q = 0.2262$ m²/s.

Measurement of the water depths and the gate opening for a sluice gate allows us to obtain the discharge through the gate. Various other structures can be used to achieve the same goal and all of these result in a rapid variation of flow depth close to the structure. In the rest of this chapter, we describe four types of structures—free overfalls, weirs, spillways, and flumes. We start with the sudden drop in the channel bed (e.g., that occurring at a canal fall) and its utility as a discharge measuring device. We then describe the broad-crested weirs since its analysis is simpler (based on critical flow conditions) followed by the sharp-crested weir, which is generally analysed using the slot-flow concept. The ogee spillway follows as a logical extension of the sharp-crested weir and its profile is discussed. Finally, convergent flumes, which create critical condition to measure the discharge, are discussed. As was the case for the sluice gate and, to some extent, for the hydraulic jump, all of these require the use of empirical coefficients.

4.4 Free Overfall

For a typical canal, the design bed slope is flatter than the available ground slope. To avoid the channel being in embankment[19] for a long reach, canal falls are provided at appropriate places. The flow near the fall is rapidly varied but may be analysed with the help of some simplifying assumptions. Since most channels have a mild slope, we would mostly discuss these (a brief discussion on steep channels is provided at the end of this section). As we saw in the previous chapter, when a mild slope channel is followed by a steep slope channel, critical depth occurs at the junction. The free overfall (Fig. 4.16) may be thought of as the same scenario but with the steep channel becoming

---

[19]A channel in embankment generally has a higher cost, more seepage, and greater chances of flooding of the surrounding area.

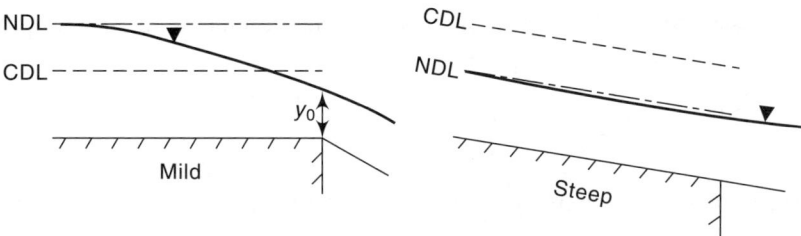

Fig. 4.16 Free overfall in mild- and steep-slope channels

vertical. The flow should, therefore, be in a critical state at the brink. This leads to the possibility of using the fall as a discharge measuring structure. However, the convex upward curvature of flow near the brink and the non-hydrostatic pressure distribution[20] at this section implies that the pressure head will be *less than* the flow depth (see Appendix B, Section B.2). Therefore, even though the specific energy at the brink is equal to the critical specific energy for that discharge, the velocity head should be *more than* that corresponding to the critical depth. Thus the depth at the overfall, y_o (called the end depth or the brink depth), would be smaller than the critical depth, y_c. Since a relation between y_c and the discharge is known for a given channel shape (Section 3.3), we can obtain the discharge if we are able to relate y_0 to y_c.

Dey (2002) provides a review of various attempts to study the free overfall. We first look at two different approaches of obtaining the end depth, one based on the momentum equation and the other on energy equation. Since simplifying assumptions have to be made in the analysis, the theoretical values are then compared with the experimental observations. For simplicity, rectangular channels are considered. Extension to other shapes is left to the reader.

4.4.1 Momentum Equation

We assume that (i) the critical depth occurs a short distance upstream of the brink so that the boundary shear over this length can be neglected, (ii) the streamlines are nearly parallel at this section such that the pressure may be taken as hydrostatic, and (iii) the pressure distribution at the brink may be approximated as giving rise to a pressure force equal to k times the hydrostatic pressure force (clearly, k would be less than 1). Applying the momentum

---

[20]Pressure will be atmospheric at the water surface as well as at the channel bottom (since a free jet exists at the brink).

equation between the brink and the upstream section having critical depth, we have

$$\frac{\rho g y_c^2}{2} - k \frac{\rho g y_0^2}{2} = \rho q^2 \left(\frac{1}{y_0} - \frac{1}{y_c} \right)$$

which can be written, using $q^2 = g y_c^3$ and the end depth ratio (EDR), $r_0 = y_0/y_c$, as

$$kr_0^3 - 3r_0 + 2 = 0 \tag{4.39}$$

Based on the observed pressure profile, the value of k is likely to be less than 0.5. It would also depend on whether the jet issuing from the brink is confined at the sides or not. At canal falls, typically the side walls extend beyond the brink and the jet is confined. In laboratory experiments, generally the side walls end at the brink and the jet is unconfined. The k value should be larger for the confined case due to this confining pressure. If we take $k = 0$ (i.e., neglect the pressure force at the brink), we get $r_0 = 0.667$ and for $k = 0.5$, we get $r_0 = 0.732$. It is expected that the actual value of EDR for a rectangular channel would lie between these two.

4.4.2 Energy Equation

We assume that (i) the critical energy, E_c, occurs at the brink, (ii) the horizontal velocity at all points at the brink section is same and is equal to the average velocity, V, (iii) the specific (per unit width) discharge at the brink is given by $q = V y_0$, and (iv) the pressure distribution at the brink is parabolic with zero pressure at the water surface and at the bed (the pressure distribution accounts for both the water surface curvature and the fact that the lower nappe is open to atmosphere). The specific energy at the brink is then written as (see Appendix A)

$$E_c = \frac{1}{y_0} \int_0^{y_0} \left(\frac{p}{\rho g} + Y \right) dY + \frac{q^2}{2 g y_0^2}$$

$$= \frac{1}{y_0} \int_0^{y_0} \left(\kappa \left(Y - \frac{Y^2}{y_0} \right) + Y \right) dY + \frac{q^2}{2 g y_0^2} \tag{4.40}$$

with κ being an empirical coefficient (note that it would be different from the coefficient, k, used in the momentum equation since k applies to the pressure force on the whole section and κ to the pressure at a point in the section). In terms of the EDR, using $E_c = 1.5\, y_c$, we may now write

$$\left(1 + \frac{\kappa}{3}\right)r_0^3 - 3r_0^2 + 1 = 0 \tag{4.41}$$

Anderson (1967) obtained a similar expression by matching the curvature of the water surface and the upper nappe beyond the brink,[21] and found that the value of κ should be equal to 1. If we take $\kappa = 0$ (i.e., neglect the pressure at the brink), we get $r_0 = 0.653$ and for $\kappa = 1$, we get $r_0 = 0.694$.

Experimental studies conducted on rectangular channels with confined nappe have shown the EDR to be about 0.715, not very different from the theoretical predictions. Similarly, for an unconfined nappe, experiments on various shapes of the channel show that the EDR is 0.705 for a rectangular channel, 0.725 for a circular channel, and 0.795 for a triangular channel (Rajaratnam and co-workers 1964, 1968). For a trapezoidal channel (Rajaratnam and Murlidhar 1970, Bhallamudi 1994), it is a function of the nondimensional critical depth, my_c/B, and increases monotonically from about 0.7 to 0.8, i.e., the rectangular to triangular value, with increase in my_c/B, achieving a value of about 0.75 at $my_c/B = 1$.

Supercritical channel slopes are not very common in practical situations. When the flow in the channel is supercritical, the flow depth at the brink should theoretically be equal to the normal depth (comparing with the flow profile on a steep channel followed by a steeper one). Again, due to the curvature of streamlines and the atmospheric exposure of the lower nappe, the pressure head is less than the hydrostatic pressure head and the velocity head should be larger than that at the normal depth. Therefore, the brink depth, y_0, is less than the normal depth, y_n. For a rectangular channel with a bed slope more than three times the critical slope, the brink depth may be taken as about 90% of the *normal* depth. For the slopes one to three times the critical slope, the EDR (y_0/y_c) varies from about 0.7 to 0.63. It should be noted, however, that the critical slope computations require the use of the Manning's n and the *unknown* discharge. If the roughness coefficient and the channel slope have to be used, it may be better to measure the normal depth a little upstream of the drop and obtain the discharge using the Manning's equation. Utility of the free overfall to estimate the discharge under supercritical flow conditions is, therefore, rather limited and we do not discuss it in detail.

---

[21]The upper nappe is assumed to follow a gravity-fall trajectory. The lower nappe does not follow a gravity-fall trajectory as it is constrained to have zero curvature at the brink. The gravity-fall trajectory is parabolic with a constant curvature.

Example 4.6   A 1 m wide rectangular tilting flume[22] ends in a free overfall where the depth of flow is measured as 0.30 m. The flow depth upstream of the drop was not measured but was observed to be considerably larger than the brink depth. What is the discharge in the channel? The discharge was then reduced till the flow in the channel becomes just supercritical and the brink depth was measured as 0.05 m. What is the discharge under this condition? If the channel is now tilted to increase the slope three-fold, keeping the discharge same, what would be the flow depth at the brink?

Solution

Since it is a rectangular channel and assuming that the nappe is unconfined, the end depth ratio would be 0.705 for subcritical flow. Since the first case has a depth smaller than the uniform flow depth, the flow would be subcritical. Therefore, the critical depth would be

$$y_c = \frac{y_0}{0.705} = 0.426 \text{ m}$$

and the discharge is

$$Q = B\sqrt{gy_c^3} = \textbf{1.74 m}^3\textbf{/s}$$

For supercritical flow, the end depth ratio is a function of the bed slope and for $S_0/S_c = 1$ (which is the case here since the flow is *nearly critical*), it is 0.7. Hence, the critical depth is 0.05/0.7 m = 0.0714 m and the discharge is **0.12 m³/s**.

Keeping the discharge same, if the bed slope is increased three fold, $S_0/S_c = 3$. The corresponding end depth ratio would be 0.63 and with a critical depth of 7.14 cm obtained above, the end depth would be **4.5 cm**.

It has been recommended (Bos 1989) that for an overfall to function as a discharge measuring device, a straight reach of about $12y_0$ length should be available before the drop to ensure a uniform velocity distribution, and the drop height should be more than 0.6 times the critical depth for the maximum discharge to eliminate the effect of submergence.

Canal drops, when they exist, are thus a good way of estimating the discharge. However, these do not occur very frequently and may not be available at the location where we want to measure the discharge. In such cases, we would have to either construct a flow-measuring structure or use a flow-control structure as a flow meter. Some of these are described in subsequent sections.

---

[22]A tilting flume has a mechanism to adjust the bed slope to a desired value.

4.5 Weirs

A weir is a small dam-like obstruction in a stream or river to raise the water level or divert its flow. Depending on the width of the crest, a weir may be classified as broad-crested or sharp-crested[23] (Fig. 4.17). While the sharp-crested weir is completely characterized by its height, h_w, and cross section, the broad-crested weir requires additional parameters such the crest-length, L_w, the slope of the upstream face, s_u, and the slope of the downstream face, s_d. We assume that the weir spans the whole width of the channel, B. If not, the streamlines would show contractions at the ends and the weir is called a *contracted weir*. (The weir which spans the whole width may be thought of as one which suppresses the end contractions and is, therefore, called a *suppressed weir*.) Its effective width would be reduced from its actual width, B_w, typically by 10% of the head above the weir crest for each contraction. We start with broad-crested weirs since these can be analysed with the already discussed concepts of critical flow and momentum equation.

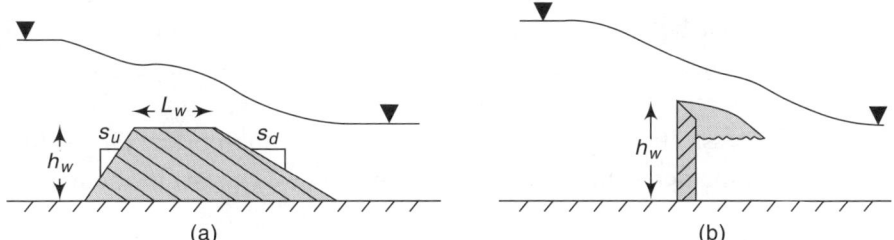

Fig. 4.17 Weirs in a channel: (a) broad-crested and (b) sharp-crested

4.5.1 Broad-Crested Weirs

To simplify the discussion, we consider a rectangular channel with a rectangular broad-crested weir with vertical upstream and downstream faces spanning the entire channel width. We assume that the upstream corner of the crest is sharp. Figure 4.18 shows the flow over a broad-crested weir which can be analysed using the energy or the momentum equation.

Energy equation

Assuming the total energy to be same, the specific energy over the weir crest will be quite small (recall that the specific energy is measured from the channel bed, which is now at the crest). For a sufficiently high weir, the specific

---

[23]Sometimes another classification, the short-crested weir, is also used to represent the situation when the crest is not very broad, but we will not discuss it.

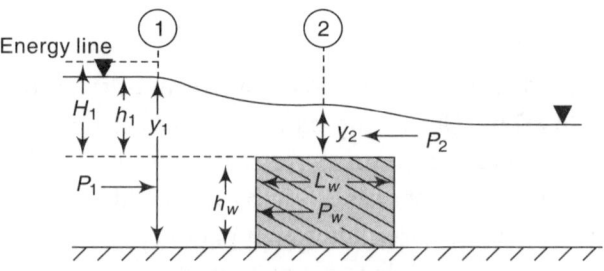

Fig. 4.18 Flow over a broad-crested weir

energy at the crest will achieve the smallest possible value, which is the critical condition. (In the next chapter, we will discuss more about the transition in bed level and establish that critical flow occurs at a 'large' hump in bed.) Since the relationship between the critical depth and the discharge is known, we can obtain the discharge from measurement of the depth at the crest. However, because of curvature of streamlines as the flow approaches the weir, the assumption of hydrostatic pressure distribution will not be valid and the relationship between the critical depth and the discharge obtained in the previous chapter may not hold good. We, therefore, want the weir crest to be long enough to produce almost parallel flow over some part of the crest. On the other hand, the crest length, L_w, should be small enough to neglect the energy loss over the crest. It is suggested (Bos 1989) that the appropriate range of L_w/H_1 (H_1 being the upstream energy head above the weir crest) is about 1.5 to 12.

Assuming that critical depth occurs somewhere over the crest with straight and parallel streamlines, a uniform velocity distribution, and neglecting the energy loss, we write the specific discharge (Q/B) as

$$q = \sqrt{gy_c^3} = \sqrt{g\left(\frac{2H_1}{3}\right)^3} = \sqrt{\frac{8g}{27}}H_1^{1.5} \tag{4.42}$$

This theoretical discharge has to be modified to obtain the actual discharge because the assumptions of no energy loss, parallel streamlines, uniform velocity distribution, and no viscous or turbulence effects, are not necessarily true. We introduce a discharge coefficient as $C_d' = \dfrac{q_{\text{actual}}}{q_{\text{theoretical}}}$, and write

$$q = C_d' \sqrt{\frac{8g}{27}}H_1^{1.5} \tag{4.43}$$

Hager and Schwalt (1994) provide the following expression for C_d' for the practical range of L/H_1 (assuming that H_1 is large enough so that viscosity and surface tension do not affect the flow. Generally this limit is about 5 cm and would be exceeded in most practical cases):

$$C_d' = 1.09\left[1 - \frac{0.222}{1 + (H_1/L_w)^4}\right] \tag{4.44}$$

Since it is inconvenient to use the energy head, H_1, and it is preferable to use the easily-measurable water depth above the weir crest, h_1, a velocity coefficient is introduced as $C_v = \left(\dfrac{H_1}{h_1}\right)^{1.5}$ to obtain

$$q = C_d' C_v \sqrt{\frac{8g}{27}} h_1^{1.5} \tag{4.45}$$

The velocity coefficient is related to the height of weir and the head over the weir. Absorbing the two empirical constants into one, we may write

$$q = C_d \sqrt{g} h_1^{1.5} \tag{4.46}$$

where C_d is another discharge coefficient.[24] Based on experiments with $h_1/h_w < 1$, where h_w is the height of the weir, Govinda Rao and Muralidhar (1963) proposed the following relation for C_d:

$$C_d = 0.491 + 0.0264 \frac{h_1}{L_w} \tag{4.47}$$

which is valid for the range $0.1 \le h_1/L_w \le 0.35$ (which covers most of the recommended range for a broad-crested weir) and results in a C_d that is nearly constant at about 0.5. Equation (4.43) or (4.46) can be used to obtain the discharge for a measured water head, h_1. Note that Eq. (4.43) needs an iterative solution since H_1 depends on the velocity head, $V_1^2/2g$, which, in turn, depends on the discharge.

Momentum equation

Assuming the force on the upstream face of the weir to be equivalent to a hydrostatic pressure distribution with water depth of y_1 (see Fig. 4.18) and neglecting the frictional forces, application of the momentum equation over the control volume between sections 1 and 2 results in

$$P_1 - P_w - P_2 = \rho q^2 \left(\frac{1}{y_2} - \frac{1}{y_1}\right) \tag{4.48}$$

---

[24]Sometimes, $\sqrt{8/27}$ is not incorporated into C_d and it is defined as $C_d = C_d' C_v$. We prefer Eq. (4.46).

If section 2 is chosen in such a way as to have parallel flow and hydrostatic pressure distribution, and if we assume that $y_2 = kh_1$, Eq. (4.48) becomes

$$\frac{1}{2}\rho g y_1^2 - \left(\frac{1}{2}\rho g y_1^2 - \frac{1}{2}\rho g h_1^2\right) - \frac{1}{2}\rho g k^2 \, h_1^2 = \rho q^2 \left(\frac{1}{kh_1} - \frac{1}{y_1}\right) \quad (4.49)$$

which can be written in the form of Eq. (4.46) with C_d given by

$$C_d = \sqrt{\frac{k(1-k^2)}{2\left(1 - \dfrac{k}{1 + \dfrac{h_w}{h_1}}\right)}} \quad (4.50)$$

since the term $\sqrt{k(1-k^2)/2}$ is close to 0.43 for the expected range of k (from 0.5 to 0.7, with observed data indicating a value closer to 0.5), we can approximately write

$$C_d = \frac{0.43}{\sqrt{1 - \dfrac{k}{1 + \dfrac{h_w}{h_1}}}} \quad (4.51)$$

The following example illustrates the estimation of discharge using a broad-crested weir.

Example 4.7   A rectangular weir with vertical faces is built in a 3 m wide channel. The weir height is 1 m, the crest length is 3 m, and the weir spans the whole width of the channel. What would be the discharge when the upstream water depth in the channel is 1.6 m?

Solution
We assume that the downstream water level is lower than the modular limit and free flow is occurring over the weir. Since $h_1/h_w = 0.6$ and $h_1/L_w = 0.2$, we use Eq. (4.47) to obtain C_d as 0.496 and the discharge is obtained from Eq. (4.46) as 0.722 m³/s per m width or a total discharge of **2.166 m³/s**. We can also use Eq. (4.51) to estimate the coefficient of discharge (with $k = 0.5$) as 0.477, which is not very different from the value computed above.[25]

Once an estimate of discharge is obtained, we can estimate the velocity head upstream of the weir as $(0.722/1.6)^2/2g = 0.01$ m. The energy head above the weir crest is, therefore, $H_1 = 0.61$ m. C_d', from Eq. (4.44), is 0.848 and the

---

[25]If we use $k = 0.6$, we get $C_d = 0.488$ and with $k = 0.7$, we get $C_d = 0.5$.

discharge, from Eq. (4.43), is 0.689 m³/s per m width, a difference of only about 5% from the previously computed value. The different approaches—one based on the water depth above the crest and the other based on the energy head above the crest—therefore, provide comparable results.[26]

Some points to be noted while using the broad-crested weir for flow measurement are as follows:

1. Since the water surface just upstream of the weir has a curvature, the measurement of h_1 should be done at a location, which is at a distance of roughly two to three times the maximum expected head on the weir. This would ensure that the correct measurement is taken since the effect of curvature would not extend to this distance.

2. To avoid side wall effects, the channel width should be more than one-fifth of the crest length.

3. To avoid effects of surface tension and viscosity, the head over the weir should be greater than the larger of 6 cm or $L_w/20$. In field measurements, this condition would be generally satisfied. However, when conducting model studies in the laboratory, the length scale used for the model may necessitate the use of a head smaller than 6 cm. The discharge obtained from the model studies in such cases cannot directly be scaled up to predict the prototype discharge since the model results would be influenced by viscosity and surface tension but the prototype would not. Some studies have been conducted to enable one to account for these effects. However, these use empirical coefficients, which vary with Reynolds number $(\rho\sqrt{gh_1^3}/\mu)$, Weber number $(\rho g h_1^2/\sigma)$, and slope of weir faces. These also show a large scatter of data and are not very well established. Therefore, these are not described here. Interested reader should refer to Ranga Raju and Asawa (1979).

4. If the weir length is more than about ten times the head over the weir, the energy loss over the crest would be appreciable. The critical condition will still occur over the crest but the upstream energy would be larger than the critical energy. Thus the critical depth would be smaller than $2H_1/3$ and the discharge coefficient would also be smaller.

---

[26] We had assumed that the velocity is uniform and equal to the discharge divided by the entire area of flow. In reality, the velocity would be highly nonuniform and we should use an energy correction factor. If we use a correction factor of 2, we get $H_1 = 0.62$ and discharge equal to 0.71 m³/s per m width.

5. If the weir length is less than about two times the head over the weir, the curvature of streamlines over the crest would be quite pronounced and parallel flow would not occur anywhere on the crest. The critical condition still occurs over the crest but the piezometric head would be less than the flow depth and the critical depth would be more than $2H_1/3$ resulting in a larger C_d value. (Taking the specific energy as $ky + (q^2/2gy^2)$ with $k < 1$, it can be shown that the minimum specific energy is equal to $3ky/2$ and the critical depth may be taken as $2H_1/3k$.)

6. Generally the upstream corner is rounded to avoid separation of streamlines and thus minimize the energy loss. Separation can also be avoided by providing an upstream sloping face with slopes of 2-to-3H:1V. Another advantage of sloping upstream face is the reduction in accumulation of debris.

7. A downstream sloping face (4-to-6H:1V) is used to reduce the effect of submergence when the downstream water depth is expected to be close to the upstream depth (for a vertical downstream face, submergence starts affecting the discharge when the d/s water level relative to the crest, h_t, is about 80% of the upstream value, h_1, and with a sloping face, the limit increases to about 90%). Since the submergence limits are so high, it is unusual for a broad-crested weir to operate beyond the modular limit. Hence, the effect of submergence is not discussed here.

8. In Example 4.7, we have conveniently used the parameters to fall within the specific ranges required for application of various empirical equations. If the head over the weir is more than the weir height, the weir faces are not vertical, the corners are rounded, or the head-length ratio is outside the range specified in the empirical equations, we would have to use different equations or charts to obtain the discharge coefficient. Bos (1989) provides comprehensive tables and charts for such cases. Swamee (1988) provides an approximate expression valid over a wide range. It should be noted that the definition of C_d is different from that used in Eq. (4.46) and a factor has to be used depending on how C_d is defined in the relevant study.

4.5.2   Sharp-Crested Weirs

If the crest length of the weir is reduced, keeping the upstream corner sharp, the streamline separating at the corner may clear the crest completely. Typically this occurs when the crest length is about two-thirds of the head, h_1. These weirs are known as the *sharp-crested weirs* and are generally in the form of a thin metal plate placed across the channel width. Due to their small thickness, sharp-crested weirs are not used to measure high discharges since a large upstream depth would result in a large force on the weir leading to possibility of structural failure. Typically these weirs are analysed by treating it as flow through a slot with the upper edge of the slot above the water level, as described further in this section.

To simplify the discussion, we consider a rectangular channel with a rectangular sharp-crested weir spanning the entire channel width. We also assume that the flow is modular so that the downstream water level does not influence the discharge. Submerged flow is discussed in a subsequent subsection. Figure 4.19 shows the flow over a sharp-crested weir that can be analysed using the rectangular slot equation. (We can use the energy or the momentum equation, but there are several unknowns, e.g., the pressure distribution on the weir face, the velocity distribution above the weir, the water depth at the weir crest, and the pressure distribution above the weir crest.)

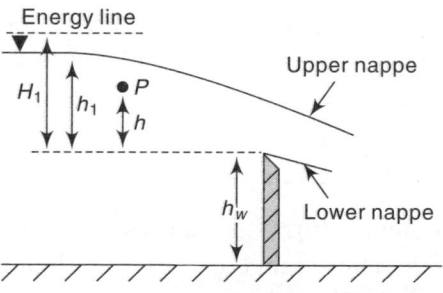

Fig. 4.19   Flow over a sharp-crested weir

Considering ideal fluid flow through a rectangular slot with the weir as its bottom plate and an imaginary upper plate, the velocity at a point P (which is at a height h from the weir crest) is given by $\sqrt{2g(H_1 - h)}$. The specific discharge, q, is obtained by integrating the velocity over the entire flow depth (we ignore the contraction due to curvature of streamlines and would account for it later through an empirical coefficient) as

$$q = \int_0^{h_1} \sqrt{2g(H_1 - h)}\,dh = \sqrt{\frac{8g}{9}}\left[(H_1 - h)^{1.5}\right]_{h_1}^0$$

$$= \sqrt{\frac{8g}{9}}\left[H_1^{1.5} - (H_1 - h_1)^{1.5}\right] \tag{4.52}$$

As was done for the broad-crested weir, we introduce a coefficient of discharge (thought of as a product of a contraction coefficient and a velocity coefficient) to account for contraction and other non-ideal conditions and to express q in terms of the water depth, h_1, in place of the energy head, H_1, as[27]

$$q = C_d \sqrt{g} h_1^{1.5} \qquad (4.53)$$

For large flow depths, i.e., when the flow is not affected by viscous and surface tension forces, the coefficient of discharge is a function of the relative head, h_1/h_w. As this ratio becomes large, the weir starts to behave like a sill and may be analysed by assuming that the critical depth occurs at the sill.[28] This condition is observed when $h_1/h_w > 20$ and gives rise to

$$q = C_d \sqrt{g} h_1^{1.5} = \sqrt{g(h_1 + h_w)^3} \qquad (4.54)$$

implying that

$$C_d = \left(1 + \frac{h_w}{h_1}\right)^{1.5} \qquad (4.55)$$

For $h_1/h_w < 5$ (with h_1 large enough to neglect the viscosity and surface tension effects), experimental determination of C_d led to the following expression (proposed by Rehbock in 1929):

$$C_d = 0.578 + 0.0707 \frac{h_1}{h_w} \qquad (4.56)$$

Swamee (1988) combined Eqs (4.55) and (4.56) to obtain an approximate expression valid for the entire range of h_1/h_w. A slightly modified form is given below and plotted in Fig. 4.20.

$$C_d = \left[\left(1 + 1.5 \frac{h_w}{h_1}\right)^{-10} + \left(0.58 + 0.07 \frac{h_1}{h_w}\right)^{-10}\right]^{-0.1} \qquad (4.57)$$

We have described a rectangular suppressed weir operating with modular flow. Field conditions sometimes dictate the use of a different arrangement. Some of these situations and some other factors to be kept in mind are described further.

---

[27]Again, commonly the factor $\sqrt{8/9}$ is not included in C_d. However, we prefer to use Eq. (4.53).

[28]Since we consider the downstream water level to be less than the submergence limit, the bed level should be much lower than the sill level. The flow would be similar to that at a drop discussed in Section 4.4. As seen for the free overfall, the critical depth occurs a little upstream of the brink. Since the head measurement is also performed a little upstream of the weir, it is not unreasonable to assume that the flow depth, $h_1 + h_w$, is equal to the critical depth.

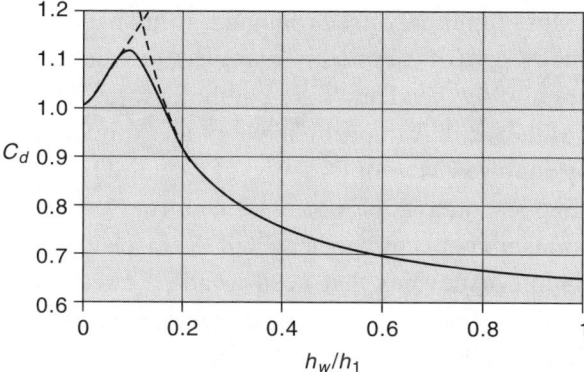

Fig. 4.20 Discharge coefficient for a sharp-crested weir [Solid line shows Eq. (4.57) and dashed lines show Eqs (4.55) and (4.56).]

1. For low discharges, the rectangular weir will have a small head over the crest. Besides difficulty in measuring these heads accurately, there may also be undesirable viscosity and surface tension effects. One solution is to use a contracted weir, spanning only part of the channel width. Other options are use of weirs of other shapes, e.g., triangular, parabolic, circular, and trapezoidal. The triangular weir, also called a V-notch (see Fig. 4.21) is the most commonly used weir for small discharges. (However, its drawback is a large head at high discharges. Sometimes, a compound weir is used with a triangular notch cut in the rectangular weir plate.) The analysis is similar to that of a rectangular weir [Eq. (4.52), except that the area of the flow strip now depends on h] and results in a discharge given by

$$Q = C_d \sqrt{g} h_1^{2.5}$$

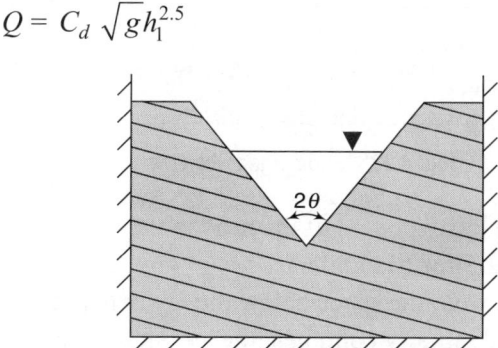

Fig. 4.21 A triangular weir or V-notch

(Note that the total discharge Q is used and the head is raised to the power 2.5.) The discharge coefficient, C_d, is a function of the bottom angle, 2θ. Most commonly used triangular weirs have a bottom angle of 90° for which $C_d = 0.44$. For a circular weir, the discharge is a complicated function of the head and for a parabolic weir, it is proportional to the square of the head. A trapezoidal weir (also known as Cipoletti weir) is generally analysed as a combination of a central rectangular weir and a surrounding V-notch. A proportional weir (or Sutro weir) provides a linear relationship between the head and the discharge and is useful for float-regulation of devices that need to add a chemical to water in a fixed proportion. However, since these are not very common, we do not describe these in detail.

2. If the tailwater level approaches the weir crest, it starts affecting the discharge. The recommended value of the tailwater level is about 5–10 cm below the crest. However, if it is unavoidable to use the weir under submerged conditions, it will involve the measurement of two heads, one upstream and the other downstream (Fig. 4.22). Since the discharge would be based on a difference of these heads, which is likely to be small, the resulting value may be subject to a large error. An estimate of the discharge may be obtained by the Villemonte formula, based on a 1947 study, as

$$Q_{\text{sumberged}} = Q_{\text{free}} \left[1 - \left(\frac{h_2}{h_1} \right)^p \right]^{0.385} \tag{4.58}$$

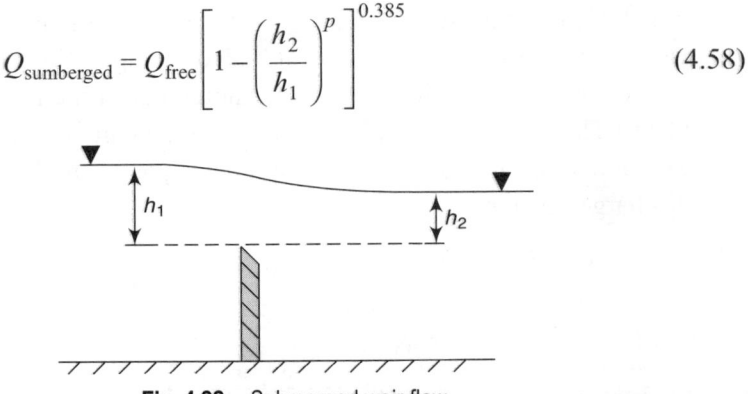

Fig. 4.22 Submerged weir flow

where p refers to the exponent of the head in the discharge equation (1.5 for rectangular, 2.5 for triangular).

3. For a suppressed weir, the space below the lower nappe is enclosed from all sides. It would initially contain some air but the water flowing above it gradually entrains the air and causes the pressure within this pocket to fall below the atmospheric pressure. This leads to an additional pull on the nappe surface thereby depressing it and causing an increase in discharge over that obtained under *standard conditions*. With time, all air

from this pocket may be removed and the nappe may cling to the weir plate. In order to be able to use the discharge coefficient obtained under standard conditions, provision of air-entry has to be made. An air-vent is normally provided to provide access to atmospheric air for the area just downstream of the weir plate. Bos (1989) provides an expression to obtain the rate of air supply needed to satisfy the aeration needs.

4. A contracted weir does not span the whole width of the channel. The jet of water passing over the weir is subjected to side contractions as well as crest contractions. An advantage of the contracted weir is that aeration below the lower nappe is not needed since this area is open to atmosphere from the sides. Kindsvater and Carter (1957) describe a technique for estimating the discharge over a contracted weir. An effective width of weir and an effective head are used to replace the actual width of the weir crest and the actual head of water above the crest, and the discharge coefficient is obtained as a function of the contraction ratio and the relative head. Details are not provided here but may be seen in USBR (1997). Another option is to use Eq. (4.53) to obtain the specific discharge with C_d of about 0.6, and multiply it by a reduced crest width (the reduction is typically 10% of the head for each contraction).

5. Some attempts have been made to design a weir shape for a given exponent in the discharge equation. For example, if the exponent is 1.5, we get a rectangular weir, for 2 we get a parabolic weir, and for 2.5 we get a triangular weir. The special case of linear weir has an exponent of 1 and a general weir shape may be designed to obtain any value of the exponent. However, we believe that these are more of theoretical exercises rather than having practical utility and so do not discuss them here.

Example 4.8 A sharp-crested weir is built across the entire width of a 3 m wide channel. The weir height is 1 m. What would be the discharge when the upstream water depth in the channel is 1.6 m?

Solution
Here, $h_1/h_w = 0.6$ and we may use Rehbock's equation, Eq. (4.56), to obtain C_d as 0.62 [we can also use Eq. (4.57) to obtain the same value of C_d]. The discharge is obtained from Eq. (4.53) as **0.903 m³/s per m width**. Note that it is almost 25% larger than the discharge for a broad-crested weir of the same height and with the same upstream water depth (see Example 4.7).

The thin-plate sharp-crested weirs may not be able to withstand the water force at large discharges. The broad-crested weirs can be used for such cases but involve a more complicated variation of C_d. If we still want to use the sharp-crested weir formula, we can strengthen the weir by filling the portion between the weir plate and the lower nappe by, say, concrete, without affecting the discharge equation. However, since the nappe profile changes with the discharge, our new weir would not behave as a sharp-crested weir at all discharges. This idea of designing the profile of the structure based on the nappe profile is utilized in design of the overflow spillway to pass the flood from a reservoir to the downstream channel. It is called the *ogee*[29] *spillway* and is discussed next (although a more appropriate place is a book on Hydraulic Structures, we briefly describe this spillway here to extend our discussion of the flow over a sharp-crested weir).

4.6 Spillways

Spillways are used to provide a controlled release of flood waters from a reservoir to a downstream area (generally the channel). Water does not normally flow over the spillway and it is used only during severe floods in order to keep the reservoir level below the maximum permissible safe level for the associated dam or levee. An ogee spillway is shaped in the form of the lower nappe of an equivalent sharp-crested weir. Although the weir can be taken as inclined and its height can be small, here we consider only vertical upstream face and a weir height quite large compared to the head over the crest (conditions usually encountered in practice). Figure 4.23 shows an ogee spillway along with the hypothetical sharp-crested rectangular weir. As mentioned in the previous section, this spillway profile corresponds to the nappe profile for a particular head, known as the design head, h_d. For heads larger than this, the nappe tends to go farther, which creates a negative pressure area on the spillway crest. It leads to an additional suction being applied to the flow, which

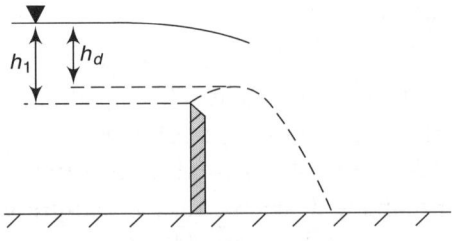

Fig. 4.23 Profile of an ogee spillway

---

[29]An *ogee* is a double curve with the shape of an elongated *S*. It consists of a concave arc flowing into a convex arc.

causes the coefficient of discharge to increase in comparison with that for a sharp-crested weir at the same head. For smaller heads, the nappe is pushed towards the weir and consequently applies a positive pressure on the spillway crest. This results in a smaller discharge over the spillway than the corresponding weir discharge. To take advantage of the increased discharge coefficient at the higher heads, generally the spillway crest profile is *underdesigned* (i.e., it is designed for a head smaller than the maximum expected head). This implies that the maximum flood would be passed with a smaller rise in water level than needed if the crest is designed for the maximum flood level. On the other hand, since negative pressures are created on the crest, there is a likelihood of cavitation if the design head is significantly smaller than the maximum head. Since the higher floods are expected less frequently, it may be more economical to underdesign the spillway. In the following discussion, we assume that the design head, h_d, is given, and the crest profile and discharge coefficient have to be determined.

Since the lower nappe of a sharp-crested weir springs higher than the weir crest due to crest contraction effect, the head over the weir, h_1, would be larger than the design head, h_d. Experimental observations indicate that the nappe rises a distance of about $0.12\,h_1$, above the weir crest. Therefore, $h_d = 0.88\,h_1$ and the specific discharge over the spillway at design head, q_d, is written as

$$q_d = C_d\,\sqrt{g}h_1^{1.5} = 1.211C_d\,\sqrt{g}h_d^{1.5} \tag{4.59}$$

in which C_d is the weir discharge coefficient already described in the previous section. Since the spillway height is generally much greater than the design head, Eq. (4.56) is used to obtain the discharge coefficient as 0.578, so that $1.211C_d$ equals 0.70. The spillway crest profile would follow the shape of lower nappe for a sharp-crested weir of height $h_d/0.88$, and is described next.

Crest profile

Extensive experiments on the lower nappe profile for sharp-crested weirs have led to empirical equations for the spillway profile, which would conform to this profile at the design head. Assuming that the surface is smooth and the friction does not affect the flow, the discharge coefficient for a free flowing sharp-crested weir has been used to obtain the design head for the given design discharge (which is smaller than the maximum expected discharge). The crest profile for the design head as recommended by the US Army Corps of Engineers (HDC 2007) consists of a power law profile downstream of the origin (see Fig. 4.23, where the origin is taken at the highest point on the crest) and a combination of three circular segments upstream of it. Obviously the profile would depend on the slope of the upstream face and the relative design head compared to the height of the weir. Here we discuss only *high spillways* with *vertical* upstream faces. For these, the downstream profile is given by

$$\eta = 0.5\xi^{1.85} \tag{4.60}$$

where η is the nondimensional vertical distance below the origin and ξ is the non-dimensional horizontal distance to the right of the origin (the design head, h_d, is used to normalize all distances). Note that the head is measured from the top of the crest to the *water level* and not energy line.

The upstream profile consists of three circular arcs with nondimensional radii of 0.5, 0.2, and 0.04, starting from the origin and going upstream. However, since this profile showed a discontinuity in curvature at the junction of different arcs, Hager (1987) proposed the following profile, which differs from the earlier profile by less than 1% of the design head, has no discontinuity in curvature, and is expected to perform better in reducing cavitation at heads higher than the design head:

$$\eta = 0.4826(\xi + 0.2818)[0.2666 + \ln(\xi + 0.2818)] + 0.136 \tag{4.61}$$

Figure 4.24 shows a plot of Eqs (4.60) and (4.61).

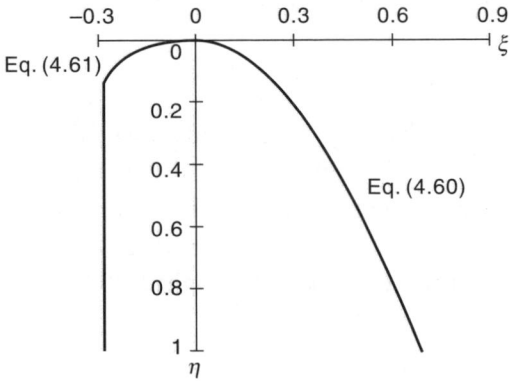

Fig. 4.24 Computed profile of an ogee spillway

For the heads higher than the design head, the nappe separates from the spillway crest and a negative pressure exists. It has been found that the minimum pressure occurs somewhere between $-0.15 \geq \xi \geq -0.20$ and it should, at maximum flood, exceed the cavitation pressure to ensure safe operation of the spillway. The coefficient of discharge at heads other than the design head is a function of the ratio of the operating head and design head and may vary from about *80% of the 'design C_d'* at low heads (*one-tenth of the design head*) to about *110% at twice the design head* and *115% at high heads* (five times the design head). Extensive discussion may be found in HDC (2007).

Example 4.9 The maximum discharge expected to be passed through an ogee spillway is 200 m³/s. The crest length is 50 m and the height of the spillway is 10 m. Find the maximum height of water above the spillway crest assuming that the design discharge is taken equal to the maximum discharge. If the spillway is underdesigned by taking the design discharge as 60% of the maximum discharge, what would be the maximum height of water above the crest?

Solution
Since the height of the spillway is large compared to the expected head, we may assume the coefficient of discharge to be 0.578 and obtain the design head from Eq. (4.59) as

$$h_d = \left(\frac{q_d}{1.211 C_d \sqrt{g}} \right)^{2/3} = 1.493 \text{ m}$$

When the design discharge is taken as 60% of the maximum discharge, i.e., 120 m³/s, the design head would be 1.062 m. Assuming that the maximum head corresponding to the discharge of 200 m³/s remains close to the value obtained earlier, the ratio of the maximum head to the design head would be about 1.4. The ratio of the coefficient of discharge at the maximum head to the 'design C_d' for this ratio is close to 1.04 (considering a linear variation of ratio from 1 at $h/h_d = 1$ to 1.1 at $h/h_d = 2$). Therefore, the maximum head would be equal to

$$h_{max} = \left(\frac{q_{max}}{1.211 \times 1.04 \times 0.578 \sqrt{g}} \right) = 1.455 \text{ m},$$

which is about 5 cm lower than the value when the spillway is designed for the maximum expected flow.

4.7 Convergent Flumes

We saw that in a broad-crested weir, critical flow occurs over the crest since the specific energy becomes small. As will be shown in the next chapter, raising the bed is not the only means of achieving critical conditions. A contraction in the channel width (or a combination of contraction and raising) may also cause the flow to become critical at the contracted section. This principle is used to design flumes, which can be used for the purpose of estimating the discharge in a channel. The flume is fitted in existing channels and, clearly, will have its width equal to the parent channel width at both ends. A converging section is generally followed by a throat of uniform width and then a diverging section to regain the original width (Fig. 4.25). Sometimes, the throat is very short (short-throated flume) or even non-existent (cut-throat or throatless flume).

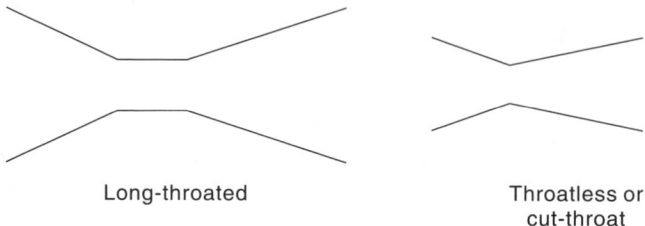

Long-throated   Throatless or
                                  cut-throat

Fig. 4.25   Convergent flumes used for flow measurment

The converging section should be sufficiently long to avoid separation of flow and generally follows a 1:3 (width:length) slope. When both the width and the bed level change, it is recommended that the transition for both of these start at the same location. The diverging section follows a slope of 1:6 to as high as 1:15. The more gradual expansion increases the modular limit slightly but is more expensive. Therefore, the 1:6 expansion is more commonly used. The head, h, is generally measured at a section that is upstream of the flume by two to three times the maximum expected head. However, for some flumes (e.g., the Parshall flume, which is commonly used in the United States, and the cut-throat flume), it is measured at a specified location in the converging section. The discharge is expressed as

$$Q = C_d B_t \sqrt{g} h^p \qquad (4.62)$$

where B_t is the width at the throat section and the power p depends on the configuration of the flume and location of the head measurement. For flumes with rectangular control section, $p = 1.5$, and a discharge coefficient close to 0.5 can be used if h is measured in the parent channel. The modular limit of these flumes is generally high (0.7 to 0.8) and they would not typically operate under submerged conditions. However, when submergence affects the flow, an additional head measurement has to be taken downstream of the throat section. Due to their higher cost compared to a weir, flumes are not very commonly used for flow measurement. Also, since the theory of these flumes is essentially similar to that of the broad-crested weir, we do not discuss it here. Bos (1989) provides greater details of these flumes along with comprehensive tables of values of the discharge coefficients for various flume geometries. Due to its popularity, we briefly describe the Parshall flume here.

Parshall flume

This flume was developed by Parshall in 1920s and is widely used in the United States. Figure 4.26 shows a Parshall flume, which has a converging section, a throat section and a diverging section. The converging section has a horizontal floor while the throat slopes down and the diverging section slopes upwards. The upstream head, h_1, is measured at a specified location and the downstream head, h_2, is measured (for estimating discharge under submerged flow conditions) at the throat. The dimensions shown in Fig. 4.26 are dependent on the throat size, B_t, but are not a linear function of B_t (i.e., various flumes are not geometrically similar). For example, the dimensions of flumes with throat widths of 0.3 m and 3 m are shown in Table 4.1 (all dimensions in meter).

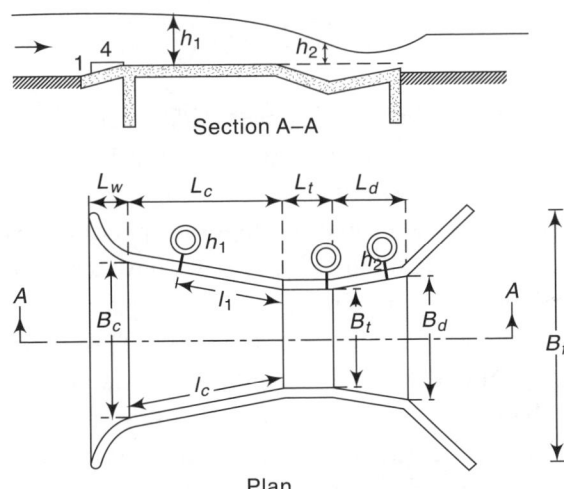

Fig. 4.26   Parshall flume

Table 4.1   Dimensions of flumes with throat widths of approximately 0.3 m, 3 m, and 15 m

B_t	B_c	B_d	B_f	L_c	L_t	L_d	L_w	l_1	l_c
0.305	0.845	0.610	1.492	1.343	0.610	0.914	0.381	0.914	1.372
3.05	4.756	3.658		4.267	0.914	1.829		1.829	
15.24	18.529	17.272		8.230	1.829	6.096		5.893	

The coefficient of discharge and the exponent p in Eq. (4.62) for different throat sizes are given in Table 4.2. The values have been found to be valid over certain ranges of the head, but we do not mention the ranges here. The interested reader may refer to Bos (1989).

Table 4.2   Coefficients of discharge and exponent p for different throat sizes

B_t(m)	0.076	0.229	0.305	0.914	1.52	2.13	3.05	4.57	7.62	12. 2
C_d	0.7420	0.7478	0.7237	0.7626	0.7818	0.7949	0.7817	0.7654	0.7517	0.7445
p	1.550	1.530	1.522	1.566	1.587	1.601	1.600	1.600	1.600	1.600

Example 4.10   A Parshall flume, with a 3.05 m throat width and having standard dimensions shown in Table 4.1, is operating within modular limit and the upstream head, h_1, was measured as 0.50 m. What is the discharge?

Solution

From Table 4.2, $C_d = 0.7817$ and $p = 1.600$. Therefore, from Eq. (4.62), we get

$$Q = 0.7817 \times 3.05 \times \sqrt{9.81} \times 0.5^{1.6} \text{ m}^3/\text{s} = \textbf{2.463 m}^3\textbf{/s}$$

The modular limit for this size of flume is 0.8, which implies that h_2 should be less than 40 cm.

SUMMARY

In this chapter, we have discussed some rapidly varied flows with an emphasis on the hydraulic jump as an energy dissipator, and on other structures as discharge measuring devices. All of these involve rapid variation in flow depth in a short distance and need some assumptions and empirical coefficients for their complete solution. In almost all cases, we have assumed that the channel geometry and the bed elevation do not change appreciably over the region of interest, or, even when it changes (e.g., the weir and the drop), the changes are rather abrupt and of considerable magnitude. In a number of practical cases, we encounter changes in the cross section, alignment, or bed elevation in the channel, which are either gradual or deliberately made gradual to reduce the energy losses. Sometimes, we come across sudden changes in the geometry or alignment that have not been discussed in this chapter since the emphasis was on energy dissipation and flow measurement. Analysis and design of these transitions is described in the next chapter.

REFERENCES

Anderson, M.V. (1967): 'Non-uniform flow in front of a free overfall,' Acta Polytech. Scand, *Civ. Eng. Constr. Ser.*, 42, 1–24.

Bhallamudi, S.M. (1994). "End depth in trapezoidal and exponential channels." J. Hydr. Res., 32(2), 219–32.

Bos, M.G. (1989): *Discharge Measurement Structures*, 3rd edn, International Institute for Land Reclamation and Improvement, Wageningen, The Netherlands.
(website: http://content.alterra.wur.nl/Internet/webdocs/ilri-publicaties/publicaties/Pub20/pub20.pdf)
Chaudhry, M.H. (1993): *Open-Channel Flow*, Prentice-Hall Inc., Englewood Cliffs, NJ, USA.
Chow, V.T. (1981): *Open-Channel Hydraulics*, McGraw-Hill International, Tokyo, Japan.
Dey, S. (2002): 'Free overfall in open channels: state-of-the-art review', *Flow Meas. Instru*, **13**(5–6), pp. 247–64.
Ead, S. A. and Rajaratnam, N. (2002): 'Hydraulic jumps on corrugated beds', *J. of Hydr. Eng.*, **128**(7), pp. 656-63.
Govinda Rao, N.S. and Muralidhar, D. (1963): 'Discharge characteristics of weirs of finite-crest width', *La Houille Blanche*, **18**(5), pp. 537-45.
Hager, W.H. (1987): 'Continuous crest profile for standard spillway', *J. of Hydr. Eng.*, **113**(11), pp. 4533-57.
Hager, W.H. and Schwalt, M. (1994): 'Broad-crested weir', *J. of Irrig. Dra. Eng.*, **120**(1), pp. 13-26.
HDC (2007): 'Hydraulic design criteria', US Army Corps of Engineers, *http://chl.erdc.usace.army.mil/Media/2/7/7/100-a.pdf* (last accessed on Feb. 17, 2007).
Kindsvater, C. E. and Carter, R.W. (1957): 'Discharge characteristics of rectangular thin-plate weirs', *J. of Hydr. Div.*, ASCE, **83**(6), pp. 1-36.
Lin, C.H., Yen, J.F., and Tsai, C.T. (2002): 'Influence of sluice gate contraction coefficient on distinguishing condition', *J. of Irrig. Dra. Eng.*, **128**(4), pp. 249-52.
Rajaratnam, N. (1967): 'Hydraulic jumps', in *Advances in Hydroscience*, Academic Press, New York, **4**, pp. 255-62.
Rajaratnam, N. and Muralidhar, D. (1964): 'End depth for circular channels', *J. of Hydr. Div.*, ASCE, **90**(2), pp. 99–119.
Rajaratnam, N. and Muralidhar, D. (1968): 'Characteristics of rectangular free overfall', *J. of Hydr. Res.*, **6**(3), pp. 233–58.
Rajaratnam, N. and Muralidhar, D. (1970): 'The trapezoidal free overfall', *J. of Hydr. Res.*, **8**(4), pp. 419–47.
Ranga Raju, K.G. (1993): *Flow through Open Channels*, Tata McGraw-Hill, New Delhi.
Ranga Raju, K.G., and Asawa, G.L. (1979): 'Comprehensive weir discharge formulae', *Proc. of IMEKO in Industry*, Tokyo, Japan.

Ranga Raju, K.G. and Visavadia, D.S. (1979): 'Discharge characteristics of a sluice gate located on a raised crest', *Proc. IMEKO in Industry*, Tokyo, Japan.

Ranga Raju, K.G. and Gopalkrishna, K.B. (1981): 'Submerged flow through a sluice gate located on a raised crest', *Proc. of Sec. Int. Sympo. on Flow*, St. Louis, USA.

Renner, J. and Naudascher, E. (1975): 'Entrainment in surface rollers', *J. of Hydr. Div.*, ASCE, **101**(HY2), pp. 325-27.

Swamee, P.K. (1992): 'Sluice-gate discharge equations', *J. of Irrig. Dra. Eng.*, **118**(1), pp. 56-60.

Swamee, P.K. and Prasad, K. (1977): 'Direct equations for hydraulic jump elements', *J. of Irrig. and Pow.*, **34**(5), pp. 503–06, Central Board of Irrigation and Power, New Delhi.

Swamee, P.K., Mishra, G.C., and Salem, A.A.S. (1996): 'Optimal design of sloping weir', *J. of Irrig. Dra. Eng.*, **122**(4), pp. 248-55 (website: *http://chl.erdc.usace.army.mil/Media/2/7/7/100-a.pdf*).

USBR (1955): 'Research studies on stilling basins, energy dissipators, and associated appurtenances', *Hydraulic Laboratory report no. Hyd-399*, US Bureau of Reclamation, *http://www.usbr.gov/pmts/hydraulics_lab/pubs/HYD/HYD-399.pdf* (last accessed on Mar. 2007).

USBR (1997): *Water Measurement Manual*, 3rd edn, US Bureau of Reclamation, *http://www.usbr.gov/pmts/hydraulics_lab/pubs/wmm/* (last accessed on Mar. 2007).

EXERCISES

4.1 A hydraulic jump is formed in a 2 m wide smooth horizontal channel carrying a discharge of 2 m³/s. The pre-jump depth is 15 cm. Find the post-jump depth and length of the jump. If the boundary is now roughened and the shear force on the boundary over the entire length of the jump is estimated as the pre-jump pressure force multiplied by $(F_1 - 1)^2$, what would be the post-jump depth?[30]   [1.09 m, 6.6 m, 0.82 m]

4.2 A trapezoidal channel has a bed width of 1 m and side slopes of 1H:1V. A hydraulic jump occurs at a location from a flow depth of 0.10 m to 0.60 m. Estimate the discharge in the channel. Compute the specific force at the pre-jump and post-jump depths and plot these points on Fig. 4.5.

---

[30]See Ead and Rajaratnam (2002) for recent discussion of hydraulic jump in channels with rough beds.

Hint: Use Fig. 4.6 or Eq. (4.10) for estimating the discharge in the channel.

[0.55 m³/s, 0.284 m³]

4.3 For the hydraulic jump on a smooth bed in Exercise 4.1, compute the energy loss within the jump and plot the profile of the jump. If the energy loss has to be increased by 50% for the same discharge, what pre-jump and post-jump depths would be required?

[1.27 m, 0.13 m, 1.19 m]

4.4 A rectangular channel has a constant top-width while a triangular channel has its top-width varying linearly with flow depth. For a general power-law geometry in which the channel coordinates at a flow depth of y are given by $\pm ay^b$, derive an expression relating the sequent depth ratio to the pre-jump Froude number and the exponent, b (note that it is not a function of a). For a parabolic channel ($b = 0.5$), the sequent depth ratio was found to be 5. What is the pre-jump Froude number?

$$\left[r^{b+2} + r^{-b-1} + \overline{F}_1^2 = 1 + \overline{F}_1^2, \text{ where } \overline{F}_1^2 = \frac{V^2}{g\overline{y}_1}; 6.0 \right]$$

4.5 In Example 4.5, what would be the flow depth just downstream of the gate when the tailwater depth is 0.7 m? If the discharge is kept same and free flow conditions have to be achieved by lowering the gate, what should be the maximum gate opening? What would be the depth upstream of the gate? [0.69 m, 5 mm, 39 m]

4.6 A laboratory flume made of glass ($n = 0.01$) is 0.6 m wide and has a bed slope of 1 in 1000. The channel ends in a free overfall where the flow depth is measured as 0.30 m. Estimate the discharge in the flume (it must be verified that the flow is subcritical). [0.522 m³/s]

4.7 A rectangular broad-crested weir is placed across the entire width of a 3 m wide rectangular channel. The crest width is 1 m and the weir height is also 1 m. At a certain time, the water level at a slightly upstream location was measured to be 0.50 m above the weir crest. What is the discharge? If the weir is contracted by reducing its width to 2.5 m (i.e., 25 cm at either end), what increase in water level would be observed for the same discharge (assume each contraction to be 10% of the head and take the head as 0.50 m)?[31] [1.56 m³/s, 8 cm]

4.8 A 1 m high sharp-crested weir spans the whole width of a 4 m wide channel. At a certain discharge, it was observed that the downstream water level is higher than the weir crest. The upstream water level was

---

[31]Although the head would increase due to the contraction, it would not make a significant difference in the computations if we use the 'old' head to estimate the contractions.

0.5 m above the crest while the downstream water level was 0.3 m above the crest. Estimate the discharge.   [2.14 m³/s]

4.9 Derive an equation similar to Eqs (4.52) and (4.53) for a triangular weir with bottom angle of 2θ and for a parabolic weir ($x = \pm k \sqrt{y}$, with origin at the crest, y being vertical height and x being the distance to the weir profile). Note the different powers of head in these equations and comment on the implications. Obtain the profile of the weir that will have its discharge proportional to the pth power of head.

4.10 It was decided to convert the sharp-crested weir of Exercise 4.8 into an ogee shape by following the profile of lower nappe when the water level is 0.5 m above the crest of the sharp-crested weir. Obtain the ogee profile and compute the discharge at the design head. Also estimate the discharge when the head over the ogee crest is twice the design head.
                    [0.679 m³/s per m; 2.11 m³/s per m]

4.11 A standard Parshall flume has a throat width of 0.3 m and the upstream head is measured as 0.2 m. What is the discharge? If the discharge is doubled, what would be the upstream head? Assume that no submergence occurs under both conditions.   [0.0587 m³/s, 0.315 m]

5 Channel Transitions

5.1 Introduction: Occurrence and Importance

Most of the discussion till now has focused on prismatic channels in which the bed slope and the cross-sectional shape do not change along the length. Natural channels, naturally, show a wide variation in the bed slope, cross-section shape, and alignment. However, even man-made channels may have to undergo some kind of transition due to engineering or economic considerations. For example, width of a canal can be reduced when it crosses another feature such as a river or a road to reduce the width of the aqueduct or the length of the bridge. The reduction in width may be accompanied by a raising or lowering of the bed to adjust the water level. Bends are also unavoidable in most canals although the location and curvature can be in the engineers' control at the design stage. We first analyse these transitions with a view to compute the flow characteristics in the transition section in absence of any energy loss. This is followed by a description of the design of the transition with emphasis on minimizing the energy loss. Since a majority of practical cases involve subcritical flow, we put greater emphasis on these. In supercritical flow, the sensitivity of the flow to small disturbances implies that large-amplitude waves may be created due to presence of surface irregularities or change in direction of flow. These waves may persist for a considerable length and would require higher sidewalls.

We first describe the effect of a change in bed elevation on the flow characteristics. A smooth step (positive or negative), in which the energy loss may be neglected, is analysed first since the specific energy principle described in Section 3.3 can be used to obtain the flow depth and velocity. As discussed in the section on broad-crested weirs (Section 4.5.1), we will show that a large step causes critical flow. The specific energy concept is then applied to the case of change in channel width, in form of gradual expansion or contraction, and then extended to cover the cases of abrupt width changes. Channel bends are analysed to obtain the energy loss as well as the rise in water level along the outer bank of a bend. Finally, transitions in supercritical flows are described in Section 5.5.

5.2   Change in Bed Elevation

Figure 5.1 shows a smooth step (or hump) which raises the bed level by an amount Δz. The flow would be rapidly varied and the assumption of hydrostatic pressure distribution may not be valid as the flow approaches the step and flows over it. However, if the step is long enough, we may assume that at section 2, the streamlines are parallel and the hydrostatic pressure distribution exists. If we want to find the flow depth at this section for given upstream conditions, application of either the energy equation between the sections 1 and 2 or the momentum equation on the control volume enclosed by these sections can be tried. However, in order to apply these equations, we would require a knowledge of either the energy loss or the force acting on the step. Assuming the step to be smooth enough to have negligible energy loss, we write the energy equation as (taking the datum at the upstream bed level)

$$y_1 + \frac{Q^2}{2\,gA_1^2} = y_2 + \frac{Q^2}{2\,gA_2^2} + \Delta z \tag{5.1}$$

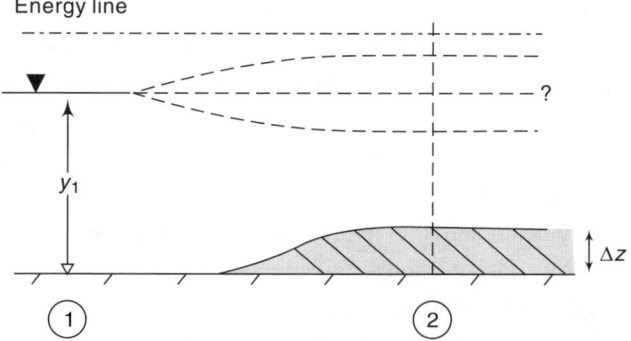

Fig. 5.1   A smooth step in a channel

which results in a nonlinear equation in y_2 which can be solved using an iterative technique. There would be multiple solutions, however, and we will have to decide which one is the 'correct' solution. (Mathematically, of course, all of these solutions would be correct but physical considerations would eliminate all but one of them.) For example, a rectangular channel has area proportional to the depth and Eq. (5.1) becomes a cubic equation in y_2. It may be shown that the three solutions comprise one negative and two positive values.

The negative solution, of course, is meaningless but there is no reason to prefer one of the remaining two over the other on mathematical grounds. Here the concept of specific energy comes in handy to analyse the physics of the problem and find the plausible solution. For simplicity, we describe the application of this concept to a rectangular channel and, as done earlier, leave the problems of non-rectangular channels to be solved by the reader.

For a rectangular channel of width B, the specific energy at a section is written as

$$E = y + \frac{q^2}{2\,gy^2} \tag{5.2}$$

It is, therefore, clear from Fig. 5.1 or Eq. (5.1) that the specific energy at section 2 is less than that at section 1 by the amount Δz. Figure 5.2 shows a typical specific energy diagram (see Section 3.3) for a given specific discharge, q. (A more convenient representation of the specific energy diagram can be obtained by nondimensionalizing the specific energy and the flow depth by the critical depth, i.e., $E_* = E/y_c$ and $y_* = y/y_c$. Equation (5.2) then collapses to a single curve for all discharges as $E_* = y_* + (1/2y_*^2)$.)

From Fig. 5.2, we can easily find the specific energy at section 1, E_1, since y_1 is known. The specific energy at section 2, E_2, is then obtained by subtracting Δz from E_1, and is shown by a dashed line in Fig. 5.2. But now we are back to the same question—which depth to choose—because there are two depths, $y_{2,\text{sub}}$ and $y_{2,\text{sup}}$, one subcritical and the other supercritical, which result in the same specific energy, E_2. Since the specific discharge is constant, all sections

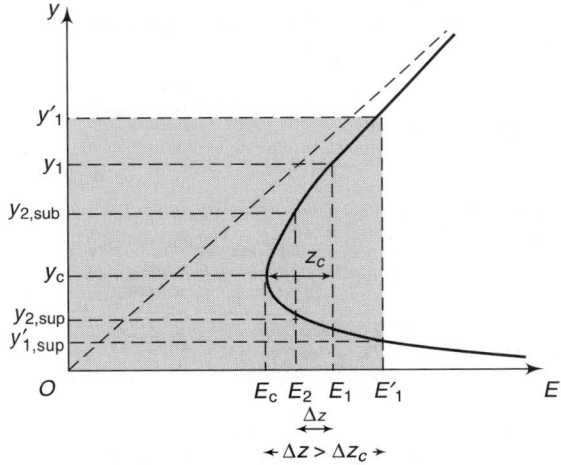

Fig. 5.2 Specific energy diagram for a rectangular channel

all sections between 1 and 2 would have conditions corresponding to the specific energy diagram shown in the figure, i.e., the flow depth and specific energy at the intermediate sections would fall on the plotted curve. If the step has a monotonic rise, the specific energy will continually decrease from section 1 to 2 (since we have assumed the losses to be negligible and total energy to be same). Therefore, depending on the shape of the step, the flow depth will gradually decrease from y_1 to $y_{2,\text{sub}}$, but will not reach the other possible depth $y_{2,\text{sup}}$ (since it would mean a decrease in specific energy till the critical point and then an increase).[1] Note that a supercritical approach depth, $y_{1,\text{sup}}$ also has the same specific energy, E_1, and will result in a depth $y_{2,\text{sup}}$ at section 2. It is also clear from Fig. 5.2 that there is a *critical height* of the step, Δz_c, which would cause the flow to become critical at section 2. Any step larger than this size would imply a specific energy less than the critical energy, which is not possible (since the critical conditions correspond to the *minimum* specific energy). This indicates that flow would not take place with specified conditions and the energy of flow must increase if the flow has to be sustained. Naturally, this increase would be the *minimum increase* required to sustain the flow and corresponds to critical condition at section 2. Therefore, for a subcritical approach flow, the flow depth at section 1 must increase to y_1' corresponding to a specific energy of $E_c + \Delta z$. Similarly, if the approach flow is supercritical, the approach flow depth must *decrease* to $y_{1,\text{sup}}'$ to achieve the required specific energy. Computations for most channel shapes would involve an iterative procedure to obtain the flow depth for a given specific energy and discharge. However, *for a rectangular channel*, the resulting cubic equation may be directly solved using the Cordano method to obtain the subcritical and supercritical flow depths for a known specific energy, E, and specific discharge, q, as follows:

$$y_{\text{sub}} = \frac{E}{3}(1 + 2\cos\theta) \tag{5.3a}$$

$$y_{\text{sup}} = \frac{E}{3}\left[1 + 2\cos\left(\frac{2\pi}{3} - \theta\right)\right] \tag{5.3b}$$

in which $\theta = \dfrac{1}{3}\cos^{-1}\left(1 - \dfrac{27\,q^2}{4\,gE^3}\right)$.

---

[1] If we want to attain the supercritical depth, $y_{2,\text{sup}}$, starting from the subcritical upstream depth, y_1, we would have to provide a step which first increases in height to Δz_c and then decreases to its final height of Δz.

Note, from the definition of θ, that the specific energy must be greater than $(27q^2/8g)^{1/3}$, i.e., $3y_c/2$, which represents the critical specific energy.

The following example illustrates the concepts of flow over a smooth step.

Example 5.1   A discharge of 16 m^3/s flows in a 4 m wide rectangular channel at a flow depth of 2 m. At a section, a smooth hump of height Δz is placed. Plot the variation of the upstream depth and the depth at the hump for Δz ranging from 0 to 1 m. Repeat the process if the upstream flow occurs at the alternate depth of 0.75 m.

Solution

The specific discharge, $q = 4$ m^2/s, the critical depth, $y_c = (q^2/g)^{1/3} = 1.18$ m, and the critical specific energy, $E_c = 1.5\, y_c = 1.77$ m.

The specific energy of the approach flow, $E_1 = y_1 + q^2/2gy_1^2 = 2.20$ m for both the approach flow depths of 2 m (greater than y_c, subcritical) and 0.75 m (supercritical). Therefore, the critical height of the hump, $\Delta z_c = (2.20 - 1.77)$ m = 0.43 m. For hump heights smaller than the critical height, the specific energy at the hump would be equal to $E_1 - \Delta z$ and the corresponding flow depth is found using the appropriate equation [Eq. (5.3a) for subcritical approach flow or Eq. (5.3b) for supercritical]. For larger hump heights, critical depth would occur at the hump and the upstream specific energy will be equal to $E_c + \Delta z$. The flow depth is again obtained using Eq. (5.3a) or Eq. (5.3b). Figures 5.3 a and b show the variation of the flow depths and clearly show the critical hump height, the critical depth, and the 'backing up' of subcritical flow for large hump heights.

For the subcritical approach flow at the hump, the water depth at the hump is shown to be smaller than the approach depth [Fig. 5.3(a)]. However, since the bed level has gone up at the hump, the net effect on the *water level* is not immediately apparent. One can conclude that the water level will go down due to the combination of the following factors:

(i) The total energy remains same.

(ii) Since the flow depth is smaller at the hump, the velocity head must be larger than the approach velocity head.

(iii) The water level will be below the total energy line by an amount equal to the velocity head.

An alternative argument can be based on the specific energy diagram (Fig. 5.2) as follows:

The upper (subcritical) limb of the specific energy diagram is asymptotic to the line $y = E$ indicating that, in the limiting case of very large flow depth,

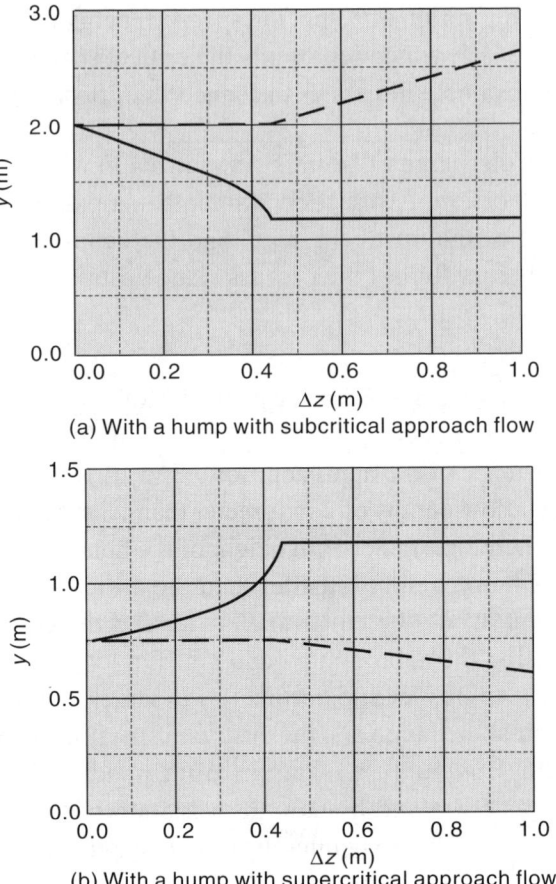

(a) With a hump with subcritical approach flow

(b) With a hump with supercritical approach flow

Fig. 5.3 Flow depth variation (solid line: at the hump, dashed line: upstream section)

the change in flow depth would be equal to the change in specific energy. It is clearly seen from Fig. 5.2 that at all points on the subcritical limb, the difference between the flow depths corresponding to two different values of specific energy would be *larger than* the difference between the specific energies.[2] Since the difference between the specific energies is Δz, $y_1 - y_2$ would be larger than Δz, i.e., y_1 would be larger than $y_2 + \Delta z$, indicating a drop in the water level.

Analysis of flow for a negative smooth step is exactly similar, except that the specific energy now increases at section 2 and there is no *critical drop* in the bed. The upstream depth will remain unaltered, and for subcritical approach

---

[2]The same conclusion may be drawn from the relation $dE/dy = 1 - F^2$, which is positive and less than 1 for subcritical flows.

flow, the downstream depth will keep on increasing with increase in the drop of bed. Figure 5.4(a) shows the variation of the downstream depth for bed drop of up to 1 m for the subcritical approach flow of Example 5.1, with an approach flow depth of 2.0 m. With a 1 m drop in bed level, the downstream depth is 3.12 m, indicating a rise in water level of 12 cm.

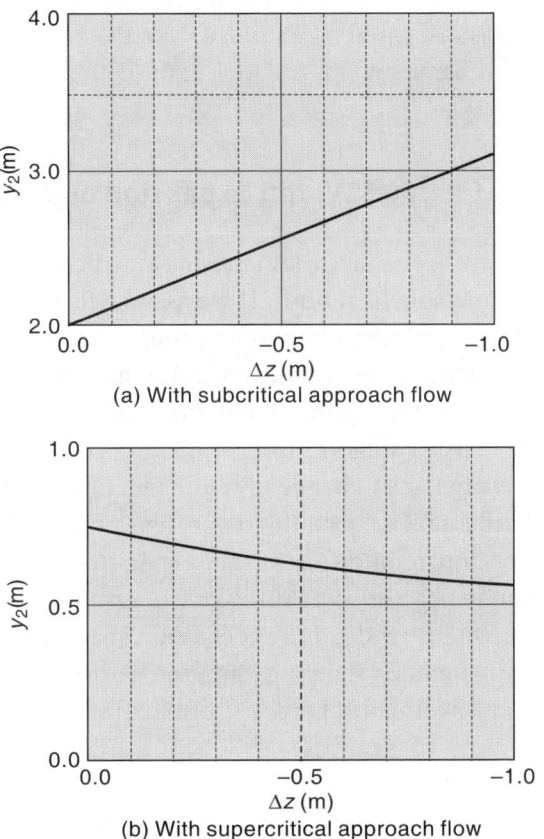

(a) With subcritical approach flow

(b) With supercritical approach flow

Fig. 5.4   Flow depth variation at a smooth drop

With supercritical approach flow, the drop in bed level will give rise to an increase in the specific energy and a corresponding decrease in the flow depth at the downstream section. Figure 5.4(b) shows the variation computed for the channel of Example 5.1 under supercritical conditions with an approach depth of 0.75 m. With a 1 m drop in bed level, the downstream depth is 0.56 m, indicating a drop in water level of 1.19 m.

For an abrupt rise or drop in the bed level, we must account for the energy loss in the transition. The usual technique is to assume this energy loss as some fraction of the approach velocity head. It is best to avoid abrupt changes in the bed level (unless we deliberately want to increase the energy loss, for example, in a hydraulic jump) and we do not discuss these further.

We have seen that if the size of a the step is large, critical flow occurs over the step. This principle has been discussed in Section 4.5.1. Similarly, if we reduce the width of the channel beyond a critical width, critical condition occurs at the contraction. Use of this principle for flow measurement using convergent flumes is discussed in Section 4.7. In the next section, we analyse a general change in channel width and look at the condition for critical flow at a contraction.

5.3   Change in Channel Width: Expansion and Contraction

As discussed earlier, the channel width may have to be changed at some locations due to economic considerations. If we assume that the bed level is kept same and the transition in width is smooth enough to neglect the energy loss, the specific energy at the upstream section and at the contracted (or expanded) section would be same. We again assume the channel to be rectangular but would not be able to use a diagram similar to Fig. 5.2 since it was drawn for a fixed specific discharge, q. A change in the channel width results in a change in the specific discharge. One can, of course, use a figure similar to Fig. 5.2, with different lines drawn for different specific discharge values. However, it is more convenient to use a *specific discharge diagram* (showing variation of specific discharge with flow depth for a constant specific energy) in place of the specific energy diagram (showing variation of specific energy with flow depth for a constant specific discharge). From Eq. (5.2), we write the specific discharge as

$$q = y\sqrt{2g(E - y)} \tag{5.4}$$

which is a multiplication of the flow depth and the velocity [note that $(E - y)$ is the velocity head]. For a given specific energy, the specific discharge would clearly be zero when either the flow depth is zero ($y = 0$) or the velocity head is zero ($y = E$). It would attain a maximum value at some depth, which can be obtained from Eq. (5.4), written for a general shape of channel, as

$$\frac{dQ}{dy} = 0 \qquad \Rightarrow \qquad \frac{dA}{dy}\sqrt{2g(E - y)} - \frac{A}{2}\sqrt{\frac{2g}{E - y}} = 0$$

$$\Rightarrow \qquad T(E - y) = \frac{A}{2}$$

$$\Rightarrow \quad \frac{Q^2 T}{2\,gA^2} = \frac{A}{2}$$

$$\Rightarrow \quad \frac{Q^2 T}{gA^3} = 1$$

which is the same condition as obtained earlier for critical flow. Hence, the critical condition also corresponds to the maximum discharge for a given specific energy. A typical specific discharge diagram[3] is shown in Fig. 5.5.

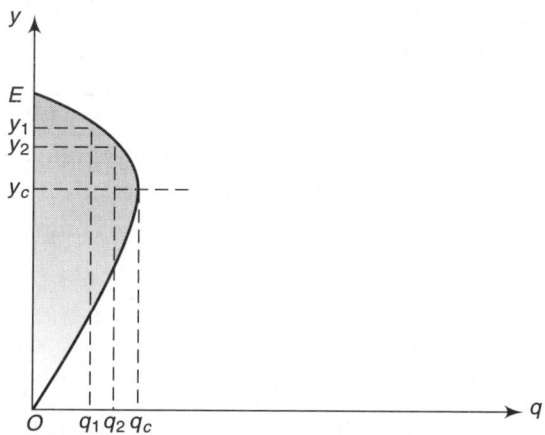

Fig. 5.5 Specific discharge diagram for a rectangular channel

For a given discharge, Q, a given approach depth, y_1, and the channel width, B_1, we may compute the specific energy, E_1 (the specific energy at section 2 would be same if the bed level is not changed), and prepare a specific discharge diagram. Figure 5.5 shows the diagram and the point representing the approach flow. As the section is contracted to a width, B_2, the specific discharge would increase from q_1 ($= Q/B_1$) to q_2 ($= Q/B_2$). The flow depth at the constriction can be obtained corresponding to this specific discharge value. It is clear from Fig. 5.4 that for subcritical flows, the flow depth increases with decrease in specific discharge for the same specific energy. Conversely, for supercritical flows, the flow depth decreases with decrease in specific discharge. There is

---

[3]As we have discussed for the specific energy diagram, we can prepare a nondimensional specific discharge diagram by using a nondimensional discharge as $q^* = q/\sqrt{2\,g\,E^{1.5}}$, and a nondimensional depth as $y^* = y/E$. Equation (5.4) becomes a single curve for all specific energy values as $q^* = y^* \sqrt{1 - y^*}$.

a critical width at contraction, B_c, which would correspond to the critical specific discharge, q_c (related to the specific energy as $q_c = \sqrt{gy_c^3} = \sqrt{8gE^3/27}$). If B_2 is reduced beyond this width, the specific discharge becomes more than q_c, and the flow would not be possible with given conditions. As we have seen in the previous section, the upstream depth would then increase (or decrease, if the approach flow is supercritical) to increase the available energy in such a way that critical conditions exist at the contraction. In other words, the specific energy of the flow would become $3q_2^{2/3}/2g^{1/3}$, where q_2 ($> q_c$) is the specific discharge at section 2. The corresponding upstream depth, y'_1, can be obtained from Eq. (5.3 a) or (5.3b). The following example illustrates the concepts of flow through a smooth contraction.

Example 5.2   A discharge of 16 m³/s flows in a 4 m wide rectangular channel at a flow depth of 2 m. At a section, the width is gradually changed to B_2. Plot the variation of the upstream depth and the depth at the contraction for values of B_2/B_1 ranging from 0.05 to 1. Repeat the computations for the upstream flow at the alternate depth of 0.75 m.[4]

Solution
As we have seen in Example 5.1, the specific energy of the approach flow, $E_1 = y_1 + [q^2/2gy_1^2] = 2.20$ m for both the approach flow depths of 2 m (greater than y_c, subcritical) and 0.75 m (supercritical). Therefore, the critical specific discharge would be

$$q_c = \sqrt{\frac{8gE^3}{27}} = 5.578 \text{ m}^2/\text{s}$$

and the critical width of contraction is

$$B_c = \frac{Q}{q_c} = 2.868 \text{ m}$$

giving a critical contraction ratio of 0.717. For contraction widths larger than the critical width, the flow depth is found by computing q and using Eq. (5.3 a) or (5.3 b). However, for widths smaller than the critical width, critical flow occurs at the contraction implying that the flow depth at the contraction, y_2, would be equal to $(q^2/g)^{1/3}$, the specific energy at the contraction (and everywhere) would be equal to $1.5y_2$ and the corresponding flow depth is found using the appropriate equation [Eq. (5.3a) for subcritical approach flow or Eq. (5.3b) for supercritical]. Figures 5.6(a) and (b) show the variation of the flow depths and clearly show the critical width, the critical depth, and the 'backing up' of subcritical flow for smaller widths

---

[4]In supercritical flows, changes in wall alignment will lead to formation of waves, which will be discussed in Section 5.5. Here we will ignore this phenomenon.

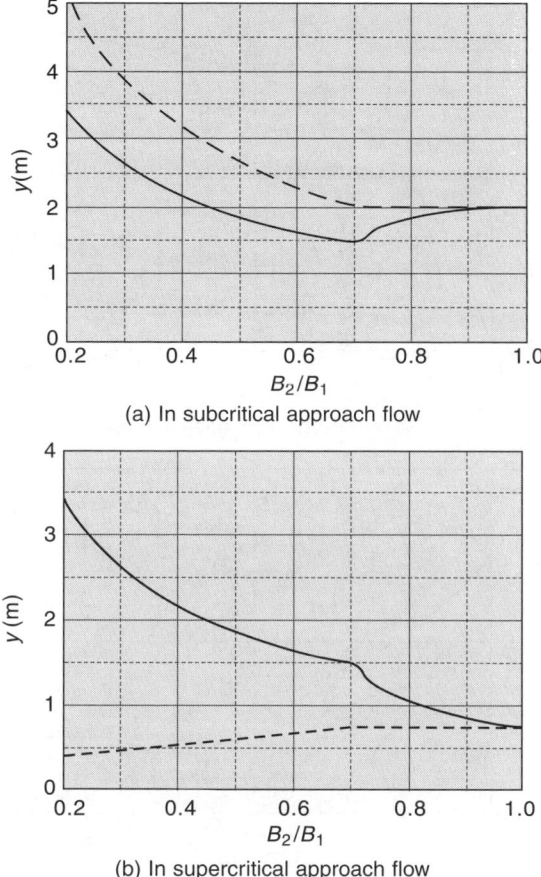

(a) In subcritical approach flow

(b) In supercritical approach flow

Fig. 5.6 Flow depth variation with a contraction (solid line: at the contraction, dashed line: upstream section)

(known as a *choked* flow). Note that unlike the flow over a hump, the critical depth is not constant but increases with decrease in the contraction width (i.e., increase in the specific discharge).

Analysis of flow for an expansion is exactly similar, except that the specific discharge now decreases at section 2 and there is no *critical width* and choking does not occur. The upstream depth will remain unaltered, and for subcritical approach flow, the downstream depth will keep on increasing with increase in the width of channel, i.e., decrease in the specific discharge. Figure 5.7(a) shows the variation of the downstream depth for expansion ratio of up to 3 for the subcritical approach flow of Example 5.2, with an approach flow depth of 2.0 m. With a three-fold increase in the channel width, the downstream depth

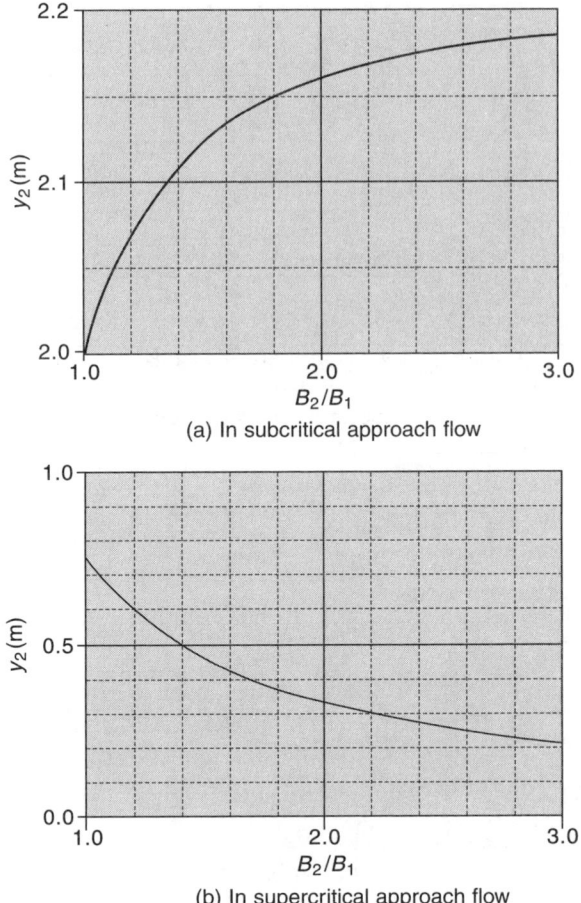

(a) In subcritical approach flow

(b) In supercritical approach flow

Fig. 5.7 Flow depth variation at an expansion

is 2.19 m, indicating a rise in water level of 19 cm. One should note that since the upstream specific energy is 2.20 m, the flow depth cannot increase beyond 2.20 m. Also, we have neglected all energy losses, which may not be a reasonable assumption for expansions.

With supercritical approach flow, the increase in channel width will give rise to a decrease in the flow depth at the downstream section. Figure 5.7(b) shows the variation computed for the channel of Example 5.2 under supercritical conditions with an approach depth of 0.75 m. With a three-fold increase in the channel width, the downstream depth is 0.21 m, indicating a fall of 54 cm in water level.

For an abrupt expansion or contraction in the bed width, we must account for the energy loss at the transition. The usual technique is to assume this energy loss as some fraction of the approach velocity head. Also, sometimes, both the width and the bed elevation may be changed simultaneously. Similar analysis is applicable to such cases also but accounting for the fact that both the specific energy and the specific discharge are changing.

While analysing the flow through expansions or contractions, we assumed that there was no energy loss between the two sections. However, depending on the profile of the variation of bed width (and side slope for trapezoidal channels), there would be some energy loss, which is generally higher in an expansion as compared to contraction. There have been a number of studies to arrive at a design of the bed width profile and side slope variation in order to achieve a low energy loss. A brief description of the design aspects is provided further.

5.3.1 Designs of Expansions

Hinds in 1928 suggested a method for the design of expansion, which assumed the water surface to be a combination of a concave upward parabola followed by a convex upward parabola of same length. The profile, therefore, has an inflection point at the midpoint of the transition and is tangential to the water surface at the beginning and end. The head loss due to expansion is assumed to be 20 to 30% of the difference in velocity head at the upstream and downstream sections and is assumed to vary linearly along the length of the transition. The head loss due to friction is normally small and, therefore, neglected. The bed level is generally assumed to vary linearly. The side slope is assumed to vary from its pre-expansion value (zero since the channel is generally rectangular) to the post-expansion value (equal to the side slope of the trapezoidal section) in a specified manner and the bed width is computed. If the bed width thus obtained shows a smooth variation, the design is accepted, otherwise a different pattern of variation of the side slope is assumed and the process is repeated. We do not discuss this method in detail since it involves an iterative procedure. Vittal and Chiranjeevi (1983) proposed a design method based on choosing the bed width on the basis of the profile of the separating streamline and the side slope to achieve a minimum energy loss. Swamee and Basak (1991,1992,1993,1994) provided comprehensive designs of both expansion and contractions by minimizing the head loss and fitting approximate expressions to the resulting bed width and side slope variations. They showed that their designs had the lowest head loss compared to the previously proposed designs. Here we present only some of the designs without detailing the methodology. Figure 5.8 shows the symbols used in the designs of expansions and contractions.

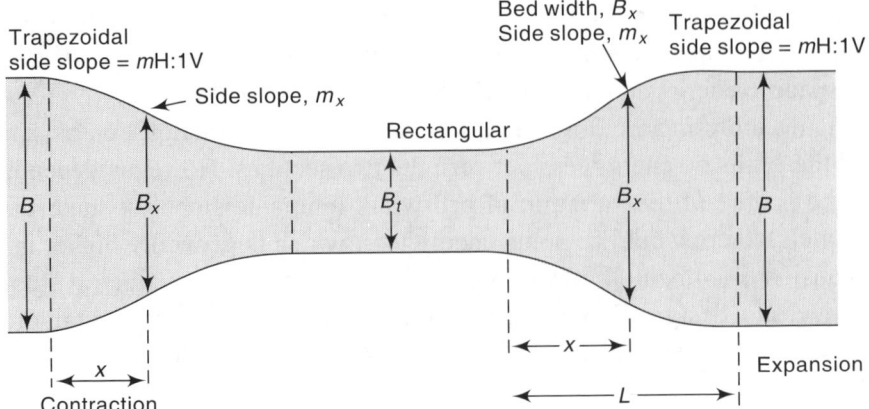

Fig. 5.8 Exapansion to and contraction from a trapezoidal channel

Vittal–Chiranjeevi's method for designing expansions

This method is based on an extensive study of flow patterns at expansion from rectangular sections of width B_t to trapezoidal sections of bed width B and side slope mH:1V.

The transition length is recommended as

$$L = 2.35(B - B_t) + 1.65my \tag{5.5}$$

The bed width profile is given by

$$B_x = B_t + (B - B_t)\frac{x}{L}\left[1 - \left(1 - \frac{x}{L}\right)^{0.80 - 0.26\sqrt{m}}\right] \tag{5.6}$$

and the side slope varies as

$$m_x = m\left(1 - \sqrt{1 - \frac{x}{L}}\right) \tag{5.7}$$

The bed elevation within the transition may be changed linearly either to keep the specific energy equal at the two ends, or to effect a specified change in the bed level. Another alternative is to change the bed elevation in such a way as to maintain a constant water depth in the transition.

Swamee–Basak's method for expansions

The head loss in an elemental width of the transition is obtained by applying the momentum and energy equations for an abrupt expansion and is then integrated over the length of transition to obtain the total head loss due to expansion. Minimization of this head loss leads to the recommended bed-width and side-slope profiles in the transition. Additional constraints were imposed to avoid separation near the entrance and eddy formation near the end of the expansion by requiring the bed-width profile to be tangential to the pre-transition and post-transition width profiles. A large number of numerical simulations were performed using a wide range of parameters. For example, the channel bed slope ranges from 10^{-4} to 10^{-2}, the side slope of the trapezoidal channel ranges from 0.5 to 3.5, the bed-width expansion ratio ranges from 1.25 to 3, nondimensional discharge ($Q/g^{1/2}B_t^{5/2}$) ranges from 0.1 to 2 and the length of transition ranges from two to eight times B_t. Since an optimum length of transition is not specified, Eq. (5.5) may be used to obtain the length. The bed-width profile is given by

$$B_x = B_t + (B - B_t)\left[1 + 2.52\left(\frac{L - x}{x}\right)^{1.35}\right]^{-0.775} \tag{5.8}$$

and the side slope varies as

$$m_x = m\left(\frac{x}{L}\right)^{1.23} \tag{5.9}$$

The head loss is about 5% lower than that in the Vittal–Chiranjeevi's design. Figure 5.9 compares the two methods for expansion from a rectangular section to a trapezoidal section of side slope 2H:1V. Example 5.3 illustrates the methodology of both types of designs.

5.3.2 Design of Contractions

Since the head loss in a contraction is generally small compared to that in expansion, most experts recommend using either a linear or elliptical width change (for small discharge) or a profile similar to that suggested for an expansion, e.g., given by Vittal and Chiranjeevi. Swamee and Basak (1994) suggested the following equations for minimizing the head loss:
The bed-width profile is given by

$$B_x = B - (B - B_t)\left[1 + 1.41\left(\frac{L - x}{x}\right)^{1.23}\right]^{-0.924} \tag{5.10}$$

and the side slope varies as

$$m_x = m\left[1 - \left(\frac{x}{L}\right)^{1.52}\right] \tag{5.11}$$

The head loss is about 15% lower than that in the Vittal–Chiranjeevi's design and slightly lower than that in the elliptic transition. An expression for

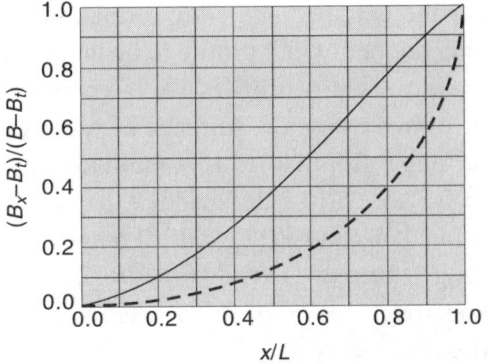

(a) Comparison of bed-width profiles

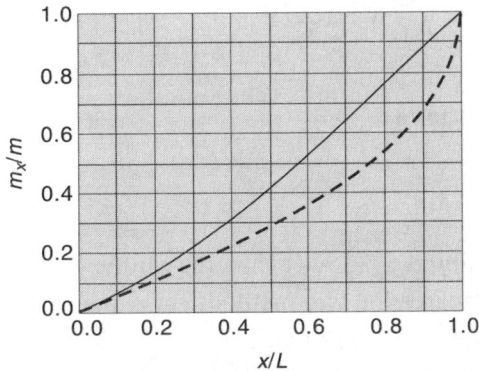

(b) Comparison of side slope variations

Fig. 5.9 Comparison of Swamee–Basak and Vittal–Chiranjeevi methods of expansions (solid lines: Swamee–Basak's method, dashed lines: Vittal–Chiranjeevi's method)

the change in bed elevation is also given. However, since the overall head loss in the contraction has a small magnitude, we believe that a linear or elliptic transition would be sufficient in the contraction.

Example 5.3 A canal, carrying a discharge of 10 m³/s, has a trapezoidal section with bed width of 6 m, side slopes of 2H:1V, and flow depth of 1.3 m. It has to be carried over a river via an aqueduct. Due to economic considerations, it was decided to adopt a 4 m wide rectangular section at the aqueduct. Design the contraction before the aqueduct and the expansion after it.

Solution

The contraction is designed by applying the Swamee–Basak's method. The length of transition is obtained from Eq. (5.5) as

$$L = 2.35 \, (5 - 3) + 1.65 \times 2 \times 1.3 \text{ m} = 9.0 \text{ m}$$

Equations (5.10) and (5.11) are used to obtain the bed width and side slope at different locations and the condition of constant flow depth (of 1.3 m) is used to obtain the bed elevation with the assumption of negligible energy loss. The following table shows the elements of the contraction using the Swamee–Basak's method:

Distance (m)	Bed width (m) Eq. (5.10)	Side slope Eq. (5.11)	Specific energy (m)	Drop in bed level (m)
0	6.00	2.00	1.341	0.000
1.8	5.73	1.83	1.346	0.005
3.6	5.34	1.50	1.357	0.016
5.4	4.87	1.08	1.377	0.036
7.2	4.38	0.58	1.415	0.074
9.0	4.00	0.00	1.488	0.148

The expansion is designed using both the Vittal–Chiranjeevi's method and the Swamee–Basak's method. The flow depth is assumed to be constant at 1.30 m and the head loss in the transition is taken as 0.3 time the difference in velocity heads at the two ends, i.e.,

$$\frac{0.3}{2 \times 9.81}\left[\left(\frac{10}{4 \times 1.30}\right)^2 - \left(\frac{10}{6 \times 1.30 + 2 \times 1.3^2}\right)^2\right] \text{ m} = 0.044 \text{ m}$$

This is assumed to be distributed linearly along the length of the expansion, i.e., about 0.009 m in each 1.8 m long segment. The following table shows the elements of the expansion using the Vittal–Chiranjeevi's method:

Distance (m)	Bed width (m) Eq. (5.6)	Side slope Eq. (5.7)	Specific energy (m)	Rise in bed level (m)
0	4.00	0.00	1.488	0.000
1.8	4.04	0.21	1.462	0.017
3.6	4.16	0.45	1.434	0.037
5.4	4.39	0.74	1.405	0.056
7.2	4.80	1.11	1.377	0.076
9.0	6.00	2.00	1.341	0.103

The Swamee–Basak's method results in the following profile:

Distance (m)	Bed width (m) Eq. (5.6)	Side slope Eq. (5.7)	Specific energy (m)	Rise in bed level (m)
0	4.00	0.00	1.488	0.000
1.8	4.22	0.28	1.444	0.036
3.6	4.54	0.65	1.404	0.067
5.4	5.00	1.07	1.374	0.088
7.2	5.55	1.52	1.353	0.100
9.0	6.00	2.00	1.341	0.103

The transitions discussed so far included change in the channel width, side slope, and/or bed elevation in a *straight* channel. However, the presence of bends in canals is unavoidable and the natural channels exhibit a meandering tendency leading to several bends in the longitudinal alignment. These bends lead to a significant energy loss and an increase in flow depth due to the centripetal force.[5] In erodible channels, an additional complication arises because of the pattern of scour and deposition induced by the bend. However, this aspect will be discussed later in Chapter 8. In the next section, we look at the flow pattern in a bend and analyse the energy loss and depth increase.

5.4 Bends

Figure 5.10 shows the flow through a bend. Clearly, the one-dimensional method of analysis that we have applied successfully till now is not applicable to this case due to the existence of complex three-dimensional flow in the bend and for a considerable distance after the bend.

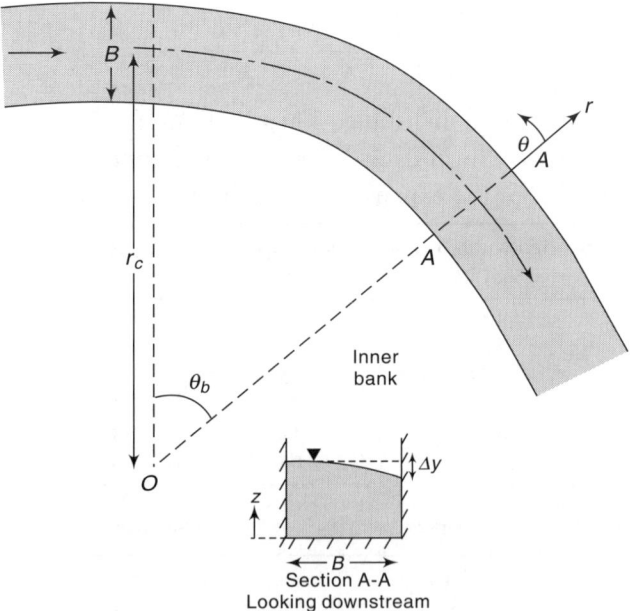

Fig. 5.10 Flow through a bend in the channel

---

[5]The term *centrifugal force* is also used sometimes and is equal and opposite to the centripetal force. While the centripetal force acts on the body, the centrifugal force is applied by the body. Many physicists argue that centrifugal force is a fictitious force since it exists only in a rotating reference frame.

Using a cylindrical coordinate system r-θ-z, as shown in Fig. 5.10, we may write the r-direction momentum equation as (see Appendix A):

$$(V \cdot \nabla)v_r - \frac{v_\theta^2}{r} = -\frac{1}{\rho}\frac{\partial(p + \rho g z)}{\partial r} + F_f \tag{5.12}$$

in which F_f is the friction force per unit mass. We take the datum at the channel bed, assuming that the channel bed is horizontal in both the transverse and the longitudinal directions, and also assume that a hydrostatic pressure distribution exists so that the piezometric head, $p/\gamma + z$, at all points of a vertical section is equal to the water depth, y. Since the radial (cross-stream) component of the velocity would be very small compared to the tangential (longitudinal) component, we neglect the first term on the left hand side of Eq. (5.12). Integrating Eq. (5.12) over the vertical dimension, we may write

$$-\beta\frac{V_\theta^2}{r} = -g\frac{\partial y}{\partial r} + \frac{\tau_{r,0}\, r d\theta dr}{\rho y r d\theta dr} \tag{5.13}$$

in which $\tau_{r,0}$ is the shear stress at the bed in the radial direction and would be related to the radial velocity, V_r. The last term in Eq. (5.13) represents the shear force acting on the bed in the radial direction divided by the mass of water within the soil column of height y. Using Manning's equation and assuming the channel to be wide, the shear stress may be expressed as $\rho g n^2 V_r^2 / y^{1/3}$. Again, since the radial velocity is small, we neglect it and obtain an equation for the water depth variation along the radial direction as

$$\frac{dy}{dr} = \beta\frac{V_\theta^2}{gr} \tag{5.14}$$

The momentum correction factor is generally close to unity and, therefore, dropped from this equation. Equation (5.14) demonstrates that the flow depth increases as r increases, i.e., as we move from the inner bank to the outer bank. This is similar to the *cant* provided on road curves and the *superelevation* provided on railway curves and is called the superelevation of the water surface. We may interpret this equation as the pressure difference between two radial distances balancing the outward acceleration of the fluid particles. However, one must realize that the tangential velocity in a vertical line is not constant but increases from zero near the bed to a maximum at or very near the water surface. Thus the fluid particles near the bed will have a 'lower than average' outward acceleration and those near the water surface will have higher acceleration. The net effect of the pressure difference due to superelevation and acceleration due to curvature of streamlines would, therefore, be an inward movement of particles near the bed and an outward movement near the surface. This secondary circulatory motion, when superimposed on the main tangential velocity, results in a spiral flow. Experimental results confirm the existence of the secondary currents and also show that at the beginning of the bend the maximum longitudinal velocity occurs near the inner bank, but slowly shifts towards the outer bank. At the end of the bend, the maximum tangential velocity occurs close to the outer bank and the velocity profile in a radial section is almost opposite of that at the beginning of the bend. Thus a single expression for V_θ as a function of r [which is required to integrate Eq. (5.14)] may not be applicable throughout the bend. However, some approximations have been suggested and have been found to work well in obtaining the superelevation of the water surface, as described next.

5.4.1 Superelevation

If we assume that the tangential velocity in Eq. (5.14) may be replaced by its average value over the cross section, V, we get

$$y = \frac{V^2}{g} \ln r + C_1 \tag{5.15}$$

Using the centreline radius, r_c, and the channel width, B, we may write the superelevation as

$$\Delta y = \frac{V^2}{g} \ln \frac{1 + (B/2 r_c)}{1 - (B/2 r_c)} \tag{5.16}$$

known as the Grashof's equation[6] obtained in 1870s.

On the other hand, if we assume that the specific energy is constant along the radial direction, and use Eq. (5.14) with $\beta = 1$, we get

$$\frac{dE}{dr} = 0 = \frac{dy}{dr} + \frac{d \frac{V_\theta^2}{2g}}{dr} = \frac{V_\theta^2}{gr} + \frac{V_\theta}{g} \frac{dV_\theta}{dr} \Rightarrow V_\theta = \frac{\Gamma}{2 \pi r} \tag{5.17}$$

which is the velocity profile for a free vortex having a circulation constant, Γ. We can approximate the circulation constant as the circulation at the centreline, i.e., $2\pi r_c V$, and write Eq. (5.14) as

$$\frac{dy}{dr} = \frac{V^2 r_c^2}{gr^3} \Rightarrow y = -\frac{V^2 r_c^2}{2 gr^2} + C_1 \tag{5.18}$$

---

[6]Clearly, the argument of the ln term can also be written as r_o/r_i. Use of the centerline radius is, however, more common.

from which the superelevation is obtained as

$$\Delta y = \frac{V^2}{g} \frac{B/r_c}{(1 - B^2/4r_c^2)^2}$$ (5.19)

Equations (5.16) and (5.19) give comparable results (within 5%) for B/r_c less than about one-third, which is valid for most practical cases.

In addition to the superelevation, another quantity of practical interest is the energy loss in the bend, which is discussed in the next subsection.

5.4.2   Energy Loss

The energy loss in the bend would be more than that in a comparable straight channel because of the transverse circulation, which creates additional friction at the bottom and increases turbulence. For sharp bends, streamline separation may occur which leads to higher loss. The *additional* energy loss due to the bend is usually expressed as

$$E_B = K_B \frac{V^2}{2g}$$ (5.20)

where K_B is a loss coefficient, which would be a function of the bend angle, θ_b, its curvature, B/r_c, and the relative flow depth, y/B. Some empirical suggestions account for the increased loss by increasing the Manning's n. A more theoretical approach[7] (Chang 1983) is based on an assumed radial and tangential velocity distribution and division of the total energy loss into longitudinal and transverse components. The transverse component is further subdivided into two components, one due to internal turbulent friction associated with transverse circulation and the other associated with the transverse boundary shear. It has been found to work well for subcritical flow in bends with moderate curvature (i.e., no separation) and can be written as

$$K_B = \frac{28.58 + 58.76 \dfrac{n\sqrt{g}}{y^{1/6}}}{5.01 + \dfrac{y^{1/6}}{n\sqrt{g}}} \frac{y}{r_c} \theta_b$$ (5.21)

in which the Manning's roughness, n, includes the effect of the bend loss and is higher than that for a straight channel. Chang (1983) used an iterative

---

[7]A similar analysis was performed earlier by Rozovskii in 1957.

technique to estimate its value starting from the roughness coefficient for the straight channel. A rough estimate of *increase* in *n* over that of the straight channel was provided by Scobey in 1930s as $0.087/r_c$ (the radius being in metre).

Example 5.4   A 5 m wide rectangular canal carries a discharge of 10 m³/s at a flow depth of 1.25 m and has a Manning's *n* of 0.02. It has a bend with centreline radius of 30 m and included angle of 45°. Find the superelevation and the energy loss due to the bend. What would be these values if the bend had a centreline radius of 15 m.

Solution
The average velocity of flow is (10/5)/1.25 m/s = 1.6 m/s. The channel width to bend radius ratio is 5/30, i.e., 1/6. Hence, either Eq. (5.16) or Eq. (5.19) can be used to obtain the superelevation. Using Eq. (5.16), we get the superelevation as 0.021 m.

Using Scobey's empirical relation, the increase in *n* due to the bend is estimated as 0.087/30 = 0.0029, resulting in *n* = 0.023 at the bend. Equation (5.21) then provides K_B = 0.055. The additional energy loss in the bend is, therefore, only about 0.007 m.

For the sharper bend with centreline radius of 15 m, we obtain the superelevation as 0.044 m and the energy loss as 0.016 m—the values almost double of those computed previously.

We have discussed various types of transitions for subcritical flow conditions in which the velocity is small and the free surface relatively calm. For supercritical flows, small perturbations created due to transitions tend to amplify and create a wavefront. The water surface becomes wavy and analysis of these waves becomes our main concern as described in the next section.

5.5   Supercritical Flow Transitions

Any transition in a flow creates disturbances in the flow. To understand the difference in the nature of subcritical and supercritical flows, we consider the perturbation created by an irregularity on the boundary such as a change in direction. To simplify the analysis, instead of considering fluid moving to the right with a velocity, *V*, and a stationary source of perturbation, we consider the equivalent case of a disturbance moving to the left with a velocity, *V*, in a stationary pool of fluid. If we assume that the amplitude of the disturbance is small compared to the flow depth, the speed (or celerity) of the wave, C, is given by $C = \sqrt{gy}$ (as will be shown later in Chapter 7). Figure 5.11 shows

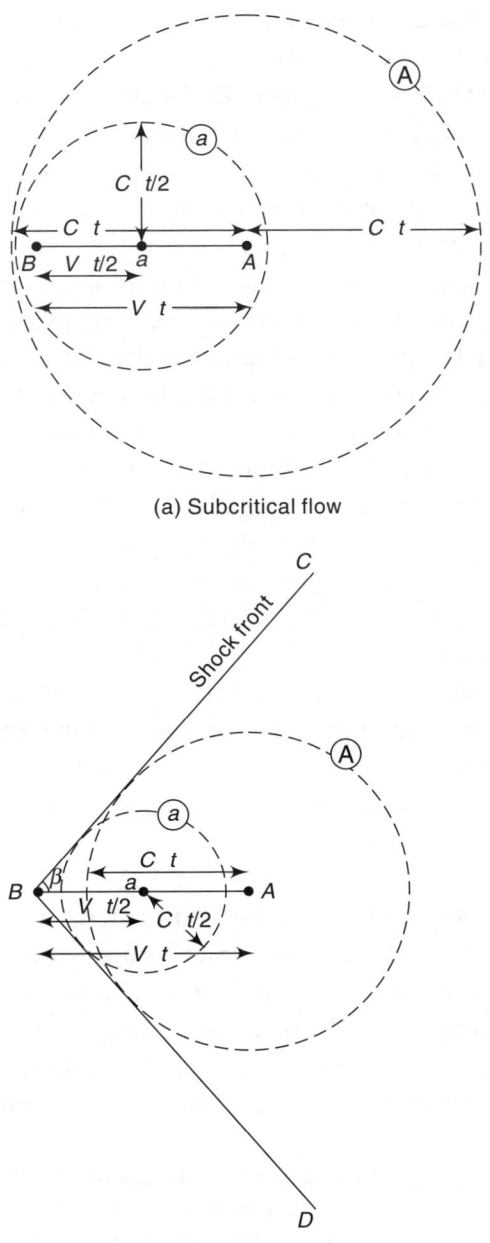

(a) Subcritical flow

(b) Supercritical flow

Fig. 5.11   Spreading of a disturbance in (a) subcritical and (b) supercritical flows

the resulting wave pattern for subcritical ($V < C$) and supercritical ($V > C$) flows.[8]

Let us consider the movement of the object representing the irregularity from its location A at a previous time $t - \Delta t$ to the location B at the *current time*, t. Obviously the distance between A and B is equal to $V\Delta t$. The object, as it moves along the line AB, generates disturbances, which spread with a speed of C in a circular pattern emanating from the location of the irregularity.

In the subcritical flow [Fig. 5.11(a)], the object travels at a speed slower than that of the wave, and by the time the object reaches B, the waves[9] generated by it when it was at A would travel farther ($C\Delta t$) than B ($V\Delta t$). The same would be true for the waves generated from all the intermediate locations between A and B. Thus at the current time, the wave pattern would be circles with centres along the line AB and radii equal to Vt^*, with t^* denoting the difference between the current time and the time when the object was at the corresponding centre. Clearly, t^* will vary linearly from 0 at B to Δt at A. Two of these waves are shown in the figure. It is, therefore, apparent that at any time there would be wavefronts ahead of the object as well as behind it, with those ahead of the body being more closely spaced. Of course, when the velocity, V, approaches zero, the wavefronts approach concentric circles and when the velocity approaches the celerity, all the wavefronts ahead of the body collapse to the location of the body.

On the other hand, for a supercritical flow [Fig. 5.11(b)], the object travels at a speed faster than that of the wave, and by the time the object reaches B, the waves generated by it when it was at A would not reach B. The same would be true for the waves generated from all the intermediate locations between A and B. Again, at the current time, the wave pattern would be circles with centres along the line AB and radii equal to Vt^*. However, these wavefronts would lie entirely behind the body. In terms of the actual flow past an irregularity, it implies that the wave front is downstream of the irregularity. Two of these waves are shown in the figure, and it is seen that at the current time, the disturbances due to the object are contained within the enveloping tangents BC and BD. We may think of these tangents as separating an area affected by the disturbance and an area that remains undisturbed. In other

---

[8]Although we are analysing a moving source of disturbance in still water, it effectively represents the flow of fluid past a stationary irregularity. The terms *subcritical* and *supercritical*, therefore, refer to the flow characteristics although we consider the irregularity moving with a velocity V.

[9]From our point of view, the waves travelling in the direction same as that of the object are more important. There would, of course, be waves spreading in all directions.

words, the disturbance propagates along these lines, which are called *shock fronts*. It is quite straightforward to show that the angle between the shock front and the flow direction, β, is given by

$$\sin \beta = \frac{C}{V} = \frac{1}{F} \tag{5.22}$$

F being the Froude number. Thus the angle decreases with increase in Froude number, a fact that would have a great bearing on the behaviour of flow near a transition in channel width.

To analyse the supercritical flow through an expansion, contraction, and bend, one must consider these shock waves and their effect on flow depth and velocity. To help us in this analysis, we first look at the effect of a small change in wall alignment of a channel having supercritical flow. Straight expansions and contractions can then be analysed by considering the deflection at the corner to be made up of a series of small deflections *at the same point*, and curved transitions or bends can be analysed by considering the curve to be made up of a series of small deflections *at different locations* along the curve. Figure 5.12 shows a small inward[10] deflection, $\Delta\theta$, of the wall and the resulting shock front.

We use the continuity and momentum equations to obtain the location of the front and the change in flow depth and velocity as follows.

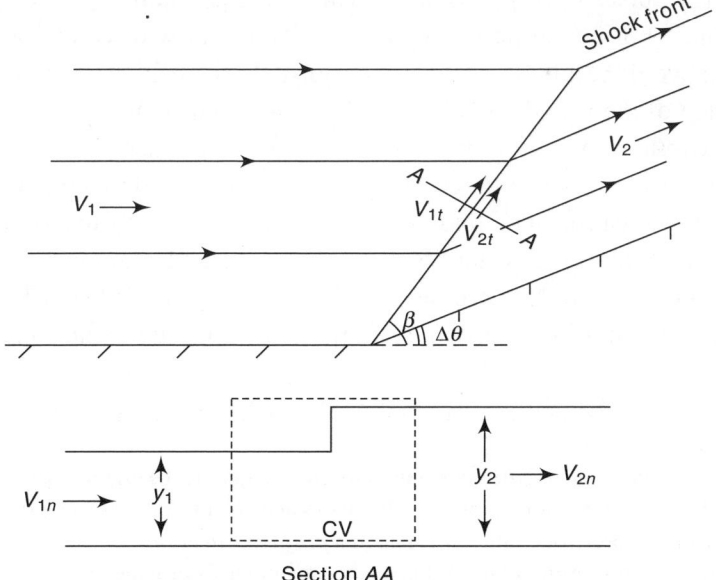

Fig. 5.12 Effect of a small change in wall alignment

[10]For outward deflection, $\Delta\theta$ would be negative.

Considering the channel to be rectangular, horizontal, and frictionless, choosing the control volume as shown in the figure, and assuming the pressure distribution to be hydrostatic on the control surface, the continuity equation results in

$$V_{1n}y_1 - V_{2n}y_2 = 0 \qquad (5.23)$$

and the momentum equations in the normal and tangential directions provide

$$\frac{1}{2}\rho g y_1^2 - \frac{1}{2}\rho g y_2^2 = \rho V_{2n}^2 y_2 - \rho V_{1n}^2 y_1 \qquad (5.24a)$$

and $\qquad 0 = \rho V_{2n}y_2 V_{2t} - \rho V_{1n}y_1 V_{1t} \qquad (5.24b)$

With the relationships $V_{1n} = V_1 \sin \beta$, $V_{1t} = V_1 \cos \beta$, $V_{2n} = V_2 \sin (\beta - \Delta\theta)$, $V_{2t} = V_2 \cos (\beta - \Delta\theta)$, we obtain the following expressions[11] using Eqs (5.23) and (5.24):

$$\frac{y_2}{y_1} = \frac{1}{2}\left(\sqrt{1 + 8 F_1^2 \sin^2 \beta} - 1\right) \Leftrightarrow \sin \beta = \frac{1}{F_1}\sqrt{\frac{1}{2}\frac{y_2}{y_1}\left(1 + \frac{y_2}{y_1}\right)}$$

$$(5.25a)$$

$$\frac{y_2}{y_1} = \frac{\tan \beta}{\tan (\beta - \Delta\theta)} \qquad (5.25b)$$

It is seen from Eq. (5.25b) that for the case of the deflection of the wall *into the flow* (i.e., positive $\Delta\theta$), y_2 is more than y_1, and it would be called a positive wavefront.[12] For a wall deflecting away from the flow direction, a negative wavefront would result. It should also be noted that a positive wavefront causes a reduction of the Froude number for the flow downstream of the front since the flow depth becomes larger and the velocity becomes smaller. On the other hand, a negative wavefront occurring when the wall is deflected away from the flow would cause an increase in the Froude number over that of the undisturbed flow. Another important fact that we emphasise again is that the angle of the wavefront would increase with decrease in Froude number [see Eq. (5.25a) and its limiting case, Eq. (5.22)].[13] For a straight expansion both the

---

[11]The methodology is similar to that used for obtaining the sequent depth ratio in a hydraulic jump.

[12]The terms *positive* and *negative* are generally used to denote the effect of a wavefront on the flow depth. At any point in the flow, if the passing wavefront causes the flow depth to increase from its pre-front value, it is called a *positive wavefront*.

[13]One should keep in mind that all this analysis is done for a wide rectangular channel such that $F = V/(gy)^{1/2}$. For other shapes of channel or when the channel width is not very large in comparison with its depth, other factors have to be considered. We do not discuss these in this book.

sides are sloping away from the flow direction, for a contraction both deflect into the flow, and for a bend the outer boundary would be deflecting into the flow and the inner boundary would be deflecting out. However, the analysis is complicated because of reflection of these waves as they hit the other boundary. Therefore, we first analyse only one side of the transition and then consider the reflections by adding the other side.

5.5.1 Straight Expansion

Figure 5.13 shows the flow near the corner of a straight expansion. The total deflection, θ, may be thought of as a series of very small deflections, $\Delta\theta$, away from the flow with the first deflection giving rise to a *negative* wavefront at an angle β_1 from the original flow direction [the angle of the front for a small deflection is given by Eq. (5.22)]. The flow now occurs at an angle $\Delta\theta$ from the original direction and has a *larger* Froude number. This flow, when subjected to another small deflection, $\Delta\theta$, at the corner would experience another shock wave at an angle β_a (which will be *smaller* than β_1) from this modified flow direction. Thus a series of fronts would be formed at angles $\beta_1, \beta_a, \beta_b, \dots,$ β_2, which are measured from the corresponding flow directions and become successively smaller. The final wavefront is at an angle β_2 measured from the downstream wall and is related to the downstream Froude number, F_2, [Eq. (5.22)].

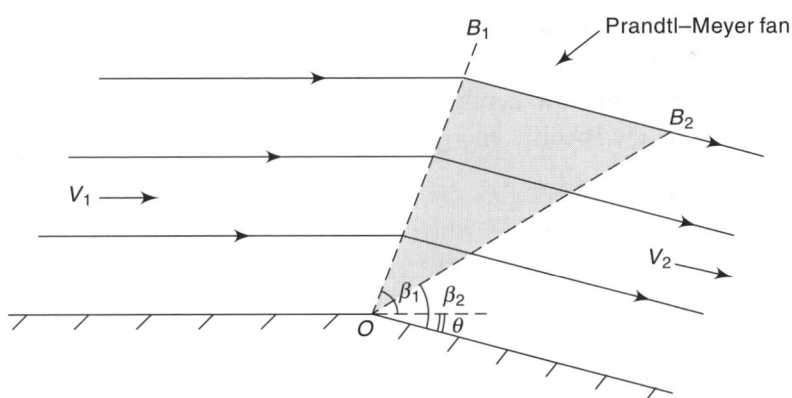

Fig. 5.13 Wavefronts near the corner of a straight expansion

The discussion in terms of the finite small deflections can now be generalized to obtain the downstream conditions for a given approach flow and the angle of transition, θ. In the limit that the finite deflections considered above become zero, we would get a continuous front spanning the area bounded by the first and the last wavefronts, i.e., B_1OB_2, with OB_1 at an angle of β_1 from the flow direction in the channel before the expansion and OB_2 at an angle of β_2 from the flow direction in the expansion. This fan-shaped area, called the Prandtl–Meyer fan (to honour the investigators who studied it in early 1900s in connection with supersonic flow of gases at a corner), would encompass the transition of the flow from an upstream Froude number, F_1, to a higher Froude number, F_2, in the expansion. (The actual behaviour in a straight expansion would, of course, be different due to the reflection of these waves from the opposite sidewall and due to the disturbance generated at that wall.) To find out the flow conditions after the fan, i.e., downstream of OB_2, Eq. (5.25b) is applied to a point within the fan, having a flow depth y, Froude number F, and a wavefront angle β, as

$$\frac{y + \Delta y}{y} = \frac{\tan \beta}{\tan(\beta - \Delta\theta)} = \frac{\tan \beta(1 + \tan \beta \tan \Delta\theta)}{\tan \beta - \tan \Delta\theta} \approx \frac{1 + \Delta\theta \tan \beta}{1 - \dfrac{\Delta\theta}{\tan \beta}}$$

(5.26)

from which, taking the limit when $\Delta\theta$ tends to zero, neglecting the multiplication of $\Delta\theta$ and Δy, and using Eq. (5.22), we get

$$\frac{dy}{d\theta} = \frac{yF^2}{\sqrt{F^2 - 1}}$$

(5.27)

However, since the Froude number involves the water depth as well as the flow velocity (both of which vary within the fan), we will have to make some assumption to relate these two in order to be able to integrate Eq. (5.27). Since the change in flow depth is gradual, we can assume that there is no energy loss and the specific energy, E, is constant. Therefore,

$$F^2 = \frac{V^2}{gy} = \frac{2(E - y)}{y}$$

(5.28)

and

$$\frac{dy}{d\theta} = \frac{2(E - y)}{\sqrt{\dfrac{2E - 3y}{y}}}$$

(5.29)

An alternative form of this equation can be written in terms of the Froude number, by using the relation $y = 2E/(2 + F^2)$ in Eq. (5.27), as

$$\frac{dF^2}{d\theta} = -\frac{F^2(2 + F^2)}{\sqrt{F^2 - 1}}$$

(5.30)

Any of these forms can be integrated to obtain the variation of either the flow depth or the Froude number as a function of θ, with the constant of integration evaluated from the condition at the start of the expansion, i.e., $y = y_1$ (or $F = F_1$) when $\theta = 0$. For the flow depth, the solution is written as[14]

$$-\theta = \sqrt{3}\ \tan^{-1} \sqrt{\frac{2E - 3y}{3y}} - \tan^{-1} \sqrt{\frac{2E - 3y}{y}} + C \qquad (5.31)$$

However, it is more convenient to write it in terms of the Froude number using the Prandtl–Meyer function[15] as

$$-\theta = v(F) - v(F_1) \qquad (5.32)$$

where $v(F)$ is the Prandtl–Meyer function defined by

$$v(F) = \sqrt{3}\ \tan^{-1} \sqrt{\frac{F^2 - 1}{3}} - \tan^{-1} \sqrt{F^2 - 1} \qquad (5.33)$$

and is shown in Fig. 5.14.

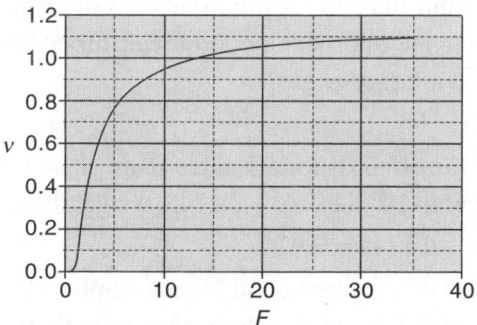

Fig. 5.14 Variation of the Prandtl–Meyer function with Froude number

It is seen that v is 0 at $F = 1$ and approaches a maximum value as F becomes very large. From Eq. (5.33), using the series expansion of *arctan* and the complementary angle property,[16] it can be readily seen that

---

[14]Since the expansion angles away from the flow direction, θ is negative. Therefore, we write the expression in terms of $-\theta$.

[15]The Prandtl–Meyer function was derived for supersonic ideal gas flow and used the Mach number and the ratio of specific heat capacities. It is equivalent to what we describe here if the Froude number replaces the Mach number and the specific heat capacity ratio is taken as 2.

[16]Series expansion of arctan(x) is given by $\tan^{-1} x = x - \dfrac{x^3}{3} + \dfrac{x^5}{5} - \dfrac{x^7}{7} + \cdots$ and the complementary angle property is $\tan^{-1} \dfrac{1}{x} = \dfrac{\pi}{2} - \tan^{-1}x$.

$$\text{As } F \rightarrow 1, \; v(F) \rightarrow \frac{2}{9}(F^2 - 1)^{3/2} \tag{5.34a}$$

$$\text{As } F \rightarrow \infty, \; v(F) \rightarrow \frac{\pi}{2}(\sqrt{3} - 1) - \frac{2}{\sqrt{F^2 - 1}} \tag{5.34b}$$

The maximum value of v is, therefore,

$$v_{max} = \frac{\pi}{2}(\sqrt{3} - 1) = 1.15 \text{ radians or } 65.9° \tag{5.35}$$

The following example illustrates the determination of the downstream Froude number, F_2, for a given approach flow Froude number, F_1, and transition angle, $-\theta$.

Example 5.5   A wide rectangular channel carries supercritical flow at a Froude number of 5 and flow depth of 0.5 m. At a corner of a straight expansion, the wall deflects away from the flow direction by 10°. What would be the Froude number and the flow depth after the fan? What is the maximum angle through which the approaching flow can turn? What happens if the transition expands at a larger angle?

Solution
From Fig. 5.14 [or Eq. (5.33)] for a Froude number of 5, $v = 0.76$. With the deflection angle equal to 10°, i.e., $\theta = -0.1745$, we get, from Eq. (5.32),

$$v(F) = 0.76 + 0.1745 = 0.9345$$

From Fig. 5.14, we get the corresponding Froude number after the fan as about **9**. Since the specific energy is assumed to be constant, the flow depth after the fan is obtained as

$$y = \frac{y_1\left(1 + \dfrac{F_1^2}{2}\right)}{\left(1 + \dfrac{F^2}{2}\right)} = \frac{0.5 \times 13.5}{41.5} \text{ m} = \mathbf{0.16 \text{ m}}$$

Since v_{max} is 1.15, the maximum angle of turn can be 1.15 − 0.76, i.e., 0.39 radian or about **22°**. If the wall angle is more than this value, in an ideal fluid, a slipstream is created at the angle of 22°, which denotes the streamline separating the stagnant water near the wall and the deflected stream. For real fluids, a shear layer is formed instead of the slipstream.

Reading the values of F for a given v from Fig. 5.14 may involve significant errors. Also, from a computational point of view, it is desirable to express F

as an explicit function of v. The following approximation is found to work well with a maximum error of less than 1% for the inverse Prandtl–Meyer function, $F(v)$:

$$F = \sqrt{1 + \left(1.6\,v^{1/3} + 1.74\,\frac{v}{v_{\max} - v}\right)^2} \tag{5.36}$$

For Example 5.5, for $v = 0.9345$, we get $F = 9.17$.

5.5.2 Straight Contraction

Figure 5.15 shows the flow near the corner of a straight contraction. The total deflection, θ, is now thought of as a series of very small deflections, $\Delta\theta$, *into the flow* with the first deflection giving rise to a *positive* wavefront at an angle β_1 from the *original* flow direction. The flow now occurs at an angle $\Delta\theta$ from the original direction and has a smaller Froude number. This flow, when subjected to another small deflection, $\Delta\theta$, at the corner would experience another shock wave at an angle β_a (which will be larger than β_1) from this modified flow direction. However, as opposed to the expansion, now the two wavefronts would not diverge from each other but would interact. Interaction of the wavefronts generated by each small deflection leads to the formation of a shockwave, which is inclined at an angle, β, from the upstream flow direction. Since the continuity and momentum equations are similar to those used for a small inward deflection [Eqs (5.23) and (5.24)], similar expressions would hold good for this case, with the transition angle, θ, replacing $\Delta\theta$.

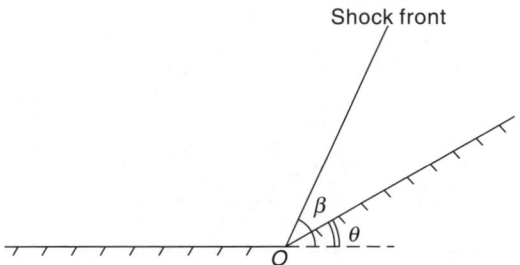

Fig. 5.15 Shockwave near the corner of a straight contraction

For known upstream Froude number, F_1, and the deflection angle, θ, Eqs (5.25 a and b) are easily manipulated to provide the following nonlinear equation in β, which can be solved by graphical or numerical techniques:

$$\tan \beta (\sqrt{1+8 F_1^2 \sin^2 \beta} - 3)$$

$$= \tan \theta (2 \tan^2 \beta + \sqrt{1+8 F_1^2 \sin^2 \beta} - 1) \qquad (5.37)$$

Figure 5.16 shows a plot of this relationship. Note that there are two possible shock angles for given approach flow and contraction angle. The larger of these corresponds to a subcritical flow ($F_2 < 1$) downstream of the shock and the smaller to either a subcritical or a supercritical flow depending on the deflection angle. Even for the lower limb of the curve, higher deflection angles may result in $F_2 < 1$. In keeping with the definition of criticality of flow based on the Froude number, we call this flow subcritical, but it should be kept in mind that $F_2 = 1$ does not correspond to a minimum specific force due to different definitions of specific force and Froude number at the shockwave (the specific force is based on a control volume perpendicular to the shock and the Froude number is based on the velocity along the deflected direction in the contraction). It is only when we consider the flow along the deflected direction that $F_2 = 1$ would correspond to a minimum specific force and $F_2 < 1$ would indicate a subcritical flow, which would be affected by the downstream conditions. However, since the subcritical flow would be affected by the downstream conditions, the assumptions used in the derivation of these equations would be violated. Hence, we consider only the supercritical

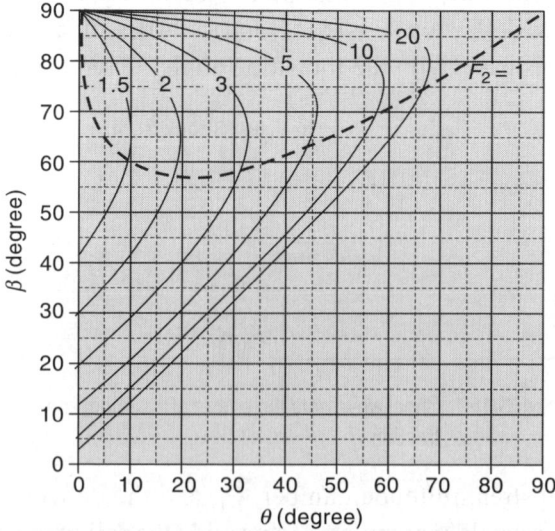

Fig. 5.16   Angle of the shockwave at a straight contraction for different values of the approach Froude number (shown on the curves)

conditions (i.e., below the dashed line showing $F_2 = 1$) and do not discuss the *oblique hydraulic jump*.[17] Once the shock angle is determined from this figure, the flow depth and velocity after the shock front can be obtained using Eqs (5.25) and (5.23), and the Froude number, F_2, can be computed. Direct determination of F_2 may also be done as

$$F_2 = F_1 \frac{\cos \beta}{\cos (\beta - \theta)} \sqrt{\frac{\tan (\beta - \theta)}{\tan \beta}} \tag{5.38}$$

The computations are illustrated in Example 5.6.

Example 5.6   A wide rectangular channel carries supercritical flow at a Froude number of 5 and flow depth of 0.5 m. At a corner of a straight contraction, the wall deflects into the flow direction by 10°. What would be Froude number and the flow depth downstream of the resulting shock?

Solution

For a Froude number of 5 and deflection angle of 10°, Fig. 5.16 provides the angle of the shockwave as $\beta = 20°$. The Froude number downstream of the shock is then obtained from Eq. (5.38) as $F_2 = 3.35$. The flow depth, y_2, is obtained from Eq. (5.25b) as 1.01 m.

5.5.3   Bends

Figure 5.17 shows the flow along a bend. Similar to the straight contraction, the outer wall of a bend may be analysed as a series of small inward deflections, the difference being that in the contraction, these small deflections were located at the same point (the corner) but in the bend, they occur at different locations along the wall. On the other hand, the inner wall of the bend may be analysed in a similar way as the straight expansion to obtain a fan, which is equivalent to the Prandtl–Meyer fan but with an inner edge comprising the entire bend length rather than only the corner. As previously discussed, inward deflections cause positive disturbances. Therefore, the water level along the outer edge of the bend would be increasing continually from its pre-bend level and that along the inner edge would be decreasing. Since the methods of analysis are identical to those discussed in the previous two subsections, we do not discuss these further.

---

[17]The shockwave with supercritical flows on both sides may also be called an oblique hydraulic jump. However, to maintain consistency from our definition of the jump from a supercritical flow to a subcritical flow, we would refer to the shockwave as an oblique jump only when the downstream flow is subcritical.

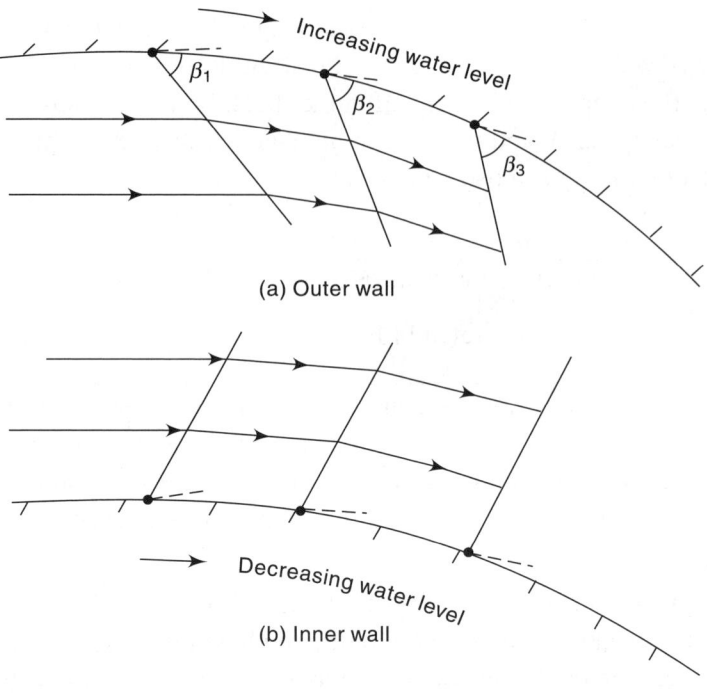

(a) Outer wall

(b) Inner wall

Fig. 5.17 Flow in a bend

5.5.4 Interaction between the Walls

While analysing the transitions in the form of expansion, contraction, and bend, we have considered only one side and ignored the interaction between the two walls. The presence of the other wall influences the analysis in the following two ways:

(i) For a finite width of channel, the definition of Froude number has to be modified to use the hydraulic depth rather than the flow depth.

(ii) The waves generated at one side would interact with the other boundary as well as with the waves generated from the other boundary.

We will not consider the first effect, i.e., we assume that even though the channel width is finite, it is still quite large compared to the flow depth (An assumption generally reasonable for supercritical flow since the flow depth is small). The interaction between a wave and a boundary and between two waves is analysed next.

Wave interaction with a boundary

We first analyse a straight contraction in which one boundary is deflected but the other one follows a straight alignment (Fig. 5.18). This would allow us to consider the interaction of the positive wave generated at the corner with the

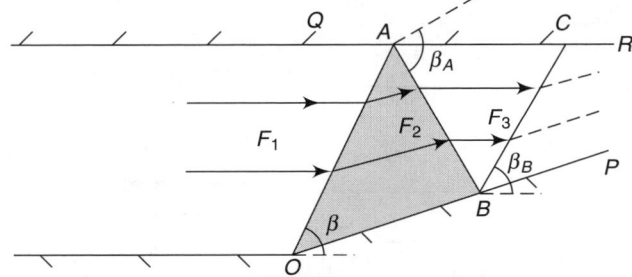

Fig. 5.18   Flow in a contraction with one deflected wall

other boundary without worrying about the wave generated at the other boundary.

As seen earlier, the corner, O, would generate a positive shockwave at an angle β, which is a function of the approach Froude number, F_1, and the degree of deflection, θ (Fig. 5.16). The post-shock flow direction would be parallel to the deflected boundary OP and the Froude number, F_2, would be smaller than F_1. Although the other boundary, QR, is straight and along the original flow direction, it would act as an *inward deflection* for the flow downstream of the first shock, OA. In other words, the point A would act as a corner with a deflection angle of θ for the flow downstream of the shock OA. This implies that another shockwave would be generated at A, at an angle β_A from the *current flow direction*, which is parallel to the deflected wall. The angle β_A would be greater than β, since $F_2 < F_1$ (see Fig. 5.16) and the flow direction after the shock AB would be parallel to QR, i.e., the approach flow direction. The flow after this *reflected* shockwave will be parallel to the deflected wall, will have a Froude number (F_3) smaller than F_2, would experience another contraction at B by an angle θ; and would generate a shockwave BC at an angle β_B, which would be larger than β_A. If the contraction is sufficiently long, this reflection will continue till the critical conditions are achieved, i.e., the post-shock Froude number becomes 1 (shown by the dashed line in Fig. 5.16). In practice, however, both the sides of the contraction would be deflected and would generate shockwaves. Interaction of the resulting waves is described next.

Wave interaction with another wave

We analyse a symmetric contraction in which both boundaries are deflected at the same angle (Fig. 5.19). Analysis of nonsymmetric contractions is similar but a little more complicated.

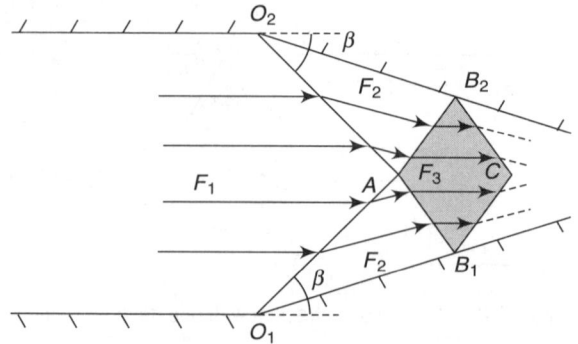

Fig. 5.19 Flow in a symmetric contraction

The corners O_1 and O_2 will generate shock waves at an angle of β from the flow direction, which will meet at the centre of the channel at point A. Downstream of O_1A and O_2A, the flow direction would be parallel to the respective walls and the Froude number would be smaller. Since A would act as a source of disturbance now, there would be shock waves generated at A (AB_1 and AB_2). The angle of these shock waves (measured from the flow direction) would be larger than β and the flow conditions in the area O_1AB_1 and O_2AB_2 would be uniform and identical. The points B_1 and B_2 generate further shocks (B_1C and B_2C) and the region AB_1CB_2 will have an even smaller Froude number and larger flow depth. This phenomenon then continues as discussed earlier under 'wave interaction with a boundary'. For a bend, however, the two waves generated at the beginning of the bend—one from the outer edge and the other from the inner side—have opposite nature (Fig. 5.20). There is a positive wave originating from the point O_2 on the outer side and a negative wave from O_1.

The positive wave O_2A causes an increase in the flow depth and corresponding reduction in the Froude number while the negative wave O_1A has an opposite effect. The point of intersection, A, would cause disturbances similar to those at the inner edge of a bend (i.e., a negative wave, AB_2) for the flow farther from the centreline of the bend and a positive wave (AB_1) for the flow near the inner side. Therefore, flow depth downstream of the shock wave AB_2 (actually, for the entire wavefront O_1B_2, since O_1B_2 is a negative shock wave[18]) would be smaller than that upstream of it. The maximum depth of flow will, therefore, occur at B_2. Similarly, O_2B_1 is a positive wave and the minimum

---

[18]Note that while O_1A is a straight line at an angle of $\sin^{-1}(1/F_1)$, AB_2 is influenced by the curvature of the bend and is concave inwards (i.e., curving towards the centre of curvature of the bend).

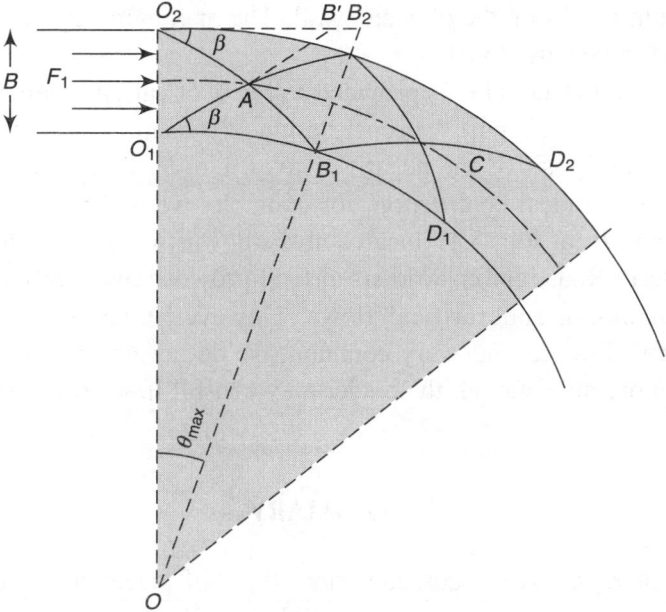

Fig. 5.20 Waves generated in supercritical flow in a bend

flow depth occurs at B_1. This wave pattern is repeated throughout the bend (except that the entire wavefront would be curved and there would be no straight portion that was observed in the first shock waves) and also persists for some distance after the bend. Starting from O_1, along the inner wall of the bend, the flow depth decreases to a minimum value at B_1, then starts increasing (due to the influence of the positive wave B_1D_2) and attains a maximum at D_1 where the reflected negative wave hits this wall. Similarly, along the outer wall, the maximum depth is achieved at B_2 and minimum at D_2. To find the location of the point of maximum depth (which would be needed in deciding the height of the sidewall), we may extend the line O_1A to intersect the tangent line at O_2 at the point B' and assume that OB' passes through B_2. The angle θ_{max} is then given by

$$\theta_{max} = \tan^{-1} \frac{B}{(r_c + B/2)\tan\beta} = \tan^{-1} \frac{B\sqrt{F_1^2 - 1}}{(r_c + B/2)} \qquad (5.39)$$

The subsequent maximum depths on the outer wall occur at locations separated by approximately $2\theta_{max}$. The maximum depth at the inner wall occurs roughly at every $2\theta_{max}$ from the starting of the bend. The points of minimum depths along one wall occur almost at the same angular location as the points of maximum depths on the opposite wall. The maximum superelevation has been found to be close to BV_1^2/gr_c.

Reinauer and Hager (1997) provided a review of various characteristics of supercritical flow in bends. A proper design of transitions aims at cancelling out the effect of positive and negative waves in order to achieve a minimum water depth. An ideal contraction, for example, would be the one with the points B_1 and B_2 of Fig. 5.19 located at the downstream end of the contraction. Similarly, Rouse and co-workers in 1951 provided empirical design curves for expansions in supercritical flows. However, since we feel that the supercritical flows are not very common, we do not describe these aspects here. For more information, the reader may consult Ippen (1949) and ASCE (1951).

SUMMARY

In this chapter, we have discussed various types of transitions encountered in practical situations such as rise or fall in bed level, expansion or contraction of channel width, and bends. The emphasis has been on subcritical flow through these transitions and a brief discussion on supercritical flows has been provided at the end. Up to this point in the book, we have discussed various situations starting from the simplest case of uniform flow in which the flow characteristics remain same everywhere, and moving to gradually and rapidly varied flows in which the flow depth and velocity change with space. However, even when the depth and velocity change spatially, the discharge has been assumed to be constant, i.e., there has been no spatial variation in discharge. In some cases, e.g., overland flow joining a river, diversion of flow through a weir in the side of a channel, the discharge in the channel would be changing spatially. Some of these situations may involve temporal changes in the discharge also (e.g., overland flow joining a river during a storm). However, we first consider steady flow with spatial variation of discharge and describe it in the next chapter. The unsteady flow is then discussed in Chapter 7.

REFERENCES

ASCE (1951): 'High velocity flow in open channels' *A symposium* on *Transactions*, ASCE, p. 116.

Chang, H.H. (1983): 'Energy expenditure in curved open channels,' *J. of Hydr. Eng.*, ASCE, **109**(7), pp. 1012-1022.

Ippen, A.T. (1949): 'Channel transitions and controls,' in *Engineering Hydraulics* (Ed: H. Rouse), John Wiley and Sons, New York.

Reinauer, R. and Hager, W.H. (1997): 'Supercritical bend flow', *J. of Hydr. Eng.*, **123** (3), pp. 208-218.

Swamee, P.K. and Basak, B.C. (1991): 'Design of rectangular open-channel expansion transitions', *J. of Irrig. and Dra. Eng.*, ASCE, **117**(6), pp. 827-838.

Swamee, P.K. and Basak, B.C. (1992): 'Design of trapezoidal expansive transitions', *J. of Irrig. and Dra. Eng.*, ASCE, **118**(1), pp. 61-73.

Swamee, P.K. and Basak, B.C. (1993): 'Comprehensive open channel expansion transition design', *J. of Irrig. and Dra. Eng.*, ASCE, **119**(1), pp. 1-17.

Swamee, P.K. and Basak, B.C. (1994): 'Design of open-channel contraction transitions', *J. of Irrig. and Dra. Eng.*, ASCE, **120**(3), pp. 660-668.

Vittal, N. and Chiranjeevi, V.V. (1983): 'Open channel transitions: rational method of design', *J. of Hydr. Eng.*, ASCE, **109**(1), pp. 99-115.

EXERCISES

5.1 In Examples 5.1 and 5.2, the channel bed level and width were changed one at a time. If the bed is raised and the width is reduced simultaneously, plot the variation of the flow depth at the downstream section with B_2/B_1 for values of $\Delta z = 0.1, 0.25$, and 0.5 m for both subcritical and supercritical approach flows.

5.2 Redo Exercise 5.1 with contraction in width but a negative step, i.e., a decrease in bed level of the same amounts (0.1, 0.25, and 0.5 m).

5.3 Design the transition for a canal carrying a discharge of 10 m³/s, when the original canal section is trapezoidal with a bed width of 10 m, side slopes of 3H:1V, and flow depth of 1.2 m, and the contracted section is an 8 m wide rectangular section. Use Vittal–Chiranjeevi's method and maintain a constant specific energy in the expansion.

5.4 Redo the previous exercise using the Swamee–Basak's method.

5.5 Find the superelevation in a 6 m wide rectangular channel at a bend with centreline radius of 12 m. The channel carries a discharge of 12 m³/s. Consider two different cases, one with a flow velocity of 1 m/s and the other with 4 m/s. [0.026 m, 0.82 m]

5.6 A wide rectangular channel carries supercritical flow at a Froude number of 6 and flow depth of 0.4 m. At a corner of a straight expansion, the wall deflects away from the flow direction by 20°. What would be the

Froude number and the flow depth after the fan? What is the maximum angle through which the approaching flow can turn?

[Maximum angle 19°]

5.7 A wide rectangular channel carries supercritical flow at a Froude number of 6 and flow depth of 0.4 m. At a corner of a straight contraction, the wall deflects into the flow direction by 20°. What would be the Froude number and the flow depth downstream of the resulting shock?

[2.84, 1.40 m]

6 | Spatially Varied Flow

6.1 Introduction: Occurrence and Importance

The discharge through a channel is an important (and probably the most important) parameter in the study of flow characteristics in a channel. So far, our discussion of various flow situations has involved spatial variation in flow depth, bed slope, cross section shape and size. We have, however, assumed that the discharge[1] remains same throughout the channel. In other words, the product of area and velocity remains constant at all sections. Sometimes, however, we encounter open channel flows in which the discharge changes along the length of the channel. Such flows are called spatially varied flow. For example, when it is not feasible to carry water from a spillway through a conventional channel, a channel is constructed parallel to the spillway crest (Fig. 6.1). This is known as a side channel spillway and the discharge in the channel increases from zero at the upstream end to its full value (equal to the total discharge over the spillway crest) at the end of the spillway crest. The lateral inflow into the side channel could be from one side [Fig. 6.1(a)] or both sides [Fig. 6.1(b)]. Similarly, a weir with its crest parallel to the flow direction is installed in the side of a channel to allow diversion of water into a side channel when the water level in the main channel rises above the weir crest. This is called a *side-weir* [Fig. 6.1(c)] and the flow in the main channel decreases from its pre-weir value to the post-weir value (equal to the difference of the pre-weir value and the discharge carried into the side channel) over the length of the weir crest. Another example of spatially varied flow (SVF) with increasing discharge is a roof gutter [Fig. 6.1(d)], which carries the rainwater from the roof. An example of spatially varied flow with decreasing discharge is a bottom rack [Fig. 6.1(e)], which diverts water away from a main channel into a trench. Some of the desired outputs of the analysis of these flows are the water surface profile (which will help in designing the

---

[1]We re-emphasize that the continuity equation states the conservation of mass. However, since water is treated as incompressible under general open channel flow conditions, volume conservation is equivalent to mass conservation.

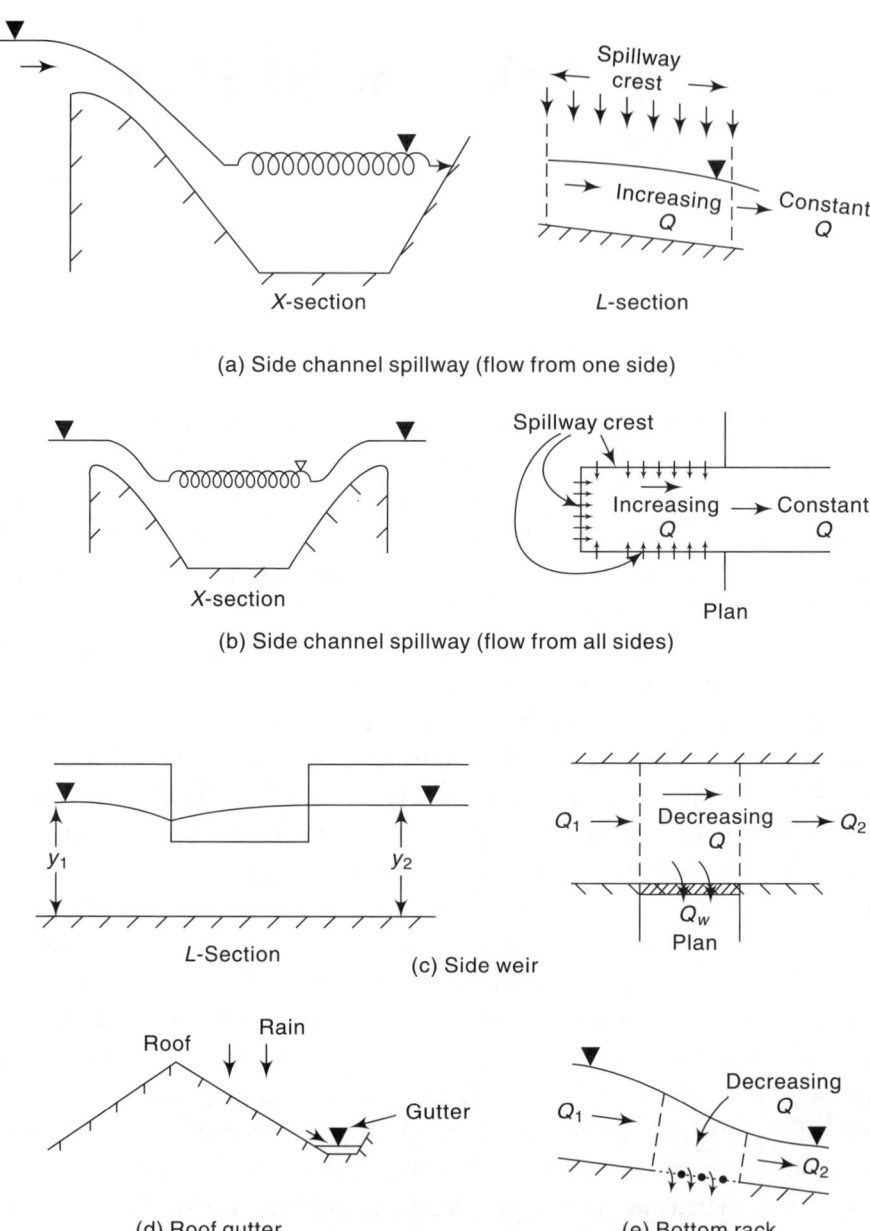

(a) Side channel spillway (flow from one side)

(b) Side channel spillway (flow from all sides)

(c) Side weir

(d) Roof gutter

(e) Bottom rack

Fig. 6.1   Examples of spatially varied flow

height of the walls in a side channel spillway), the variation of Froude number within the channel (to decide whether the flow is subcritical or supercritical or, if it changes from subcritical to supercritical, to locate the critical section), and the variation of discharge in the channel (to decide the length of a side weir that will be able to divert a pre-determined quantity of water under certain conditions). Due to the variation of discharge in the longitudinal direction, the analysis of these flows becomes more complicated than of those previously described. Additional complications arise in the analysis since SVF with lateral inflows, i.e., in which the discharge increases along the length, generally has a significant energy loss due to impact of the incoming water; and SVF with lateral outflow typically has a momentum efflux, which depends on the local velocity and is, therefore, unknown. In the next section, we look at the application of the basic equations of continuity, momentum, and energy to flows involving varying discharge and consider their relative suitability for various flow conditions.

6.2 Continuity, Momentum, and Energy Equations

6.2.1 Continuity Equation

Figure 6.2 shows the longitudinal section of an open channel along with a control volume (CV) and the cross sections at either ends of this CV. As stated earlier, we perform a one-dimensional analysis by using the cross-sectional average velocity [Eq. (2.1)]. The flow is steady but the discharge is changing spatially because of lateral inflows or outflows. We denote the lateral inflow/outflow by q^* (positive for inflow and negative for outflow), which represents the volume per unit channel length per unit time and, in general, would be a function of longitudinal distance.[2] The continuity equation is then written as

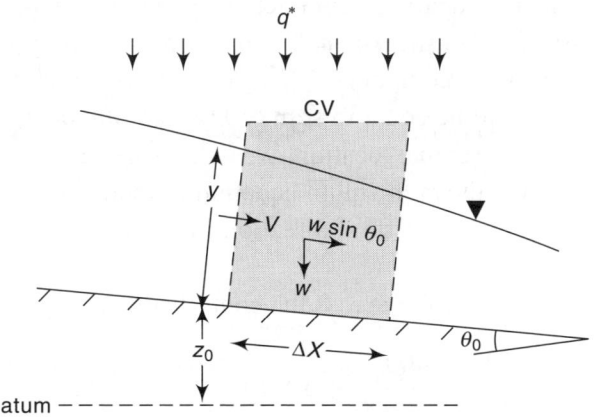

Fig. 6.2 Longitudinal section of a channel with spatially varied flow

[2]As we will see in a little while, for some cases, the lateral inflow rate may be constant, thus simplifying the analysis. We do not consider the cases in which there is a simultaneous lateral inflow and outflow, although the treatment of such cases is quite similar.

$$VA + q^*\Delta X - \left[VA + \frac{\partial(VA)}{\partial X}\Delta X\right] = 0 \quad \Rightarrow \quad \frac{\partial Q}{\partial X} = q^* \qquad (6.1)$$

Inflow rate	Outflow rate	Rate of change
into CV	from CV	within CV

implying the obvious fact that the discharge in the channel will increase along the length in the presence of lateral inflow.

6.2.2 Momentum Equation

Application of the X-momentum equation over the CV shown in Fig. 6.2 requires information about the forces acting on the CV and the net efflux of momentum across the control surface (CS). The forces acting on the CV are due to the pressure at the cross sections, shear stress at the boundary, weight of water within the CV, and air resistance at the water surface [see Eqs (3.2) to (3.13)]. We assume that the air resistance is negligible, the curvature of streamlines is small, and the average shear stress at the boundary is given by an expression similar to that used for uniform flow with the bed slope replaced by a shear slope, S_T. Once all the forces are known, the net force in the X-direction can be computed and set equal to the net momentum efflux. As compared to the GVF [Eq. (3.12)], however, we have the additional complexity due to the lateral flow. We assume that the lateral flow, as it enters (or leaves) the control surface, has a velocity component V_X^* in the X-direction. The lateral inflow generally takes place perpendicular to the main channel flow direction and does not contribute to the momentum flux in the longitudinal direction. Thus application of momentum equation is more suited to such cases. However, lateral outflows may be analysed by making suitable assumption about the velocity component, V_X^*. For example, we may assume this velocity component to be equal to some *correction factor* times the average velocity in the channel at that location. We will not describe this methodology here and assume that the momentum equation is applied to cases with lateral inflow perpendicular to the flow direction in the channel, thereby making $V_X^* = 0$.

The net momentum efflux across the control surface is obtained as

$$M_X = \beta\rho\left\{\left[\frac{Q^2}{A} + \frac{\partial(Q^2/A)}{\partial X}\Delta X\right] - \frac{Q^2}{A}\right\} - \rho q^*\Delta X V_X^*$$

$$= \rho\Delta X\left[BQ^2\frac{\partial(1/A)}{\partial X} + \beta\frac{1}{A}\frac{\partial Q^2}{\partial X} - q^* V_X^*\right] \qquad (6.2)$$

$$= \rho \Delta X \left(-\frac{\beta Q^2 T}{A^2} \frac{\partial Y}{\partial X} + \frac{2\beta Q q^*}{A} - q^* V_X^* \right)$$

Note that the momentum correction factor, β, has been assumed to be constant indicating that the velocity profiles at all sections are similar and we have used the continuity equation, Eq. (6.1), to eliminate the spatial derivative of the discharge.

Application of the momentum equation to the CV results in

$$\rho g A \Delta X S_0 + \rho g A \bar{y} \cos \theta_0 - \rho g \cos \theta_0 \left[A\bar{y} + \frac{\partial (A\bar{y})}{\partial X} \Delta X \right] - \rho g A \Delta X S \tau$$

$$= \rho \Delta X \left(-\frac{\beta Q^2 T}{A^2} \frac{\partial Y}{\partial X} + \frac{2\beta Q q^*}{A} - q^* V_X^* \right) \tag{6.3}$$

which may be simplified as [compare with Eq. (3.15)]

$$\frac{dy}{dX} = \frac{S_0 - S_\tau - (2\beta Q q^* / gA^2) + (q^* V_X^* / gA)}{\cos \theta_0 - \dfrac{\beta Q^2 T}{gA^3}} \tag{6.4}$$

This equation relates the rate of change of the flow depth in the longitudinal direction to the flow and channel properties and is called the *spatially varied flow equation* (since it is derived from the equation of motion, it is also called the *dynamic equation* of SVF). In the next subsection, we derive a similar equation by applying the mechanical energy equation.

6.2.3 Energy Equation

The mechanical energy equation applied to the CV shown in Fig. 6.2 results in the following equation [see (Eq. 3.21)]:

$$z_0 + y \cos \theta_0 + \alpha \frac{V^2}{2g} = z_0 + \frac{dz_0}{dX} \Delta X + \left(y + \frac{dy}{dX} \Delta X \right) \cos \theta_0$$

$$+ \frac{\alpha}{2g} \left(V^2 + \frac{dV^2}{dX} \Delta X \right) + h_L \tag{6.5}$$

where h_L is the head loss (i.e., loss of mechanical energy per unit weight), which includes the losses due to the interaction of the lateral inflow/outflow. Generally, lateral inflow not only brings in additional energy (since it typically has higher energy per unit weight) but also causes high energy loss due to its impact with the channel flow. On the other hand, lateral outflow flows out smoothly and does not cause significant energy loss. One should keep in mind that the *loss* represents loss of energy per unit weight and even though some mass (and, therefore, energy) is leaving the control volume, the energy *per unit weight* in the channel remains nearly constant. Thus energy equation is more suited for application to cases with lateral outflow. However, lateral inflows may be analysed by making suitable assumption about the energy balance. We do not describe it here and use the momentum equation for cases where there is substantial energy loss or gain. Note that we have assumed that the velocity profiles at various sections are similar and, therefore, the energy correction factor, α, is a constant at all sections. Using the relationships

$$\frac{dV^2}{dX} = -\frac{2Q^2 T}{A^3}\frac{dy}{dX} + \frac{2Qq^*}{A^2} \quad \text{and} \quad \frac{h_L}{\Delta X} = S_e$$

where S_e is the slope of the energy line (representing the loss of mechanical energy per unit weight of fluid per unit length of channel), Eq. (6.5) becomes

$$\frac{dy}{dX} = \frac{S_0 - S_e - (\alpha Qq^*/gA^2)}{\cos\theta_0 - \dfrac{\alpha Q^2 T}{gA^3}} \tag{6.6}$$

As before, we will assume that the energy slope is equal to that under uniform flow conditions at a flow depth of y, if the energy loss due to the lateral flow is neglected. When there is substantial energy loss or gain due to lateral flow, the energy slope has to be modified appropriately.

As discussed for the GVF equations obtained by the application of the momentum and energy equations, the two major differences between Eqs (6.4) and (6.6) are: (i) the use of *shear slope* and *energy slope* in the numerator and (ii) the use of *momentum* and *energy* correction factors in the numerator and denominator. [The other differences are the presence of the momentum efflux term in the numerator of Eq. (6.4) and the coefficient of the $Qq*/gA^2$ term. However, we will ignore the momentum efflux term by stipulating that the momentum equation would be applied to the case of lateral inflow perpendicular to the X-direction. Similarly, the difference between the coefficients is accounted for by using coefficients c_1 and c_2 in Eq. (6.7)]. Again, we will make some simplifying assumptions so that both these equations are written in the same form and then proceed towards a solution of this differential equation. The assumptions are as follows:

(i) The shear slope and energy slope are equal to the friction slope, S_f, and are obtained by using the uniform flow resistance formula[3] (Manning's equation).

(ii) The bed slope of the channel is very small[4] so that cosine of the angle is nearly equal to unity.

Equations (6.4) and (6.6) may then be written in a common framework as

$$\frac{dy}{dX} = \frac{S_0 - S_f - (c_1 Q q^*/gA^2)}{1 - \dfrac{c_2 Q^2 T}{gA^3}} \tag{6.7}$$

in which $c_1 = 2\beta$ and $c_2 = \beta$ when the momentum equation is used (for cases with lateral inflow perpendicular to the flow direction in the channel) and $c_1 = c_2 = \alpha$ when the energy equation is used (for cases with the lateral outflow that does not change the energy per unit weight in the channel). Yen and Wenzel (1970) may be consulted for advanced discussion of the dynamic equation for SVF.

6.3 Features of Water Surface Profile

As seen in Section 3.4, analysis of the flow profile for gradually varied flow can be done on the basis of the relationship between the actual flow depth, normal flow depth, and the critical depth. For spatially varied flow, however, the discharge variation along the longitudinal direction complicates the analysis. While for the GVF, the normal depth line (NDL) and the critical depth line (CDL) were straight lines parallel to the channel bed, in the SVF with lateral inflow, the critical depth will increase in the longitudinal direction as the discharge increases. Similarly, for flows with lateral outflows, the critical depth would decrease. A normal depth is not even defined for SVF since the varying discharge gives rise to changing flow depth. However, a *pseudo-normal* (pseudo-uniform, quasi-uniform, or fictitious uniform) *depth* can be defined such that the numerator of Eq. (6.7) becomes zero at a particular location for that flow depth. The *critical depth*, of course, is defined in the same way as before, i.e., it makes the denominator of Eq. (6.7) equal to zero. It should also be obvious that both the critical depth and the pseudo-normal

---

[3]It is known that the frictional loss in an SVF is a little larger than that in a uniform flow at the same flow depth. However, to simplify the analysis, we assume these to be same.

[4]Strictly speaking, we do not need this assumption. However, since most cases of open channel flow have small bed slope, we will make this simplifying assumption.

depth change along the length of the channel since the discharge is changing. Therefore, while the NDL is always below or always above the CDL in the GVF in a channel with a constant bed slope, for a SVF, both the pseudo-NDL and the CDL are curved and the pseudo-NDL may be below, above, or even partly above and partly below the CDL depending on how the discharge is changing. In the last case (partly above and partly below), there would be a point of intersection of the pseudo-NDL and the CDL, which is known as the *transitional depth* (see Example 3.5) and at which the pseudo-uniform flow will occur under critical conditions.

Since we assume that the friction slope for SVF is equal to that for uniform flow taking place at the same flow depth, S_f can be obtained from the Manning's equation. The pseudo-normal depth can then be obtained from the condition that the numerator of Eq. (6.7) becomes zero, or

$$\frac{n^2 Q^2}{A^2 R^{4/3}} + \frac{c_1 Q q^*}{gA^2} = S_0 \tag{6.8}$$

If we assume that the lateral inflow rate, q^*, is constant and the origin of X-axis is at the beginning of the channel, i.e., $Q = q^* x$, the pseudo-normal depth at any location will be obtained by solving the following nonlinear equation:

$$\frac{n^2 X^2}{A^2 R^{4/3}} + \frac{c_1 X}{gA^2} = \frac{S_0}{q^{*2}} \tag{6.9}$$

Similarly, the critical depth is obtained from the following equation:

$$\frac{A^3}{T} = \frac{c_2 q^{*2} X^2}{g} \tag{6.10}$$

and the equation for the transitional depth is obtained by combining Eqs (6.9) and (6.10) as

$$\frac{n^2 X^2}{A^2 R^{4/3}} + \frac{c_1 X}{gA^2} = \frac{S_0 T c_2 X^2}{gA^3}$$

$$\Rightarrow \quad X = \frac{1}{\dfrac{c_2 S_0 T}{c_1 A} - \dfrac{n^2 g}{c_1 R^{4/3}}} \tag{6.11}$$

Note that the transitional profile is independent of the discharge.
Example 6.1 illustrates the computations of the pseudo-normal depth line, critical depth line, and the transitional profile.

Example 6.1 A trapezoidal channel has bed width of 5 m, side slopes of 1H:1V, bed slope of 0.1, and Manning's roughness of 0.01. It is 80 m long and carries water away from a spillway whose crest is parallel to the channel. Plot the pseudo-NDL and CDL when the discharge per unit length of the spillway is 2 m^3/s/m and 3 m^3/s/m. Also plot the transitional depth profile and compute the transitional depth and its location for these two discharges. Assume the momentum correction factor to be 1.2.

Solution

Since the outflow from the spillway is perpendicular to the channel flow, there is no momentum contribution in the X-direction. Hence, we use the momentum equation, Eq. (6.4), implying that $c_1 = 2\beta = 2.4$ and $c_2 = \beta = 1.2$ in Eq. (6.7).

Since the determination of pseudo-normal depth and critical depth at a given location involves the solution of nonlinear equations, it is more convenient to assume some water depth and compute the distance, X_n, at which it would be a pseudo-normal depth [solution of a quadratic equation, see Eq. (6.9)] and the distance, X_c, at which it would be a critical depth [direct solution, see Eq. (6.10)]. The following table shows the computations and Fig. 6.3 shows the plots. From the figure, the transitional depths for the two discharges are seen as 2.7 m and 4.0 m, respectively, and the locations at which these occur are 40 m and 56 m, respectively, from the upstream end of the channel.

$y(m)$	$A(m^2)$	$P(m)$	$R(m)$	$T(m)$	$X_n(m)$	$X_c(m)$	$X_t(m)$
0.01	0.05	5.03	0.01	5.02	0.00	0.01	0.21
0.10	0.51	5.28	0.10	5.20	0.03	0.23	2.00
0.20	1.04	5.57	0.19	5.40	0.11	0.65	3.91
0.50	2.75	6.41	0.43	6.00	0.77	2.66	9.27
1.00	6.00	7.83	0.77	7.00	3.67	7.94	17.32
1.50	9.75	9.24	1.05	8.00	9.68	15.39	24.60
2.00	14.00	10.66	1.31	9.00	19.92	24.96	31.39
2.50	18.75	12.07	1.55	10.00	35.64	36.70	37.82
3.00	24.00	13.49	1.78	11.00	58.22	50.68	44.00
3.50	29.75	14.90	2.00	12.00	89.15	66.97	49.99
4.00	36.00	16.31	2.21	13.00	130.03	85.64	55.82
5.00	50.00	19.14	2.61	15.00	248.45	130.50	67.18
6.00	66.00	21.97	3.00	17.00	427.87	185.91	78.22

The transitional profile can be seen from the table and figure to be independent of the discharge.

Similar to the GVF zones based on the NDL and CDL, an SVF also may be thought of as having three different zones based on the pseudo-NDL and CDL—zone 1 above the upper line, zone 2 between the two lines, and zone 3

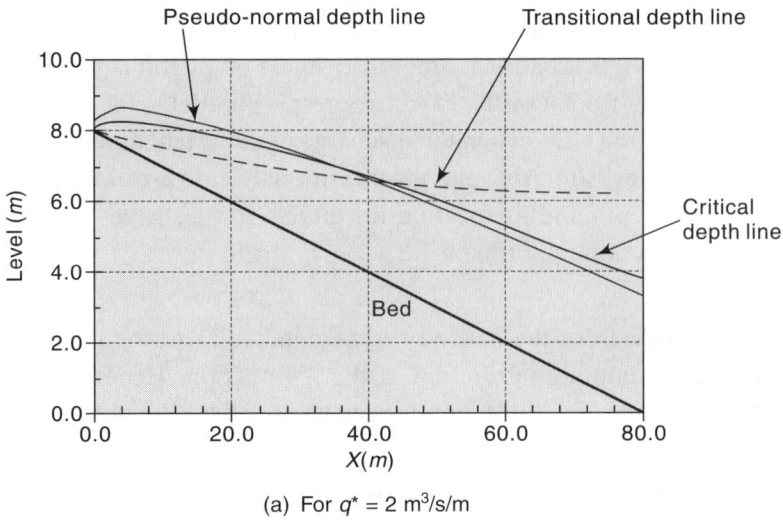

(a) For $q^* = 2$ m³/s/m

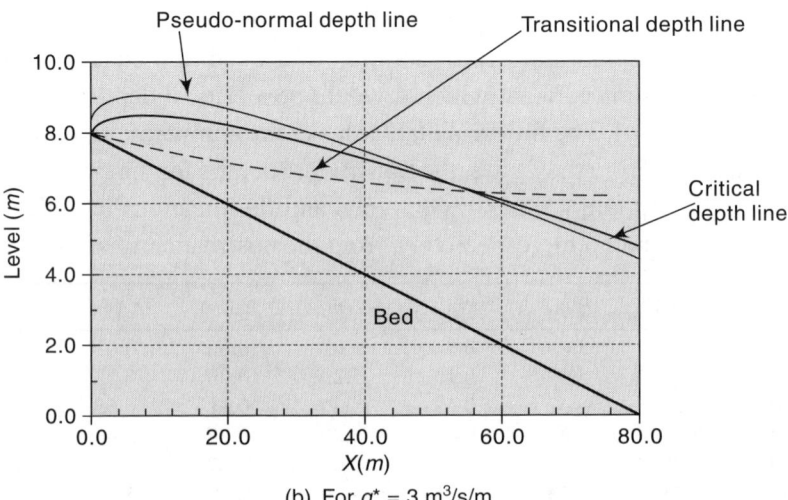

(b) For $q^* = 3$ m³/s/m

Fig. 6.3 Plots of critical depth, pseudo-normal depth, and transitional depth

below the lower line. Backwater curves (dy/dX positive) would occur in zones 1 and 3 and drawdown curves (dy/dX negative) in zone 2, as explained for the GVF. Similarly, other inferences drawn for the GVF profiles (see Section 3.4) would also hold good for the SVF profiles. However, as the flow depth approaches the transitional depth, dy/dX becomes indeterminate (of the form 0/0, since the numerator is zero at the pseudo-normal depth and the denominator is zero at the critical depth) and would have to be obtained by using the L'Hospital's rule. Analysis of flow profiles for SVF based on these zones is not very common and different classification methodologies have been used, as described in the next sections. Since the SVF with lateral inflow and the SVF with lateral outflow are inherently different, we describe each in a separate section.

6.4   SVF with Lateral Inflow

Computation of the SVF profile is generally done using numerical methods. However, if we make some simplifying assumptions, analytical solution may be obtained. For example, considering a horizontal, frictionless, rectangular side-channel spillway of length L, ending in a free overfall, and assuming $\beta = 1$, Eq. (6.7) may be simplified as

$$\frac{dy}{dX} = \frac{-2\,yXq^{*2}}{gB^2\,y^3 - q^{*2}\,X^2} \quad \Rightarrow \quad \frac{dX^2}{dy} - \frac{X^2}{y} = -\frac{gB^2\,y^2}{q^{*2}}$$

$$\Rightarrow \quad \frac{d(X^2/y)}{dy} = -\frac{gB^2\,y}{q^{*2}} \qquad (6.12)$$

If the channel ends in a free overfall, critical depth may be assumed to occur[5] at $X = L$. This critical depth is equal to $(q^{*2}L^2/gB^2)^{1/3}$ and the solution of Eq. (6.12) is readily obtained as

$$\left(\frac{X}{L}\right)^2 = 1.5\frac{y}{y_c} - 0.5\left(\frac{y}{y_c}\right)^3 \qquad (6.13)$$

On the other hand, if the tailwater level at the outlet of the channel is higher than the critical depth, it will influence the profile of the SVF within the channel. For an end depth of y_L (which is greater than y_c), and defining an outlet Froude number, F_L, as $F_L = q^* L/B \sqrt{gy_L^3}$, the flow profile can be shown to be given by

$$\left(\frac{X}{L}\right)^2 = \left(1 + \frac{0.5}{F_L^2}\right)\frac{y}{y_L} - \frac{0.5}{F_L^2}\left(\frac{y}{y_L}\right)^3 \qquad (6.14)$$

Note that for any given X, there would be two positive and one negative solutions for y/y_L, the physically meaningful solution being the one with $y/y_L > 1$ and the Froude number being less than 1 throughout the channel. At the

---

[5]As we have seen earlier in Chapter 4, the critical depth occurs a little upstream of the brink. However, we ignore this fact here.

start of the channel ($X = 0$), the flow depth is obtained from Eq. (6.14) as

$$y_0 = y_L \sqrt{1 + 2 F_L^2}.$$

After computing the flow depth at any location, the local Froude number at that point, defined by $F = q^* X / B \sqrt{gy^3}$, can be easily obtained. Figures 6.4 and 6.5 show the flow depth variation and the Froude number variation respectively in a horizontal channel for two different values of the outlet Froude number.

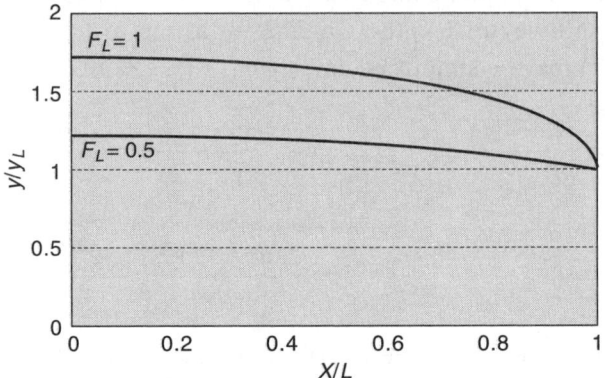

Fig. 6.4 Flow depth variation in a horizontal channel with two different outlet Froude numbers

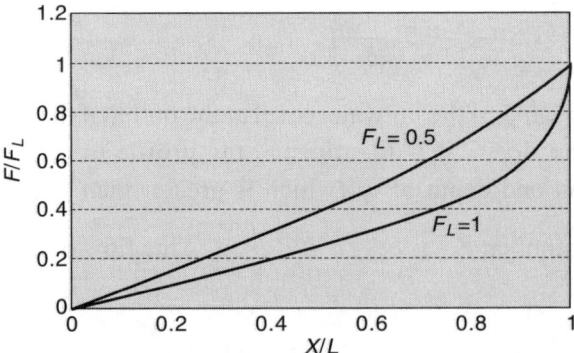

Fig. 6.5 Froude number variation in a horizontal channel with two different outlet Froude numbers

From Fig. 6.5, we see that the Froude number starts at zero and shows a monotonic increase to the outlet Froude number. For a sloping bed, however, this is not always the case. The SVF profile for a sloping bed is not amenable to an analytical solution and numerical integration has to be resorted to. Hinds in 1926 was one of the early investigators of the flow through side channel spillways. A more recent review is provided by Hager (1983). W.H. Li in 1955 performed extensive numerical simulations and classified the SVF profiles into four types based on how the Froude number varies along the length of the channel. These may be visualized by considering a side-channel spillway in which the outflow rate from the spillway, q^*, is kept constant and the downstream flow depth, y_L, is also held fixed. (We assume that y_L is greater than the critical depth at the outlet, y_c. If it is smaller, the depth at the outlet would be equal to the critical depth.) The outlet Froude number, F_L, is thus fixed. Starting with a horizontal channel, the channel slope is gradually increased and the flow profile variation as well as the Froude number variation are computed. The following classification for a *frictionless channel* covers the entire range of bed slope variation, represented by a slope parameter, $G = S_0 L/y_L$, which relates the fall in channel bed over its length to the flow depth at the outlet.

Type A. Similar to the channel discussed above, in this type of channel, the Froude number increases monotonically from zero to F_L. The flow remains subcritical throughout the channel. This condition occurs for small bed slopes and the maximum water depth occurs at the upstream end. This depth, y_0, will govern the design of the channel.

Type B. As the bed slope is increased further and becomes larger than a limiting value given by the relationship $G_{AB} = (2/3)(1 + 2F_L^2)$, the Froude number starts showing a non-monotonic behaviour. Starting from its value of zero at the upstream end, it first increases to a maximum value, which is greater than the outlet Froude number, F_L, (but less than 1, so that the flow remains subcritical) and then decreases to F_L at the outlet.

Type C. Further increase in bed slope causes the flow to become supercritical when the bed slope becomes larger than a limiting value given approximately by the relationship $G_{BC} = 1 + F_L$. The Froude number again has a monotonic behaviour, but it becomes more than 1 at some location within the channel and then increases further to cause supercritical conditions. A hydraulic jump is formed to move the flow depth from this supercritical value to the subcritical outlet depth. Location of the jump, however, is not very easy to compute due to the discharge variation. When the jump forms, the outlet conditions do not affect the flow profile upstream of the jump. This profile may be computed by locating the critical section and the critical depth, which is not described here (interested reader may refer to Chow 1981).

Type D. As the bed slope increases further, the supercritical flow near the outlet attains very high Froude number and the flow depth at the outlet would become insufficient to form a hydraulic jump. The Froude number again has a monotonic behaviour and supercritical conditions exist throughout the channel. The outlet Froude number in this case is not dependent on the water depth y_L since the supercritical flow sweeps away into the downstream pool.

The above classification provides a clear understanding of various flow situations. However, since it is based on the assumption of frictionless channel,[6] its practical utility is limited. It is recommended that the maximum depth obtained from Li's analysis be increased by about 10% to account for friction. However, we believe that a numerical method would be the best to determine the flow profile and obtain the maximum flow depth for designing the channel.

6.5 SVF with Lateral Outflow

For SVF with lateral outflow, the energy equation is used and the governing differential equation is Eq. (6.7) with $c_1 = c_2 = \alpha$. As we have seen in the previous section, analytical solutions to the SVF equations are possible for some very simple cases. In this section, we consider the case of a side weir and make some simplifying assumptions to enable us to obtain an analytical solution of the differential equation. Since mild slope channels are more common, we consider only the cases where the flow in the main channel is subcritical as it approaches the side weir.

Figure 6.6 shows a typical side weir, which diverts water from the main channel to a side channel when the water level in the main channel rises above the weir crest. The flow depth far upstream of the weir will be equal to the normal depth corresponding to the pre-weir discharge, Q_1, and the flow depth far downstream will be equal to the normal depth corresponding to the smaller post-weir discharge of Q_2. Experimental observations indicate that there is no significant energy loss due to the weir and the specific energy of the flow remains constant. The flow depth rises as we move from the upstream end of the weir to the downstream end if the flow is subcritical and it falls if the flow is supercritical (if the weir crest is at a very low height,[7] supercritical flow will occur in the main channel from $X = 0$ to $X = L$, see Fig. 6.7).

The analysis of flow at a side weir involves the computation of the surface profile in the main channel and the discharge in the side channel. As compared

---

[6]Also, the limiting relationships between G and F_L are based on the assumption that the momentum correction factor is equal to 1.

[7]To attain supercritical flow, the weir crest should be below the critical depth at the start of the weir. Note that the critical depth would be decreasing because the discharge in the main channel is decreasing.

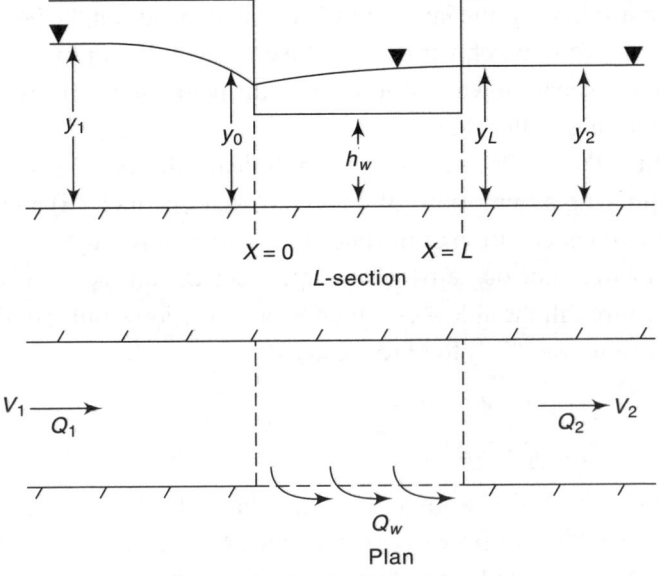

Fig. 6.6   Side weir in a channel

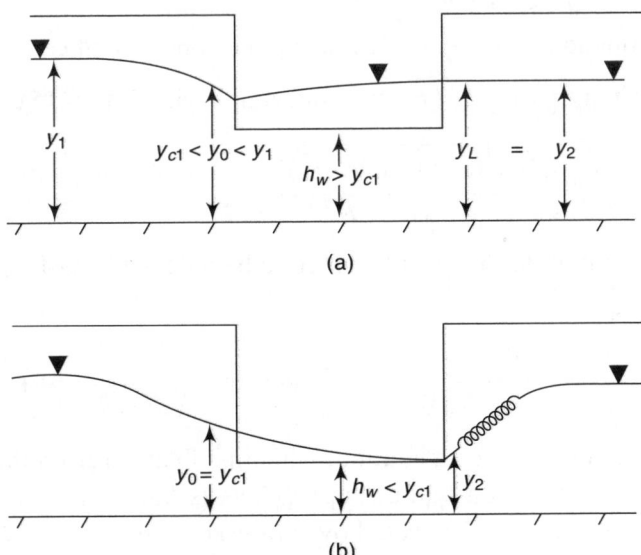

Fig. 6.7   Water surface profile at a side weir for subcritical approach flow

to the SVF with increasing discharge, this analysis is made more complicated due to the fact that the rate of lateral flow is not constant. In a side-channel spillway, the lateral inflow remains constant along the channel length as it depends on the head over the spillway crest, which is same at all locations.

However, for a side weir, the lateral outflow rate depends on the head over the crest of the side weir, which varies over the crest length. For example, in Fig. 6.7a, we should expect a lower outflow rate at the beginning of the weir because the water depth is smaller there.

De Marchi in the 1930s attempted an analytical solution of this problem by making simplifying assumptions of (i) rectangular channel, (ii) constant specific energy, (iii) cancellation of the bed slope and energy slope terms ($S_0 = S_f$, which may or may not be zero), (iv) energy correction factor of unity, and (v) discharge through the side weir given by an equation similar to that over a normal weir. Equation (6.7) then reduces to

$$\frac{dy}{dX} = \frac{-Q(dQ/dX)/gB^2 y^2}{1 - (Q^2/gB^2 y^3)}$$ (6.15)

The rate of change of discharge in the main channel, dQ/dX, would be equal and opposite to the rate of flow over the side weir, which is written as [see Eqs (4.46) and (4.53) for broad- and sharp-crested weirs]

$$dQ_w = C_d\sqrt{g}\,(y - h_w)^{1.5}\,dX$$ (6.16)

Using this equation to obtain dQ/dX, using the condition of constant specific energy to write $Q = By\sqrt{2g(E - y)}$ and simplifying Eq. (6.15), we get

$$\frac{dy}{dX} = \frac{C_d\sqrt{2}}{B}\frac{\sqrt{(E - y)(y - h_w)^3}}{3y - 2E}$$ (6.17)

If the coefficient of discharge is assumed to be independent of X, integration of Eq. (6.17) leads to

$$X = \frac{\sqrt{2}B}{C_d}\left(\frac{2E - 3h_w}{E - h_w}\sqrt{\frac{E - y}{y - h_w}} - 3\sin^{-1}\sqrt{\frac{E - y}{E - h_w}}\right) + C_1$$ (6.18)

where C_1 is a constant of integration, which is obtained from the boundary condition. The expression within the brackets is known as the *De Marchi varied flow function* and is denoted by Φ_M. It is a function of E/h_w and y/h_w and is plotted in Fig. 6.8.

The only thing which remains now is to relate the coefficient of discharge to the flow properties in the main channel. De Marchi recommended a C_d value of about 0.6 for a sharp-crested side weir (which is same as that for a normal weir). Some other investigators (e.g., Subramanya and Awasthy 1972, Ranga Raju, et al. 1979) suggest that C_d for a sharp- or broad-crested side weir depends on the Froude number, F_0, of the main channel flow at the beginning of the weir crest. For example, with a subcritical approach flow, C_d

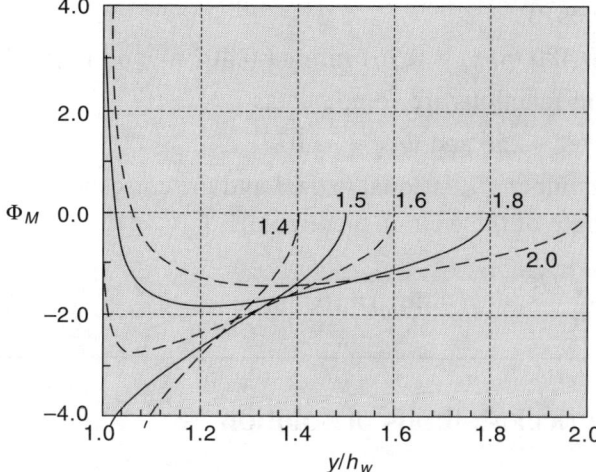

Fig. 6.8 Variation of the De Marchi function with y/h_w (Numbers near the curve represent the value of E/h_w.)

for a sharp-crested weir decreases with F_0 as $C_d = 0.59 - 0.21 F_0$. Example 6.2 describes the application of the De Marchi method for analysing the flow at a side weir.

Example 6.2 A rectangular channel has bed width of 3 m, bed slope of 1 in 1000, and Manning's roughness of 0.015. Design a side weir such that it becomes effective when the discharge in the channel is 1 m³/s and diverts 0.2 m³/s of the flow when the discharge is 1.5 m³/s.

Solution
The crest level of the side weir is fixed based on the flow depth at which it becomes active. The normal depth of flow for a discharge of 1 m³/s is obtained as 0.361 m. Therefore, the weir crest would be located at a height of 0.361 m from the channel bed.

For the discharge of 1.5 m³/s, the normal depth is 0.470 m and the critical depth is 0.294 m. Since the weir crest is above the critical depth, subcritical flow will prevail in the main channel. The flow depth in the main channel at the end of the weir crest (and beyond it) would be dictated by the normal depth corresponding to a discharge of $(1.5 - 0.2)$ m³/s, i.e., 1.3 m³/s. This is obtained as $y_L = 0.428$ m and the corresponding specific energy is obtained as 0.480 m. Since the specific energy is constant, the depth at the beginning of the weir is obtained as follows:

$$y_0 + \frac{1.5^2}{3^3\, y_0^2\, 2g} = 0.480 \quad \Rightarrow \quad y_0 = 0.402 \text{ m}$$

With the values of

$$E = 0.480 \text{ m}, \ h_w = 0.361 \text{ m}, \ y_0 = 0.402 \text{ m}, \text{ and } y_L = 0.428 \text{ m}$$

the De Marchi functions are computed as

$$\Phi_{M0} = -4.256 \text{ and } \Phi_{ML} = -3.077$$

The Froude number, F_0, is equal to 0.63 and the corresponding C_d is equal to 0.46. The length of the weir is, therefore, [Eq. (6.18)]

$$L = \frac{\sqrt{2}B}{C_d} (\Phi_{ML} - \Phi_{M0}) = 10.9 \text{ m}$$

6.6 Numerical Methods of Solution

All the numerical methods described earlier for the GVF profile computations can be used for solving the differential equation of SVF, Eq. (6.7). Due to the approximations involved in the derivation of the equation, and the uncertainty associated with the momentum and energy correction factors, it would not be prudent to use a very high accuracy scheme. Use of a forward Euler's method (see Appendix C) with a small length step should work for most cases. For higher accuracy, a modified Euler's method, involving a predictor step and a corrector step, may be used. The solution of the equation $dy/dX = f(X, y)$ using this scheme is written as

$$y_{i+1}^p = y_i + \Delta X f(X_i, y_i) \tag{6.19a}$$

$$y_{i+1} = y_i + \Delta X \frac{f(X_i, y_i) + f(X_{i+1}, y_{i+1}^p)}{2} \tag{6.19b}$$

Other methods with still higher order of accuracy, e.g., the fourth-order Runge–Kutta method, can also be used. Examples 6.3 and 6.4 illustrate the application of the forward Euler's scheme [Eq. (6.19a)] to obtain the surface profile in a side channel spillway.

Example 6.3 A rectangular channel has bed width of 10 m, bed slope of 0.05, and Manning's roughness of 0.01. It is 50 m long and carries water away from a spillway whose crest is parallel to the channel. The discharge per unit length of the spillway is 1 m³/s/m. The channel outlet is submerged such that the water depth above the bed is 2 m. Plot the water surface profile. Assume the momentum correction factor to be 1.2.

Solution

The following table shows the steps of computation and Figs 6.9(a) and (b) show the plot of water surface profile as well as the variation of the Froude number. The computations are done with a step size, ΔX, equal to 1 m. However, even with a 5 m step size, the results are comparable. The table shows some selected steps of the computations.

X (m)	y_i (m)	Q (m³/s)	A (m²)	P (m)	R (m)	S_f	dy/dX	y_{i+1} (m)	F
50	2.00	50	20.00	14.00	1.43	0.000388	0.0308	1.97	0.564
49	1.97	49	19.69	13.94	1.41	0.000391	0.0304	1.94	0.566
48	1.94	48	19.39	13.88	1.40	0.000392	0.0299	1.91	0.568
47	1.91	47	19.09	13.82	1.38	0.000394	0.0295	1.88	0.569
46	1.88	46	18.79	13.76	1.37	0.000395	0.0291	1.85	0.570
45	1.85	45	18.50	13.70	1.35	0.000396	0.0286	1.82	0.571
⋮	⋮	⋮	⋮	⋮	⋮	⋮	⋮	⋮	⋮
40	1.71	40	17.11	13.42	1.27	0.000395	0.0266	1.68	0.570
⋮	⋮	⋮	⋮	⋮	⋮	⋮	⋮	⋮	⋮
35	1.58	35	15.82	13.16	1.20	0.000383	0.0248	1.56	0.561
⋮	⋮	⋮	⋮	⋮	⋮	⋮	⋮	⋮	⋮
30	1.46	30	14.61	12.92	1.13	0.000358	0.0236	1.44	0.542
⋮	⋮	⋮	⋮	⋮	⋮	⋮	⋮	⋮	⋮
20	1.23	20	12.28	12.46	0.99	0.000270	0.0235	1.20	0.469
⋮	⋮	⋮	⋮	⋮	⋮	⋮	⋮	⋮	⋮
10	0.98	10	9.78	11.96	0.82	0.000137	0.0280	0.95	0.330
⋮	⋮	⋮	⋮	⋮	⋮	⋮	⋮	⋮	⋮
5	0.83	5	8.28	11.66	0.71	0.000057	0.0339	0.79	0.212
4	0.79	4	7.94	11.59	0.69	0.000042	0.0359	0.76	0.180
3	0.76	3	7.59	11.52	0.66	0.000027	0.0382	0.72	0.145
2	0.72	2	7.20	11.44	0.63	0.000014	0.0411	0.68	0.104
1	0.68	1	6.79	11.36	0.60	0.000004	0.0449	0.63	0.057
0	0.63	0	6.34	11.27	0.56	0.000000	0.0500	–	0.000

From Fig. 6.9(b), it is seen that the Froude number first increases, reaches a maximum value of about 0.57 at about 40 m, and then decreases to its outlet value of about 0.56. It is, therefore, a type *B* profile. *Although the momentum correction factor is not equal to 1 and the channel is not frictionless*, we may try to compare the slope parameter, G, and the outlet Froude number, F_L, to see whether it matches Li's criterion. For this case, $G = S_0 L/y_L = 1.25$. The transition from type *A* to type *B* occurs at $G_{AB} = (2/3)(1 + 2F_L^2) = 1.09$ and the transition from type *B* to type *C* occurs at $G_{BC} = 1 + F_L = 1.56$. Since G is between these two values, it indicates a type *B* profile, as observed from the numerical solution.

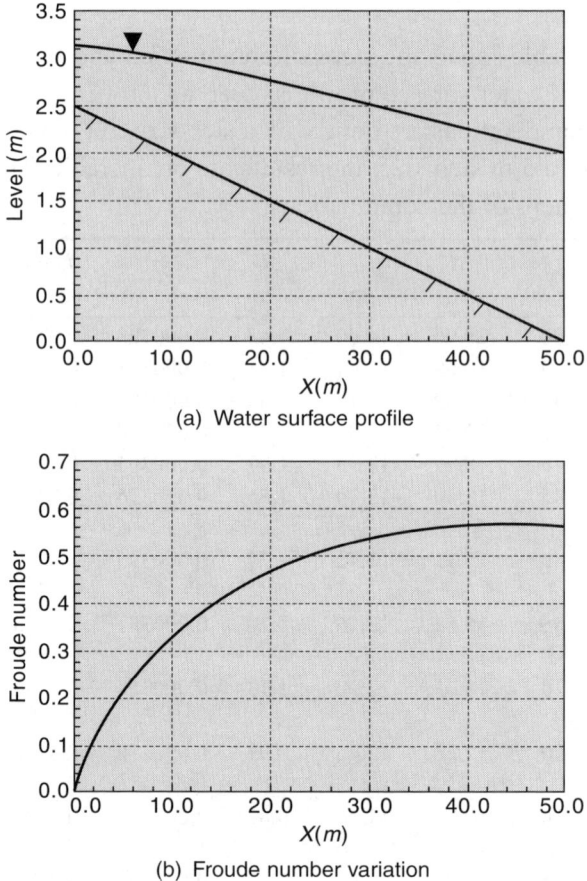

(a) Water surface profile

(b) Froude number variation

Fig. 6.9   Water surface profile and Froude number variation
in the side channel of a spillway

Application of numerical method to the case of SVF with decreasing discharge is illustrated in Example 6.4.

Example 6.4   Solve Example 6.2 using numerical integration.

Solution

The following table shows the computations using the Euler's method, starting from the downstream end of the weir where the flow depth is 0.428 m and the discharge is 1.3 m³/s. We use step length of 0.2 m and proceed upstream till the discharge in the channel becomes equal to 1.5 m³/s (and the discharge over the weir becomes equal to the desired diversion of 0.2 m³/s). The crest level of the side weir has been fixed in Example 6.2 at a height of 0.361 m from the channel bed and the coefficient of discharge is also determined as

0.46. The specific discharge dQ/dX is obtained from the normal weir equation with $C_d = 0.46$ and head equal to the local water depth above the weir crest.

X (m)	Q (m^3/s)	y_i (m)	S_f	$-dQ/dX$ (m^3/s)	dy/dX	y_{i+1} (m)	E (m)
0	1.300	0.428	0.000999	0.0250	0.0027	0.427	0.480
−0.2	1.305	0.427	0.001011	0.0247	0.0026	0.427	0.480
−0.4	1.310	0.427	0.001022	0.0244	0.0026	0.426	0.480
−0.6	1.315	0.426	0.001034	0.0241	0.0026	0.426	0.480
−0.8	1.320	0.426	0.001045	0.0238	0.0026	0.425	0.480
−1	1.324	0.425	0.001057	0.0235	0.0026	0.425	0.480
⋮	⋮	⋮	⋮	⋮		⋮	⋮
−2	1.347	0.423	0.001114	0.0222	0.0024	0.422	0.480
⋮	⋮	⋮	⋮	⋮		⋮	⋮
−4	1.389	0.418	0.001225	0.0197	0.0022	0.418	0.481
⋮	⋮	⋮	⋮	⋮		⋮	⋮
−6	1.427	0.414	0.001332	0.0176	0.0020	0.414	0.481
⋮	⋮	⋮	⋮	⋮		⋮	⋮
−8	1.460	0.410	0.001434	0.0158	0.0017	0.410	0.482
⋮	⋮	⋮	⋮	⋮		⋮	⋮
−9.8	1.488	0.407	0.001521	0.0144	0.0015	0.407	0.483
−10.0	1.491	0.407	0.001530	0.0143	0.0015	0.407	0.483
−10.2	1.493	0.407	0.001539	0.0141	0.0014	0.407	0.483
−10.4	1.496	0.407	0.001548	0.0140	0.0014	0.406	0.483
−10.6	1.499	0.406	0.001558	0.0139	0.0014	0.406	0.483
−10.8	**1.502**	0.406	0.001567	0.0137	0.0014	0.406	0.483

It is seen that the length of the weir is obtained as 10.8 m, almost identical to that obtained in Example 6.2. Also note that the specific energy remains nearly same throughout the length of the weir in accordance with the assumption made in the De Marchi's analysis.

Other spatially varied flow cases can be analysed similarly. For example, a roof gutter can be analysed in a similar manner as a side channel spillway and analysis of a bottom rack is similar to that of a side weir. For a bottom rack, however, the discharge through the rack would be obtained by considering it as an orifice (and not a weir as was done for side weir) and using an effective head equal to the depth of flow or the specific energy of flow (see Exercises 6.6 and 6.7). Since the methodology is identical to that used for the side weir, we do not describe it in detail.

Another fact which must be considered in the application of the numerical methods is the location of a control section. In Examples 6.3 and 6.4, we have assumed the flow to be subcritical and the control section is at the downstream end. If the flow changes from subcritical to supercritical within the channel, we would have to first locate the critical section and then proceed upstream from this point to analyse the subcritical flow and downstream to analyse the supercritical flow. It is left as an exercise to the reader (see Exercise 6.5).

SUMMARY

In this chapter, we have discussed various types of spatially varied flow situations. These include flow through a side weir, flow through a side-channel spillway, and several other cases in which the discharge changes along the longitudinal direction. The differential equations governing such flows have been derived based on the continuity and momentum or energy equations. Characteristics of the water surface profiles have been discussed and different methods of finding solution to the dynamic equation of SVF have been described. We have now covered various flows involving spatial variation of flow depth, velocity, and discharge. However, we have assumed that the parameters do not change with time. A number of open channel flow situations involve variation with respect to time also. For example, overland flow joining a river during a storm gives rise to spatially and temporally varying flow depth, discharge, etc. Sudden opening or closing of a gate gives rise to waves travelling in the channel, which affect the flow characteristics at different locations at different times. All these cases of unsteady flows are discussed in the next chapter.

REFERENCES

Beecham, S., Khiadani, M.H., and Kandasamy, J. (2005): 'Friction factors for spatially varied flow with increasing discharge', *J. of Hydr. Eng.*, **131**(9), pp. 792–99.

Chow, V.T. (1981): *Open-Channel Hydraulics*, McGraw-Hill International, Tokyo, Japan.

Hager, W.H.(1983): 'Open channel hydraulics of flows with increasing discharge', *J. of Hydr. Res.*, **21**(3), pp. 177–93.

Khiadani, M.H., Beecham, S., Kandasamy, J., and Sivakumar, S. (2005): 'Boundary shear stress in spatially varied flow with increasing discharge', *J. of Hydr. Eng.*, **131**(8), pp. 705–14.

Ranga Raju, K.G., Prasad, B., and Gupta, S.K. (1979): 'Side weirs in rectangular channel', *J. of Hydr. Div.*, ASCE, **105**(5), pp. 547–54.

Subramanya, K. and Awasthy, S.C. (1972): 'Spatially varied flow over side weirs', *J. of Hydr. Div.*, ASCE, **98**(1), pp. 1–10.

Yen, B.C. and Wenzel, H.G. (1970): 'Dynamic equations for steady spatially varied flow', *J. of Hydr. Div.*, ASCE, **96**(3), pp. 801–14.

EXERCISES

6.1 Prepare a transitional profile for the channel of Example 6.1 except that the bed slope of the channel is half of that used in the example.

6.2 A rectangular channel has bed width of 5 m, bed slope of 0.1, and may be assumed to be frictionless. It is 40 m long and carries water away from a spillway whose crest is parallel to the channel. The discharge per unit length of the spillway is 0.5 m^3/s/m. The channel outlet is submerged such that the water depth above the bed is 2 m. Plot the water surface profile. Assume the momentum correction factor to be 1.

6.3 A 3 m wide rectangular channel has a sharp-crested side weir in one of its sides. The weir crest is at a height of 1 m from the bed and has to be designed to divert 1 m^3/s in the side channel when the discharge in the main channel is 6 m^3/s. What should be the length of the weir? Assume that the downstream rating curve is given as $Q = 2\,y^{1.5}$ with Q in m^3/s and y in m.

6.4 Solve by numerical integration Examples 6.3 and 6.4 using the modified Euler's scheme. Use a step length twice those used in the examples and compare the results.

6.5 Analyse the case of Example 6.1 with a numerical method for the discharge of 2 m^3/s/m.

6.6 A bottom rack is used to divert part of the flow in a channel. When the rack is made of parallel bars in the bed of the channel, the effective head may be taken as equal to the specific energy. Assuming the specific energy to be constant over the rack, and considering the rack to behave as an orifice, show that the water surface profile over the rack is given

by $X = \dfrac{A_r E}{A_o C_d}\left(\dfrac{y_1}{E}\sqrt{1 - \dfrac{y_1}{E}} - \dfrac{y}{E}\sqrt{1 - \dfrac{y}{E}} \right)$, in which A_r is the area of the

rack, A_o is the area of rack opening, E is the specific energy, C_d is the coefficient of discharge for the orifice flow, y_1 is the depth at the upstream end of the rack, and y is the flow depth over the rack at a distance of X from the upstream end.

6.7 When the bottom rack is in the form of a perforated plate in the bed of the channel, the effective head may be taken as equal to the flow depth. Assuming the specific energy to be constant over the rack, show that the water surface profile over the rack is given by

$$X = \frac{A_r E}{2 A_o C_d}$$

$$\left\{ \cos^{-1} \sqrt{\frac{y}{E}} - \cos^{-1} \sqrt{\frac{y_1}{E}} + 3 \left[\sqrt{\frac{y_1}{E}\left(1 - \frac{y_1}{E}\right)} - \sqrt{\frac{y}{E}\left(1 - \frac{y}{E}\right)} \right] \right\}$$

in which C_d is the coefficient of discharge for the perforated plate flow.

7 | Unsteady Flow

7.1 Introduction

Unsteady flow denotes the flows that change with time. Strictly speaking, all flows in nature are unsteady because it is almost impossible to maintain all parameters invariant with time. Even for seemingly steady flows, where the *mean* properties do not change with time, there would be turbulent fluctuations, though generally of a much smaller magnitude than the mean. In the cases discussed so far, we have assumed that the flow was steady or, at least, could be approximated as steady. However, a number of open channel flow situations involve variations with respect to time, which may not be neglected. For example, a flood wave travelling through a river, a tidal bore[1] in an estuary,[2] waves generated due to sudden opening or closing of gates, all show significant temporal variation in velocity, flow depth, and discharge at a cross section. All these cases of unsteady flows are discussed in this chapter.

Some of the unsteady flow problems can be simplified to an equivalent steady flow problem by a simple change of the reference frame. For example, a wave travelling at a uniform velocity and maintaining its shape would be unsteady with respect to a stationary frame of reference since the flow depth and velocity at a cross section would be different before, during, and after the passage of the wave. However, if we move with the wave at the same speed as that of the wave, it would appear to be a steady flow with respect to this moving frame of reference. This is probably the easiest method of analysis for unsteady flow and will be discussed first. However, it is not always possible to convert an unsteady flow into an equivalent steady flow. For example, when a sluice gate is suddenly opened, the wave generated upstream of the gate changes its shape continuously and different fluid particles on the wavefront move with different speeds. Under some assumptions, however, it is possible to take a small elementary strip on the wavefront and analyse it as

---

[1] A *bore* (also called *eagre*) is a high wave caused by a tide or by colliding tidal currents.
[2] An *estuary* is the arm of the sea that extends inland to meet the mouth of a river.

an equivalent steady-state problem to derive the governing differential equation. Integration of this equation with appropriate boundary conditions would lead to the determination of the characteristics of this negative wave.[3] Analysis of most practical cases, however, involves the use of the continuity, momentum, and/or energy equations, which also consider the temporal variation. We will, therefore, derive these equations and then look at their applicability to practical cases and different methods of solving these equations. These equations would now be partial differential equations since the flow characteristics depend on the longitudinal coordinate, X, as well as time, t. (In steady flow also, we may get partial differential equations, if we consider the variation in the Y- and Z-directions. However, we have used the one-dimensional analysis to obtain cross-sectional average parameters, which enables us to use an ordinary differential equation.)

7.2   Conversion to Equivalent Steady Flow

If a disturbance in the channel is moving with a constant velocity and retains its shape, it can be converted to a steady-state situation with respect to a moving frame of reference. Depending on the nature of variation of depth, we can call it a *wave* (showing a gradual variation) or a *surge* (abrupt change in depth). The wave can be a small-amplitude wave in which its amplitude is very small compared to the flow depth or it can be a finite-amplitude wave. On the other hand, a surge can be a positive surge, causing an increase in flow depth or a negative surge, causing a decrease in the flow depth. Waves and surges are created in open channels due to various causes, e.g., sudden opening or closing of gates, lateral inflow due to runoff, tidal action in an estuary, discharge variation in a hydropower canal. Most commonly encountered waves in open channels comprise a single wavefront, while reservoirs and oceans can have a *train of waves*. One of the earliest observations of a wave moving at a constant velocity and maintaining its shape was done by the English scientist J.S. Russell, when he observed the wave generated in a narrow channel on stopping a moving boat. Laboratory experiments on a wave generated by dropping a weight in a channel confirmed this observation. In this section, we will analyse the waves and surges by using the equivalent steady state approach.

7.2.1   Small-Amplitude Wave

Figure 7.1 shows a channel in which a single wave is generated, say, by dropping a weight. This is a *positive wave* since it increases the flow depth as it passes a section (as opposed to a negative wave, which decreases the flow

---

[3]*Negative wave* implies that the depth of flow reduces as the wave passes through a section.

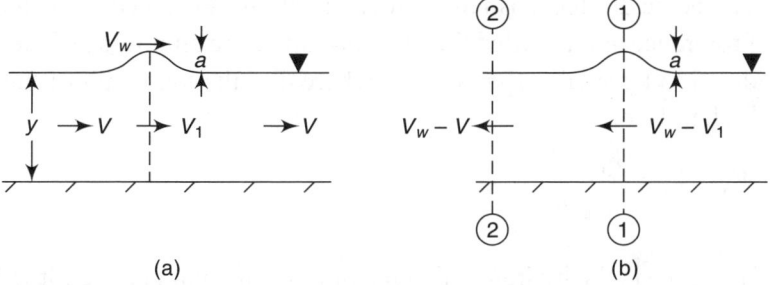

Fig. 7.1 A single small-amplitude wave in a channel

depth when it passes a section, e.g., slowly lowering a sluice gate generates a negative wave downstream of the gate and positive wave on the upstream side). We assume that the amplitude of the wave, a, is small such that the assumptions made in the theory of gradually varied flow can be applied. The piezometric head is, therefore, equal to the flow depth. We also assume the channel bed to be horizontal and neglect the energy loss due to friction (or, as done earlier, we may think of it as the bed slope and energy slope being nearly same).

The flow velocity in the channel is V and the flow depth is y with the corresponding flow area being A. The amplitude of the wave is a and the velocity of the wave is V_w. Note that the wave velocity is measured with respect to the stationary bank. It is more convenient in the analysis of waves to define a *wave celerity*, C_w, which is the velocity of the wave relative to the water,[4] i.e., $C_w = V_w - V$. Assuming that the wave velocity, V_w, is constant, we can use a frame of reference moving at the velocity of V_w to convert the problem into that of a steady-state flow.[5] In effect, a velocity of $-V_w$ is superimposed on the flow (Fig. 7.1b), and the continuity and the energy equations[6] are applied to obtain the celerity of the wave as follows.

Application of continuity equation

Denoting the velocity of flow at the section through the wave crest as V_1 and applying the continuity equation between the sections 1 and 2, we get

$$(V_w - V_1)A_1 = (V_w - V)A = Q_r \qquad (7.1)$$

---

[4]Generally the water velocity is different before and after the passage of wave. The undisturbed water velocity (i.e., before the arrival of the wave) is used to define the celerity.

[5]One of the requirements for the conversion of an unsteady state into a steady state is that the moving frame of reference must have a constant speed. If not, additional terms have to be incorporated in the basic equations. We do not discuss these.

[6]The momentum equation can also be applied. However, we will derive it in a later section.

in which Q_r is the discharge (in the –ive X-direction)[7] observed in the moving frame of reference and is called the *overrun*. Since the amplitude of the wave is considered to be small, $A_1 = A + Ta$, where T is the top width of the flow section.[8] Therefore,

$$V_w - V_1 = \frac{V_w - V}{1 + \dfrac{a}{D}} \tag{7.2}$$

where $D(= A/T)$ is the hydraulic depth (for a rectangular channel, it is equal to the flow depth).

Application of energy equation

Application of the energy equation, with the assumption of hydrostatic pressure distribution and neglecting the energy correction factor, leads to

$$y + a + \frac{(V_w - V_1)^2}{2g} = y + \frac{(V_w - V)^2}{2g} \tag{7.3}$$

Using Eq. (7.2) and writing in terms of celerity, we get

$$\frac{C_w^2}{2g}\left[1 - \frac{1}{\left(1 + \dfrac{a}{D}\right)^2}\right] = a \quad \Rightarrow \quad C_w = \pm\sqrt{gD}\,\frac{1 + (a/D)}{\sqrt{1 + (a/2D)}} \tag{7.4}$$

The $\pm$ sign for a general case indicates the celerity of the wave in the downstream direction (+ive) and the upstream direction (–ive). For example, if a wave is generated by dropping a weight in still water, it would travel in both directions,[9] X and $-X$. However, if we generate a wave by pushing a paddle in a water tank, it would generate a wave only in one direction. For a wave of very small amplitude ($a/D \ll 1$), the celerity is equal to $\sqrt{gD}$, as used in previous chapters without proof (this relationship was first derived by Lagrange in 1780s). For such cases, the wave velocity, as observed from the bank, would be given by $V_w = V + C_w = V \pm \sqrt{gD}$. This clearly shows that (again, as stated earlier) if the Froude number, $F\left(=V/\sqrt{gD}\right)$, is less than 1, the wave will move

---

[7]The wave velocity is typically larger than the flow velocity. That is why Eq. (7.1) is written in terms of discharge in the –ive X-direction.

[8]Since a is small, the top width corresponding to a flow depth of y and that at $y + a$ would be nearly same. If we want to be mathematically rigorous, we can take T at $y + (a/2)$.

[9]In fact, the wave travels in all directions. However, since we are using a one-dimensional approximation, we talk of only two directions. For a general case, the wave velocity would be obtained by a vector addition of the flow velocity and wave celertity.

both upstream and downstream (of course, with a higher velocity in the down-stream direction). For supercritical flows, however, $V > \sqrt{gD}$, and both the values of V_w would be positive, indicating only downstream movement.

We next consider a train of waves, which may or may not have a small ampli-tude. Figure 7.2 shows such a wave profile and defines its wavelength, L, and height, H.

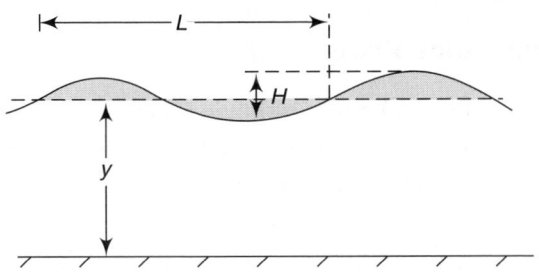

Fig. 7.2 Wave profile showing the wave height and wavelength

When a wave moves forward, it is called a *progressive wave* [Fig. 7.3(a)]. Sometimes, a progressive wave may be reflected in such a way that superposition of the incident and reflected waves gives rise to *standing waves* (particles moving only up and down) as shown in Fig. 7.3(b). Based on the transport of fluid particles by waves, we may classify waves as *oscillatory* (no net transport of fluid, e.g., a small-amplitude solitary wave) and *translatory* (net transport of fluid takes place, e.g., a flood wave).

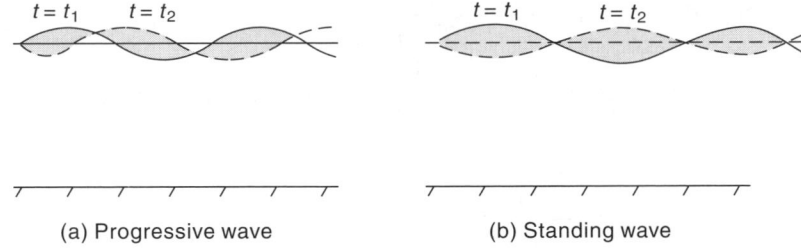

(a) Progressive wave   (b) Standing wave

Fig. 7.3   Classification of waves based on their movements

Although the above classifications are helpful in describing the nature of the wave, a more relevant classification, from the point of view of analysis of the waves, is that based on the size of the waves. This classification allows various simplifications to be made based on the *scale* of the wave. For exam-ple, waves with very small wave height to flow depth ratio, H/y, are called *small-amplitude waves* and for these, the convective acceleration term may be ignored in comparison with the inertial acceleration. For *finite-amplitude waves* (*H/y* larger than about 0.1), the terms in the solution of the governing equation involving higher powers of *H* are not negligible and the analysis becomes more complicated. It is still possible to convert it into a steady-state problem because the wave moves at a constant celerity. However, the solution is quite involved. We do not go into the details of solution and provide only a brief description of the solution process.

7.2.2 Finite-Amplitude Waves

Assuming the wave field to be a two-dimensional, inviscid, and irrotational flow in an otherwise motionless body of water, the velocity potential and the stream function satisfy the Laplace equation. Solution of the governing equation should satisfy the following boundary conditions:

  (i) Constant pressure (equal to atmospheric) on the water surface (may not be true for large-amplitude waves in presence of high wind velocities)
 (ii) No flow through the bottom boundary
(iii) The *kinematic* condition stating that a fluid particle on the surface remains on it even when the surface deforms

If we further assume that the height of wave is small in comparison with the depth of water (*short waves*, also called *Stokes waves*), the solution can be written as a power series in terms of the amplitude. The *linear wave theory*, proposed by G.B. Airy in 1840s, retains only the first term of this power series and results in a sinusoidal wave profile given by

$$\eta(X, t) = \frac{H}{2} \sin\left(\frac{2\pi X}{L} - \frac{2\pi t}{T_p} \right) \tag{7.5}$$

where T_p is the time period of the wave. The celerity of the wave, $C_w (= L/T_p)$, is given by

$$C_w = \sqrt{ \frac{gL}{2\pi} \tanh \frac{2\pi y}{L} } \tag{7.6}$$

For deep water waves (y/L greater than about 0.5), the tanh term tends to unity and the celerity is equal to $\sqrt{gL/2\pi}$. For shallow water waves or long water waves ($y/L < 0.05$), $\tanh 2\pi y/L \approx 2\pi y/L$ and the celerity is equal to $\sqrt{gy}$ (same as shown before for a single small-amplitude wave, which may be thought of as having an infinite wavelength). For intermediate waves,

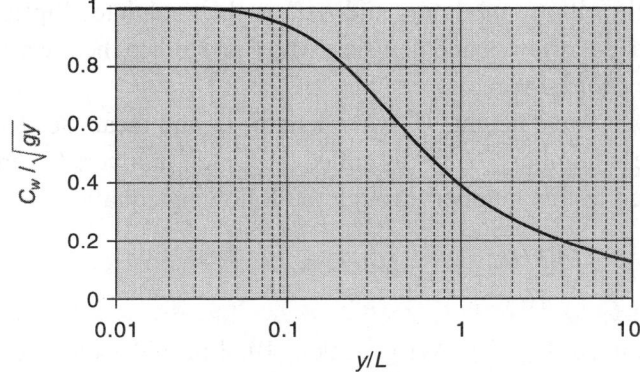

Fig. 7.4 Wave celerity as a function of the relative flow depth for small wave height

Eq. (7.6) is used to obtain the celerity. Figure 7.4 shows a plot of this equation in terms of a dimensionless celerity, $C_w/\sqrt{gy}$.

If the wave amplitude is not small, the celerity would also be a function of H/y. The water surface profile is described by (see Fig. 7.5) the Jacobian elliptical function, cn (u, k), as[10]

$$y_w = y_{\min} + H \operatorname{cn}^2\left[2K(k)\left(\frac{X_w}{L} - \frac{t}{T}\right), k\right] \tag{7.7}$$

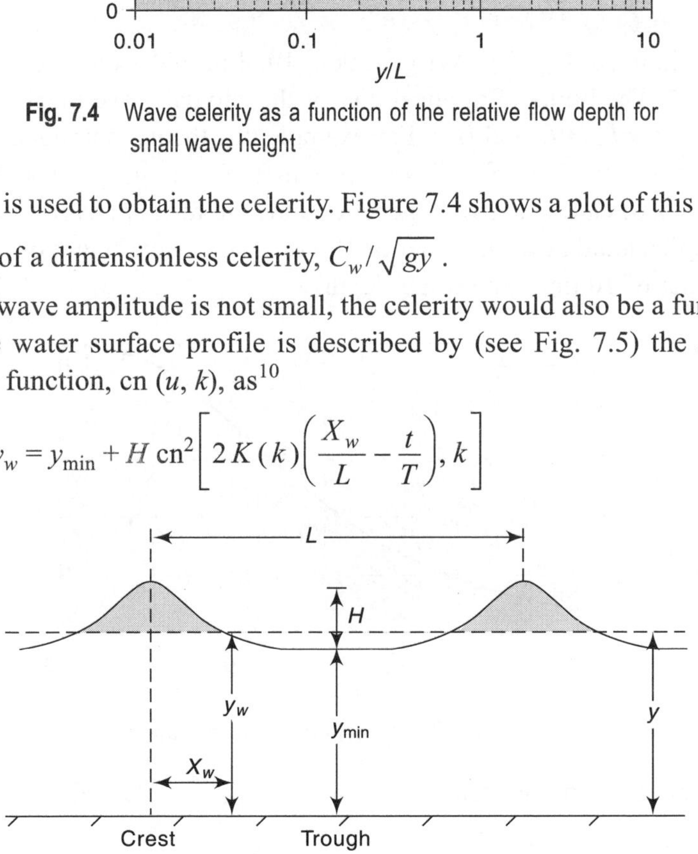

Fig. 7.5 Definition sketch for a cnoidal wave

---

[10]The function $\operatorname{cn}(u, k)$ is defined by the implicit relation $u = \int_0^{\cos^{-1}[\operatorname{cn}(u,k)]} d\theta/\sqrt{1 - k^2 \sin^2\theta}$. The variable k, which lies between 0 and 1, is called the *elliptic modulus* and the upper limit of the integral is called the *Jacobi amplitude*. The function $\operatorname{cn}(u, k)$ is a solution of the differential equation $(d^2\xi/du^2) + (1 + k^2)\xi - 2k^2\xi^3 = 0$ and reduces to the regular cosine function for $k = 0$.

where k is the elliptic modulus, and $K(k)$ is the complete elliptic integral of the first kind.[11] Korteweg and de Vries, who first studied these waves in 1890s, called these *cnoidal* waves.

For a given wave height, H, wave length, L, and mean flow depth, y, the dimensionless number, HL^2/y^3, is called the *Ursell number*, U_r, and the value of k is related to this number through the following equation:

$$kK(k) = \sqrt{\frac{3\,HL^2}{16\,y^3}} \tag{7.8}$$

which is shown in Fig. 7.6 (Wiegel 1960, 1964 provides plots for other variables also). The limits of applicability of the cnoidal wave theory are based on the ratios y/L, H/L, and H/y. For example, the theory is applicable for y/L less than about 0.15 and for U_r greater than 25 (for larger values of y/L, Eqs (7.5) and (7.6) can be used). A general guideline for applicability is that the wavelength should be of the order of 100 times the wave height or larger, and of the order of 10 times the water depth.

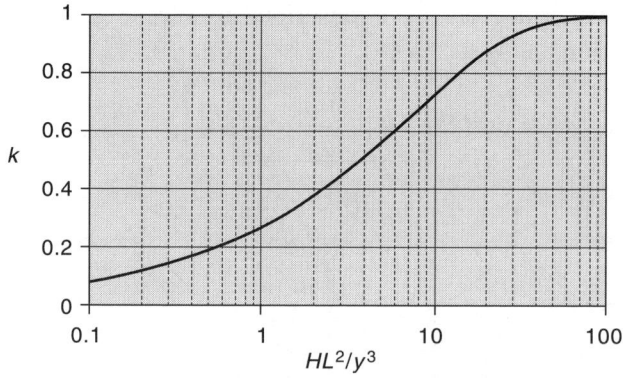

Fig. 7.6 Elliptic modulus of a cnoidal wave

The flow depth at the trough of the wave, y_{min}, is obtained from

$$y_{min} = y + \frac{16\,y^3}{3L^2}K(k)\big[K(k)-E(k)\big]-H \tag{7.9}$$

where $E(k)$ is the complete elliptic integral of the second kind. The celerity of a cnoidal wave can be expressed in terms of the depth at the trough or the mean flow depth as

---

[11]The complete elliptic integrals of first kind, $K(k)$, and second kind, $E(k)$, are defined by

$$K(k) = \int_0^{\pi/2} \frac{d\theta}{\sqrt{1-k^2\sin^2\theta}} \quad \text{and} \quad E(k) = \int_0^{\pi/2} \sqrt{1-k^2\sin^2\theta}\;d\theta$$

$$C_w = \sqrt{g\,y_{\min}}\left[1 + \frac{H}{y_{\min}\,k^2}\left(\frac{1}{2} - \frac{E(k)}{K(k)}\right)\right] \tag{7.10a}$$

or
$$C_w^2 = g\,y\left\{1 + \frac{H}{y}\left[-1 + \frac{1}{k^2}\left(2 - 3\frac{E(k)}{K(k)}\right)\right]\right\} \tag{7.10b}$$

The elliptic integrals, K and E, are readily available in a number of references (e.g., Abramowitz and Stegun 1964, available on the internet at *http://www.math.sfu.ca/~cbm/aands/toc.htm*, which also provides tabular values and approximate expressions for these) and websites (e.g., *http://mathworld.wolfram.com/topics/EllipticIntegrals.html*). These elliptic integrals are shown in Fig. 7.7.

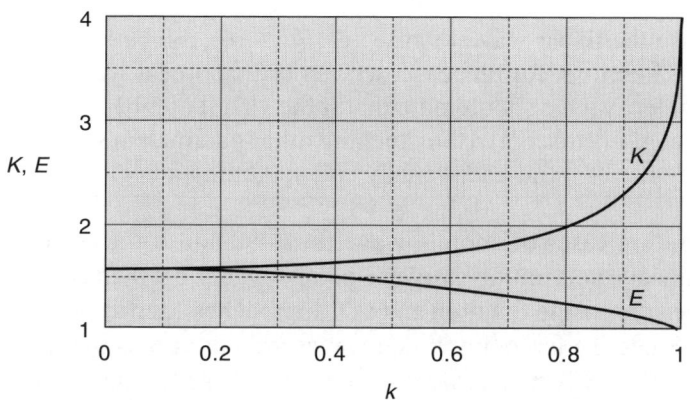

Fig. 7.7 Complete elliptic integrals of the first and second kinds

The following approximations (Hastings 1955) are quite useful in the computations of these elliptic integrals:

$$K(k) = 1.38629 + 0.1119723k_1 + 0.0725296k_1^2 - (0.5 + 0.1213478k_1 + 0.0288729k_1^2)\ln k_1$$

$$E(k) = 1 + 0.4630151k_1 + 0.1077812k_1^2 - (0.2452727k_1 + 0.0412496k_1^2)\ln k_1$$

in which $k_1 = 1 - k^2$. Note that for very small k, both $K(k)$ and $E(k)$ approach a value of $\pi/2$ and for k approaching 1, $K(k)$ becomes very large and $E(k)$ approaches 1.

Once the elliptic modulus is obtained from Fig. 7.6, and the flow depth at the trough is obtained using Eq. (7.9) and Fig. 7.7, the celerity is readily computed using Eq. (7.10a). The alternative equation for celerity, Eq. (7.10b) does not require the computation of the trough depth and provides identical results. The following example demonstrates these steps.

Example 7.1 The wave generated in a 5 m deep pool of water was found to have a wavelength of 50 m and a wave height of 1 m. Find the celerity of the wave using the cnoidal wave theory. What would be the celerity if we use the linear wave theory?

Solution
With the given data, $y = 5$ m, $L = 50$ m, and $H = 1$ m, we get $HL^2/y^3 = 20$. The Ursell number is less than 25 and y/L is 0.1, so we may be able to use the cnoidal theory. From Fig. 7.5, we get the elliptic modulus, k, as 0.88. The corresponding values of the elliptic integrals may be read from Fig. 7.6. To get more accurate values, however, we use the approximations listed above, and obtain $K(k) = 2.203$ and $E(k) = 1.195$.

The depth at the trough, y_{min}, is obtained from Eq. (7.9) as 4.59 m and the celerity obtained from Eq. (7.10a) is **6.63 m/s** [using Eq. (7.10b), we get the same value].

If we use the linear wave theory, i.e., Eq. (7.6), the celerity is obtained as **6.59 m/s**. Therefore, for this case, we can use the linear wave theory to estimate the celerity. One should note that if the Ursell number is smaller (e.g., if the wavelength is taken as 10 m keeping other parameters same), the cnoidal theory may not be valid and we may end up getting a negative celerity.

The limiting cases of cnoidal wave for $k = 0$ and $k = 1$ correspond to the sinusoidal wave and solitary wave respectively. For $k = 0$, the elliptic function cn becomes the cosine function and as k approaches 1, cn approaches the sech function. Since the large-amplitude waves are not very common in practical open channel flow scenario, we do not discuss these further. Interested reader may refer to USACE (2007).

We now look at another unsteady flow situation that can be converted into steady state. However, instead of a gradually varied flow condition, we have a rapidly varied flow with significant energy loss, necessitating the use of the momentum equation.

7.2.3 Surge

Figure 7.8 shows the formation of waves when a sluice gate is suddenly raised or lowered. When the gate is raised (Fig. 7.8a), it generates, upstream of the gate, a negative wave that moves upstream and, downstream of the gate, a positive wave that travels downstream. On the other hand, lowering of the gate [Fig. 7.8(b)] leads to formation of a positive wave (moving upstream) on the upstream side and a negative wave (moving downstream) downstream of the gate. If the opening or closing of the gate takes place over a very short interval, the wave amplitude becomes very large. Due to the large amplitude, these waves are commonly called *surges*. A consequence of the large change

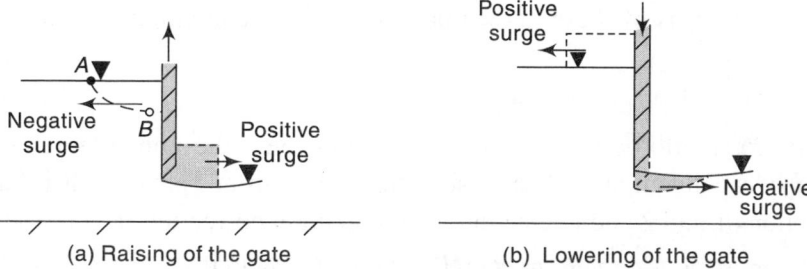

(a) Raising of the gate   (b) Lowering of the gate

Fig. 7.8 Positive and negative surges at a sluice gate

in depth is that different flow depths are associated with different particles on the wavefront. As discussed in the previous subsection, the celerity of these *particles* would increase as the flow depth increases. Thus the point A on the waves [Fig. 7.8(a)] will have a higher celerity than the point B. It implies that a negative wave will have a continuous distortion of its shape but a positive wave would maintain an almost vertical front (more discussion follows in a subsequent section). The positive wave can, therefore, be analysed by converting it into a steady-state equivalent.

Positive surge

Figure 7.9(a) shows a positive surge travelling downstream with velocity V_w and Fig. 7.9(b) shows the equivalent steady flow. Due to the large (and unknown) energy loss at the surge, the energy equation would not be helpful. The continuity and momentum equations are applied as discussed below.

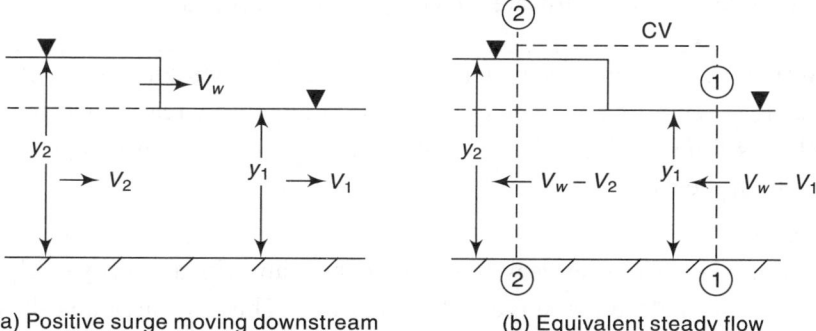

(a) Positive surge moving downstream   (b) Equivalent steady flow

Fig. 7.9 Analysis of a positive surge moving downstream

Application of continuity equation   To maintain consistency of notation with the hydraulic jump, we take section 1 at the lower depth of flow (i.e., before the arrival of the surge) and 2 at the higher depth (after the surge). Note that generally the flow conditions at section 1 would be known (since it corresponds to the undisturbed flow) and we would want to express the wave velocity in terms of these known parameters.[12] The continuity equation results in

$$(V_w - V_1)A_1 = (V_w - V_2)A_2 = Q_r \tag{7.11}$$

Application of momentum equation Application of the momentum equation in the −ive X-direction, with the assumption of hydrostatic pressure distribution at sections 1 and 2 and neglecting the momentum correction factor, leads to

$$\rho g A_1 \bar{y}_1 - \rho g A_2 \bar{y}_2 = \rho Q_r[(V_w - V_2) - (V_w - V_1)] \tag{7.12}$$

where, as before, $\bar{y}$ is the depth of the centroid of the area from the water surface (for a rectangular channel, it would be half the flow depth). Using Eq. (7.11) and writing in terms of celerity (recall that $C_w = V_w - V_1$), we get

$$C_w^2 A_1 \left(1 - \frac{A_1}{A_2}\right) = g\,(A_2\,\bar{y}_2 - A_1\,\bar{y}_1)$$

$$\Rightarrow \quad C_w = \pm\sqrt{g\,\frac{A_2}{A_1}\,\frac{A_2\,\bar{y}_2 - A_1\,\bar{y}_1}{A_2 - A_1}} \tag{7.13}$$

For this case, since the wave is travelling downstream, we will use the +ive sign. For a surge moving upstream, the −ive sign has to be used. Equation (7.13) is useful when the water depth after the surge, y_2, is measured. After obtaining C_w (and, therefore, V_w) from Eq. (7.13), Eq. (7.11) can be used to obtain the flow velocity, V_2. However, if the wave velocity, V_w, or the post-surge velocity, V_2, is known, the solution in general would involve a trial procedure since Eq. (7.13) involves nonlinear function of the post-surge depth, y_2. Examples 7.2–7.5 illustrate the computational procedure.

Example 7.2 A 2 m wide rectangular channel[13] carries a discharge of 1 m³/s at a flow depth of 1 m. The discharge is suddenly increased and the flow depth was found to increase to 1.4 m. Find the increased discharge and the velocity of the surge.

Solution
In this case, a positive surge will travel downstream with a celerity of C_w. The flow velocity before the surge is $V_1 = 0.5$ m/s. The flow area and depth to centroid at the two sections are:

$$A_1 = By_1 = 2 \text{ m}^2, \quad \bar{y}_1 = 0.5 \text{ m}, \quad A_2 = 2.8 \text{ m}^2, \quad \bar{y}_2 = 0.7 \text{ m}$$

---

[12]Since there are three unknowns (V_2, y_2, and V_w) and only two equations (continuity and momentum), one more parameter out of these three needs to be specified. For example, if we close a gate completely, V_2 would be equal to zero. Sometimes, the water depth, y_2, or the wave velocity, V_w, is measured.

[13]Although we do not explicitly mention, we assume that the channel is horizontal and frictionless.

From Eq. (7.13), therefore, $C_w = +4.06$ m/s and the velocity of the surge is

$$V_w = V_1 + C_w = \textbf{4.56 m/s}$$

From Eq. (7.11), the post-surge velocity is $V_2 = 1.66$ m/s. The increased discharge is, therefore, $Q_2 = BV_2y_2 = \textbf{4.65 m}^3\textbf{/s}$.

Example 7.3   A 2 m wide rectangular channel carries a discharge of 1 m³/s at a flow depth of 1 m. A sluice gate located in the channel is suddenly lowered and it is desired to produce a 0.4 m high surge upstream of the gate. Find the velocity of the surge and the flow velocity at a section after the surge has passed.

Solution
In this case, a positive surge will travel upstream and the celerity C_w will be negative. The flow velocity before the surge is $V_1 = 0.5$ m/s. The flow area and depth to centroid at the two sections are:

$$A_1 = By_1 = 2 \text{ m}^2, \quad \bar{y}_1 = 0.5 \text{ m}, \quad A_2 = 2.8 \text{ m}^2, \quad \bar{y}_2 = 0.7 \text{ m}$$

From Eq. (7.13), therefore, $C_w = -4.06$ m/s and the velocity of the surge is

$$V_w = V_1 + C_w = \textbf{--3.56 m/s}$$

From Eq. (7.11), the post-surge velocity is $V_2 = -0.66$ m/s. The negative velocity is not meaningful. This indicates that a 0.4 m high surge cannot be produced with the given flow conditions. (The next example clarifies this point.)

Example 7.4   A 2 m wide rectangular channel carries a discharge of 1 m³/s at a flow depth of 1 m. A sluice gate located in the channel is suddenly and completely closed. Find the velocity of the resulting surge and the flow depth at a section after the surge has passed.

Solution
In this case, a positive surge will travel upstream and the celerity C_w will be negative. The flow velocity before the surge is $V_1 = 0.5$ m/s and the velocity after the surge has passed is $V_2 = 0$ m/s since the gate is completely closed. Let the depth of flow after the surge has passed a section be y_2. The continuity equation, Eq. (7.11), is written as

$$(V_w - 0.5)\, 2 \times 1 = V_w \times 2 \times y_2 \quad \Rightarrow \quad V_w = -\frac{1}{2\,y_2 - 2}$$

and Eq. (7.13) is written as

$$V_w - 0.5 = -\sqrt{g\,\frac{y_2}{1}\,\frac{y_2 + 1}{2}}$$

Note that a negative sign has been used since the wave is moving upstream. Substituting V_w in terms of y_2 from the continuity equation, we get a nonlinear equation in y_2, which can be readily solved to obtain $y_2 = \textbf{1.166 m}$. The wave velocity is then obtained as **–3.01 m/s** and the celerity of the surge is –3.51 m/s. Thus the maximum surge height that can be created by lowering the gate is 0.166 m and the data given in the previous example with a surge height of 0.4 m was not realistic.

Example 7.5　A 2 m wide rectangular channel carries a discharge of 1 m³/s at a flow depth of 1 m. A sluice gate located in the channel is suddenly lowered and a surge was observed travelling upstream with a velocity (relative to the bank) of 3 m/s. Find the height of the surge and the flow velocity at a section after the surge has passed.

Solution

In this case, since the wave velocity is known ($V_w = -3$ m/s), Eq. (7.13) results in a quadratic equation in the flow depth, y_2, which can be directly solved to obtain the depth. The equation is written as

$$-3 - 0.5 = -\sqrt{g \frac{y_2}{1} \frac{y_2 + 1}{2}} \quad \Rightarrow \quad y_2 = \frac{\sqrt{1 + 4\dfrac{2 \times 3.5^2}{g}} - 1}{2} = 1.158 \text{ m}$$

The flow velocity after the passage of the surge is obtained from Eq. (7.11) as

$$V_2 = V_w - \frac{A_1}{A_2}(V_w - V_1) = -3 - \frac{1}{1.158}(-3 - 0.5) \text{ m/s} = 0.022 \text{ m/s}$$

which is very close to zero, as it should since the wave velocity is very close to that obtained at complete closure in the previous example. Of course, a wave velocity higher than 3.01 m/s cannot be attained under these conditions.

Negative surge

As discussed earlier, in a negative surge, different portions travel at different velocities and it is not possible to obtain an equivalent steady state. However, taking a small strip on the wave (Fig. 7.10) and considering a rectangular channel, we may write the continuity and momentum equations (in the –*X*-direction) as

$$(V_w - V)y = \left(V_w - V - \frac{dV}{dy}\delta y\right)(y + \delta y)$$

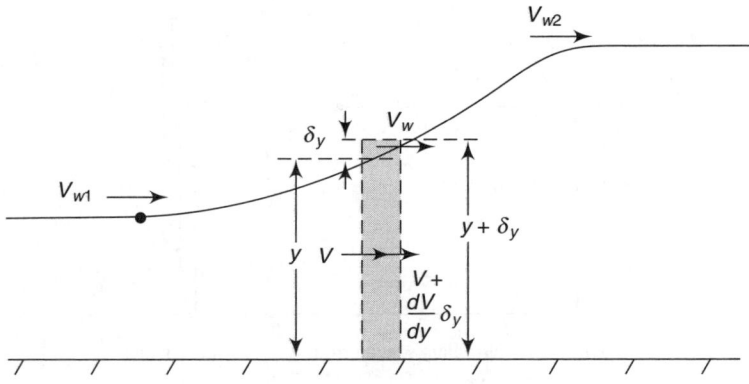

Fig. 7.10 Negative wave in a channel

$$\Rightarrow \quad \frac{dV}{dy} = \frac{V_w - V}{y} \tag{7.14}$$

$$\rho g \frac{(y + \delta y)^2}{2} - \rho g \frac{y^2}{2}$$

$$= \rho (V_w - V) y \left[(V_w - V) - \left(V_w - V - \frac{dV}{dy} \delta y \right) \right]$$

$$\Rightarrow \quad \frac{dV}{dy} = \frac{g}{V_w - V} \tag{7.15}$$

in which V_w represents the velocity of wave at a location where the water depth is y. From these two equations, we get the celerity of the wave at a water depth of y as $C_w = V_w - V = \pm\sqrt{gy}$, the same as that obtained earlier for a small-amplitude wave [Eq. (7.4) with $a = 0$ and $D = y$].

The differential equation for the negative wave is then written, using either Eq. (7.14) or Eq. (7.15), as

$$\frac{dV}{dy} = \pm \sqrt{\frac{g}{y}} \tag{7.16}$$

Integration of this equation, with the undisturbed conditions providing the boundary condition, enables us to obtain the wave velocity and profile. For example, if flow is taking place in a rectangular channel (with a sluice gate at its downstream end) at a flow depth of y_0 and velocity V_0, and the gate is suddenly raised, a negative wave will travel upstream from the gate (Fig. 7.11).

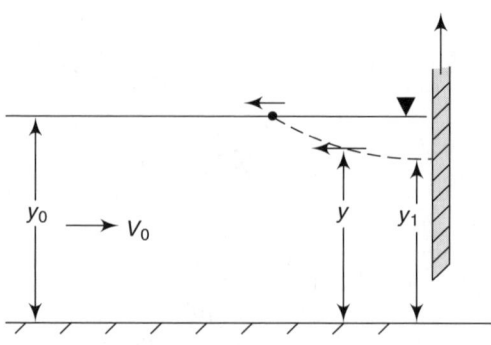

Fig. 7.11   Negative wave created by raising a sluice gate

Integration of Eq. (7.16), with the –ive sign since the wave is travelling upstream, results in

$$V = -2\sqrt{gy} + C_I \qquad (7.17)$$

where C_I is a constant of integration, which is obtained from the undisturbed condition, $V = V_0$ at $y = y_0$. The flow velocity corresponding to a water depth of y is, therefore, obtained as

$$V = V_0 + 2\sqrt{gy_0} - 2\sqrt{gy} \qquad (7.18)$$

and the wave velocity (with respect to the banks) is obtained as

$$V_w = V + C_w = V_0 + 2\sqrt{gy_0} - 3\sqrt{gy} \qquad (7.19)$$

where C_w is taken as $-\sqrt{gy}$, since the wave is moving upstream. Note that V_w is negative at $y = y_0$ since the undisturbed flow upstream of the gate is subcritical ($V_0 < \sqrt{gy_0}$) and its magnitude decreases as y becomes smaller. Thus the top of the wave moves farther upstream compared to lower portions, resulting in a continuous lengthening of the wave. If we assume that the gate is raised to such an extent that the post-surge depth, y_1, keeps V_w negative [i.e., y_1 is more than $\left(V_0 + 2\sqrt{gy_0}\right)^2/9g$, see Eq. (7.19)], the length of the wave at any time would be given by $(V_0 + 2\sqrt{gy_0} - 3\sqrt{gy_1} - V_0$ $- 2\sqrt{gy_0} + 3\sqrt{gy_0}) = 3t\left(\sqrt{gy_0} - \sqrt{gy_1}\right)$. A useful application of the negative wave analysis is in the *dam-break problem*, in which water impounded to a depth y_0 behind a dam is released due to failure of the dam. A negative wave is generated upstream of the dam and a positive wave downstream of it. This case is described in more details in Section 7.4. Example 7.6 shows the analysis of a negative wave at a sluice gate.

Example 7.6 A 2 m wide rectangular channel carries a discharge of 1 m³/s at a flow depth of 1 m. A sluice gate located in the channel is suddenly raised such that the depth of flow just upstream of the gate becomes 0.7 m. Find the flow velocity after the passage of the resulting negative surge and also the wave velocity at the topmost and bottommost points of the surge.

Solution

The undisturbed flow characteristics are defined by $y_0 = 1$ m and $V_0 = 0.5$ m/s. Using Eq. (7.19), we find that the velocity of the topmost negative wave (moving at the depth of 1 m) is

$$V_w = 0.5 + 2\sqrt{g} - 3\sqrt{g} = -2.63 \text{ m/s}$$

and the velocity of the bottommost front (moving at the depth of 0.7 m) is

$$V_w = 0.5 + 2\sqrt{g} - 3\sqrt{0.7g} = -1.10 \text{ m/s}$$

The flow velocity after the passage of the negative wave, i.e., when the depth has become 0.7 m, is obtained from Eq. (7.18) as

$$V = V_0 + 2\sqrt{gy_0} - 2\sqrt{0.7g} = 1.52 \text{ m/s}$$

The discharge has risen to 2.13 m³/s due to the raising of the gate from its initial value of 1 m³/s. Similar to Example 7.4, we note that there is a limit to the decrease in water level on raising the gate. This limit corresponds to the complete opening of the gate and can be obtained by equating the wave velocity to zero as

$$V_w = 0.5 + 2\sqrt{g} - 3\sqrt{gy} = 0 \quad \Rightarrow \quad y = 0.52 \text{ m}$$

which results in a flow velocity of 2.25 m/s and the maximum possible discharge of 2.34 m³/s.

In this section, we have looked at some unsteady flow situations that can be converted into their steady-state equivalents. We have seen that analysis of these require only the continuity, momentum, and energy equations already described in previous chapters for steady flow conditions. The theories described in this chapter till now have been based on a number of simplifying assumptions, the most notable being that of frictionless horizontal channel. Presence of the bed slope and friction slope terms in the governing equations makes the analysis quite complicated and analytical solutions become almost impossible. Numerical solution then becomes the method of choice. In the next section, we will derive the basic equations (continuity and momentum), as applied to unsteady flow situations, which will then be used to analyse cases of channels with sloping bed in presence of friction forces.

7.3 Continuity and Momentum Equations

For unsteady flows, the variation in flow characteristics with time has to be accounted for in the mass conservation and momentum equations. This implies that both these equations will have additional terms involving temporal derivatives as derived next.

7.3.1 Continuity Equation

Figure 7.12 shows the longitudinal section of an open channel along with a control volume (CV). As before, we perform a one-dimensional analysis by using the cross-sectional average velocity. The flow is unsteady implying that the water surface at two different times may be different. The continuity equation under these conditions states that the rate of change of volume of water within the CV would be equal to the net inflow rate over the control surface:

$$\underbrace{Q + q^* \Delta X}_{\substack{\text{Inflow rate} \\ \text{into CV}}} - \underbrace{\left[Q + \frac{\partial Q}{\partial X} \Delta X \right]}_{\substack{\text{Outflow rate} \\ \text{from CV}}} = \underbrace{\frac{\partial (A \Delta X)}{\partial t}}_{\substack{\text{Rate of change} \\ \text{within CV}}}$$

$$\Rightarrow \quad \frac{\partial Q}{\partial X} + \frac{\partial A}{\partial t} - q^* = 0 \tag{7.20}$$

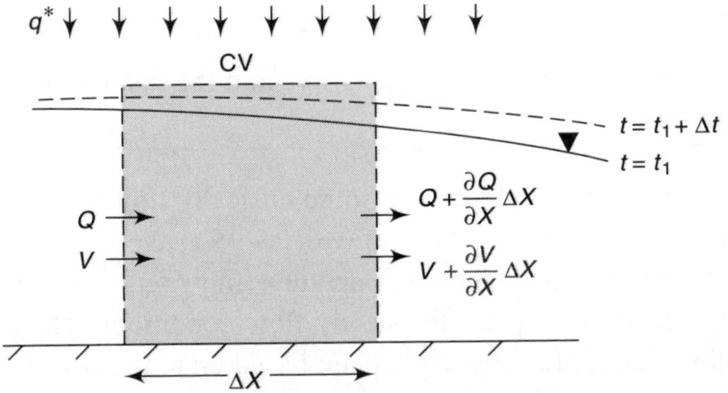

Fig. 7.12 Longitudinal section of a channel with unsteady flow

where q^* is the lateral inflow rate. Generally one uses the flow depth and velocity as the dependent variables (and not the discharge). Therefore, the continuity equation is written as

$$A \frac{\partial V}{\partial X} + V \frac{\partial A}{\partial X} + \frac{\partial A}{\partial t} - q^* = 0 \tag{7.21}$$

If we assume that there is no lateral inflow or outflow ($q* = 0$) and the channel is prismatic[14] $\left(\dfrac{\partial A}{\partial X} = T \dfrac{\partial y}{\partial X} \right)$, the above equation becomes

$$D\frac{\partial V}{\partial X} + V\frac{\partial y}{\partial X} + \frac{\partial y}{\partial t} = 0 \tag{7.22}$$

where D is the hydraulic depth ($= A/T$).

7.3.2 Momentum Equation

Application of the X-momentum equation to the CV shown in Fig. 7.12 requires information about the forces acting on the CV, the net efflux of momentum across the control surface (CS), and (the additional information in unsteady flow) *the rate of change of the X-momentum within the* CV. We assume that the air resistance is negligible, the curvature of streamlines is small, and the average shear stress at the boundary is given by an expression similar to that used for uniform flow with the bed slope replaced by the friction slope, S_f. Once all the forces are known, the net force in the X-direction can be computed and set equal to the net momentum efflux plus the rate of change of momentum within the CV. If we assume that there is no lateral flow and take the momentum correction factor as unity, this momentum term is written as

$$\begin{aligned}
M_X &= \rho\left[QV + \frac{\partial(QV)}{\partial X}\Delta X \right] - \rho QV + \frac{(\rho A \Delta X V)}{\partial t} \\
&= \rho\Delta X\left[\frac{\partial(QV)}{\partial X} + \frac{\partial Q}{\partial t} \right] \\
&= \rho\Delta X\left[Q\frac{\partial V}{\partial X} + A\frac{\partial V}{\partial t} + V\left(\frac{\partial Q}{\partial X} + \frac{\partial A}{\partial t} \right) \right]
\end{aligned} \tag{7.23}$$

From Eq. (7.20), the term in the last bracket is zero if there is no lateral inflow or outflow. (In presence of lateral flow of $q*$, this term would be equal to $q*$.) Application of the momentum equation to the CV, assuming the bed slope to be small, results in

---

[14]For nonprismatic channels, A would be a function of X, y, and t. Hence, $\partial A/\partial X$, holding the time constant, would comprise two terms, $(\partial A/\partial X)_{y=\text{constant}}$ and $(\partial A/\partial y)_{X=\text{constant}}$ $(\partial y/\partial X)$. The first term would depend on the variation of cross-section shape at different locations and the second term is equal to $T(\partial y/\partial X)$.

$$\rho g A \Delta X S_0 + \rho g A \,\bar{y} - \rho g \left[A\bar{y} + \frac{\partial (A\bar{y})}{\partial X} \Delta X \right] - \rho g A \Delta X S_f$$

$$= \rho \Delta X \left[AV \frac{\partial V}{\partial X} + A \frac{\partial V}{\partial t} \right] \tag{7.24}$$

which may be simplified as[15] [using the relationship derived earlier, $\partial(A\,\bar{y})/\partial y = A$]

$$\frac{\partial y}{\partial X} = S_0 - S_f - \frac{1}{g} \left[V \frac{\partial V}{\partial X} + \frac{\partial V}{\partial t} \right] \tag{7.25}$$

The continuity equation, Eq. (7.22), and the momentum equation [or the equation of motion, Eq. (7.25)] are known as the Saint-Venant's equations to honour the French engineer A.J.C. Barré de Saint-Venant, who first developed these in late 1800s. Obviously, for steady-state flows, the last term in both these equations drop out and we get the GVF equations. One may also note that the last term in Eq. (7.25) represents the ratio of the total acceleration, including the convective acceleration $V(\partial V/\partial X)$ and the local acceleration $\partial V/\partial t$, to the gravitational acceleration. For commonly encountered open channel flows, the friction slope (S_f) is of the same order as the bed slope (S_0), the hydrostatic pressure term ($\partial y/\partial X$) is quite small compared to the bed slope, and the acceleration terms are even smaller.

The Saint-Venant's equations involve a set of two partial differential equations in two unknowns—flow depth and velocity—which have to be solved with given initial and boundary conditions. (Alternative formulations that use the two dependent variables as depth and discharge, celerity and discharge, or celerity and velocity, may be easily derived.) Barring a few simple cases, analytical solutions are not possible (note that the friction slope is a nonlinear function of flow depth). The analytical or numerical solution techniques may be applied directly to partial differential equations (PDEs). However, similar to the technique used for converting an unsteady flow problem into a steady flow problem, the method of characteristics[16] can be used to convert these into "easier to solve" ordinary differential equations (ODEs) as described below.

7.3.3 Conversion of PDEs into ODEs

As seen earlier, an unsteady flow involving a disturbance moving with a con-

---

[15]Note that we have assumed hydrostatic pressure distribution in writing the forces [Eq. (7.24)]. Therefore, this equation is valid for *gradually varied* unsteady flow only.

[16]The characteristics (sometimes called characteristic curves) of a partial differential equation are the lines (or planes) along which it becomes an ordinary differential equation. The concept was introduced by Riemann, and the characteristic curve represents the direction in which discontinuities propagate.

stant speed along (or opposite to) the flow direction is converted into an equivalent steady flow by moving along the disturbance with a velocity equal to that of the disturbance. In a general case, the disturbance may move in both directions[17] with a velocity (with respect to the banks) of $C + V$ in the downstream direction, say X, and $C - V$ in the upstream direction, $-X$ (we have dropped the subscript w from the wave celerity and write C instead of C_w from here onwards). In other words, if we move with velocities of $V \pm C$, the partial derivatives with respect to space and time can be replaced by the total derivative with respect to time. This is illustrated mathematically below.

Multiplying Eq. (7.25) by λ and adding to Eq. (7.22), we get

$$D\frac{\partial V}{\partial X} + V\frac{\partial y}{\partial X} + \frac{\partial y}{\partial t} + \lambda\left[\frac{\partial y}{\partial X} - S_0 + S_f + \frac{1}{g}\left(V\frac{\partial V}{\partial X} + \frac{\partial V}{\partial t}\right)\right] = 0 \quad (7.26)$$

which may be written as

$$\frac{g}{\lambda}\left[\frac{\partial y}{\partial t} + (V + \lambda)\frac{\partial y}{\partial X}\right] + \left[\frac{\partial V}{\partial t} + \left(V + \frac{gD}{\lambda}\right)\frac{\partial V}{\partial X}\right] = g(S_0 - S_f) \quad (7.27)$$

In order that the quantities in square brackets become total derivatives, $V + \lambda = V + (gD/\lambda) = dX/dt$, which implies $\lambda = \pm\sqrt{gD}$, which is the wave celerity, C. Thus along the lines

$$\frac{dX}{dt} = V \pm C \quad (7.28)$$

the Saint-Venant's equations get converted into ordinary differential equations

$$\pm\frac{g}{C}\frac{Dy}{Dt} + \frac{DV}{Dt} = g(S_0 - S_f) \quad (7.29)$$

which can be solved using either analytical (for simple cases) or numerical methods. To avoid using both y and C (which is a function of y), we assume that the hydraulic depth is linearly related to the flow depth[18] as $D = y/\eta$ and modify Eq. (7.29) as

$$\frac{D(V \pm 2\eta C)}{Dt} = g(S_0 - S_f) \quad (7.30)$$

---

[17]For supercritical flow, the disturbance moves in the downstream direction only.

[18]For most channel shapes, it would be true. For example, rectangular channels have $D = y$; triangular, $D = y/2$; parabolic $D = 2y/3$. For channels that do not fall under this category (e.g., trapezoidal, circular), we may define a new variable as $C^* = \int_0^y \sqrt{g/d}\ dy$. The reader may derive the equivalent of Eq. (7.30) with this variable and verify its applicability for rectangular channels.

For rectangular channels, $\eta = 1$, and Eq. (7.30) states that the change in the quantities $(V \pm 2C)$ along the characteristic $dX/dt = V \pm C$ is equal to the difference of the bed slope and the friction slope, multiplied by the gravitational acceleration. For cases where these slopes can be ignored (e.g., a horizontal frictionless channel), the quantities $(V \pm 2C)$ are conserved along the corresponding characteristic and are known as the *Riemann invariants*. The two characteristics are distinguished by calling them the positive (or forward) characteristic $(dX/dt = V + C)$ and the negative (or backward) characteristic $(dX/dt = V - C)$ and are represented by C^+ and C^-, respectively. As is obvious from these equations, C^+ lines will have a positive slope[19] for both subcritical and supercritical flows (note that we assume the X-axis to be positive in the flow direction) and C^- lines will have a positive slope for supercritical flow ($V > C$) and a negative slope for subcritical flow ($V < C$). Under critical flow conditions, the slope of the characteristic lines would be zero [a zero slope ($dX/dt = 0$) indicates a vertical line in the X-t plane]. It again points to the fact mentioned earlier that a disturbance can travel both upstream and downstream in a subcritical flow but only downstream in a supercritical flow. Figure 7.13 shows a few typical characteristics for subcritical and supercritical flows, with the flow taking place in the + ive X-direction.

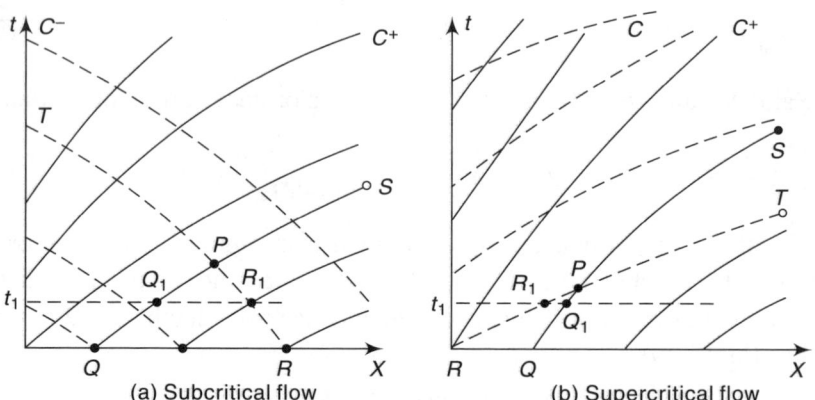

(a) Subcritical flow   (b) Supercritical flow

Fig. 7.13 Characteristic lines for subcritical and supercritical flows

From this figure, it is seen that for any point, $P(X, t)$, in the flow domain, there would be two characteristics (QP and RP) meeting there and any disturbance outside the region PQR would not influence the conditions at P. In

---

[19]The term *slope* here indicates dX/dt. However, as we will see a little later, the usual method of analysing the flow involves a plot with distance, X, on the horizontal axis and the time, t, on the vertical axis. Therefore, we should keep in mind that the slope of the characteristics is the inverse of the usual connotation of the slope of a line in a x-y graph.

other words, at initial time ($t = 0$) only the changes in flow conditions occurring between the length QR would affect the flow properties at P. Similarly, at time t_1, only disturbances created in the channel between Q_1R_1 would affect P. The region PQR is, therefore, called the *domain of dependence* of P indicating that the conditions at P depend only on conditions within this region. In an analogous manner, there is the *region of influence* of P, which is bounded by the characteristics originating at P (i.e, extension of the characteristics QP and QR, unless a change at P changes the direction of the characteristic lines), PST and the (unbounded) region beyond it. Any change at P would only influence the flow conditions in this region. It is also apparent from these figures that, in addition to the initial conditions, the two boundary conditions required for the solution of the problem are one at the upstream boundary and other at the downstream boundary for subcritical flow. For a supercritical flow, however, both boundary conditions have to be specified at the upstream boundary since both the C^+ and C^- characteristics have positive slope and the information is carried only in the downstream direction.

7.4 Method of Characteristics of Solving PDEs

In the previous section, we have looked at the derivation of the continuity and momentum equations applicable to unsteady flows. Since the partial differential equations are more difficult to solve compared to ordinary differential equations, these were converted into ordinary differential equations. In this section, we will discuss the method of solving the equations obtained using the method of characteristics. We will first consider some simple cases, for which analytical or graphical solutions are obtainable and then look at the numerical techniques.

7.4.1 Graphical Solution

We first discuss simple cases that can be solved analytically, e.g., a horizontal frictionless channel. A graphical description is provided as it helps in visualizing the flow. The aim of the solution is to determine the flow velocity and celerity in the entire X-t domain since it will enable us to obtain the flow depth and the discharge at every location and at all times. The following example illustrates the method for a negative wave generated upstream of a sluice gate by raising the gate.

Example 7.7 A 2 m wide rectangular channel carries a discharge of 1 m³/s at a flow depth of 1 m. A sluice gate located in the channel is suddenly raised such that the depth of flow just upstream of the gate becomes 0.7 m. Find the flow velocity after the passage of the resulting negative surge using the method of characteristics.

Solution

This problem has been solved earlier in Example 7.6. Figure 7.14 shows the X-t plane. (Since we have assumed X to be positive in the flow direction, we take $X = 0$ at the gate and negative before it. Some authors find it more convenient to plot X as positive in the upstream direction and use a negative velocity.)

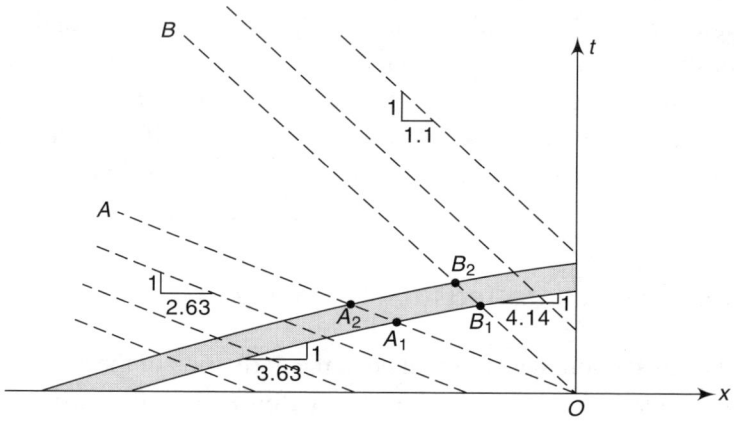

Fig. 7.14   The X-t plane for Example 7.7

At the gate ($X = 0$), the velocity before the raising of the gate is 0.5 m/s and the flow depth is 1 m. Denoting these undisturbed conditions with subscript 0, we have

$$V_0 = 0.5 \text{ m/s}, \quad C_0 = \sqrt{9.81 \times 1} \text{ m/s} = 3.13 \text{ m/s}$$

Thus the slope of the C^- lines would be -2.63 m/s and the slope of the C^+ lines would be 3.63 m/s. A few of these lines are shown in the figure. At $t = 0$ s, the depth decreases to 0.7 m, indicating a celerity of 2.62 m/s. But we cannot draw the C^- line emanating from the point ($X = 0$, $t = 0^+$) since the flow velocity is not known. Also, we do not know whether this line would be a straight line or not, i.e., whether $V - C$ along the C^- line from $(0, 0^+)$ would be constant or not. We make use of the following argument to show that the C^- lines are straight lines[20] and to obtain their slope:

---

[20]We have assumed the initial flow depth and velocity to be uniform. If it is not, the characteristics would not be straight since the velocity and celerity would be different from one location to another even in the undisturbed zone.

Let OB be the C^- line from $(0, 0^+)$,[21] which may or may not be a straight line. Since the region below OA is undisturbed (known as the zone of quiet or region of constant state), all C^+ and C^- lines would be straight lines (with slopes of 3.63 m/s and –2.63 m/s respectively). Let two of the C^+ lines cut the C^- line OA at A_1 and A_2 respectively and their extensions (which may or may not be straight) cut the characteristic OB at B_1 and B_2 respectively. Since A_1 and A_2 are in the undisturbed zone,[22] the velocity and celerity would be equal to V_0 and C_0 respectively. Also, since A_1B_1 and A_2B_2 are C^+ lines and $V + 2C$ is invariant, we have

$$V_{B_1} + 2C_{B_1} = V_{B_2} + 2C_{B_2} = V_0 + 2C_0$$

Moreover, since B_1 and B_2 are on the same C^- line, OB, and $V - 2C$ is invariant, we have

$$V_{B_1} - 2C_{B_1} = V_{B_2} - 2C_{B_2}$$

These two equations clearly show that $V_{B_1} = V_{B_2}$ and $C_{B_1} = C_{B_2}$, implying that $V - C$ would be constant along OB. Similar argument may be applied to all C^- lines indicating that all of them would be straight lines[23] and, along every one of these lines, $V + 2C = V_0 + 2C_0$.

Thus OB would be a straight line and since C is equal to 2.62 m/s along OB, the flow velocity is easily obtained as $V = V_0 + 2C_0 - 2C = 1.52$ m/s (the same as in Example 7.6). The slope of OB is $V - C = -1.1$ m/s while the slope of OA is –2.63 m/s, indicating that the length of the wave is increasing at a rate of 1.53 m/s. It is easy to see that the slope of the C^+ lines in the region above OB would be $V + C = 4.14$ m/s, which is different from the value of 3.63 m/s in the zone of quiet. This implies that A_1B_1 is NOT a straight line as its slope changes from 3.63 m/s at A_1 to 4.14 m/s at B_1. Beyond B_1, of course, the slope remains constant since everywhere above OB the depth is 0.7 m and the velocity is 1.52 m/s. It should also be mentioned that the specification of depth at the gate after it is raised meant that the celerity is known. Even if the

---

[21]We assume that OB is above OA since we know that a negative wave is created and its top moves at a higher speed than the bottom. Here OA represents the movement of the wave with the original water depth of 1 m and OB represents that with the reduced water depth of 0.7 m. Note that there would be an infinite number of characteristics originating from O (since the depth changes suddenly from 1 m to 0.7 m) spanning the region AOB.

[22]We assume that there is no sharp front (known to be true for the negative wave being discussed). We will shortly consider the case of a surge.

[23]The wave in which all positive characteristics (or all negative characteristics) are straight lines is known as a *simple wave*.

velocity is specified, we can obtain the depth using the invariance of $V + 2C$ along the C^+ line. However, if the discharge at the gate is specified, we would have to obtain V and C from the known values of $V + 2C$ ($= V_0 + 2C_0$) and VC^2/g ($= q$).

In Example 7.7, a negative wave occurs and the various disturbances at the gate, starting from the one at initial depth of 1 m and stopping at the final depth of 0.7 m, travel at increasingly smaller celerities and the C^- lines do not intersect one another. Let us now look at a positive surge where some C^- lines (or C^+ lines) may intersect one another. Intersection of two lines of the same family (positive or negative) of characteristics implies the availability of three equations—two from the intersecting lines of the same family and one from the opposite characteristic passing through that point—for the two unknowns, flow depth (or celerity) and velocity, and gives rise to discontinuous solutions.

We consider the flow in a channel in which the flow is initially uniform and then the depth at the upstream boundary is increased as a specified function of time. Figure 7.15 shows the X-t plane for this problem.

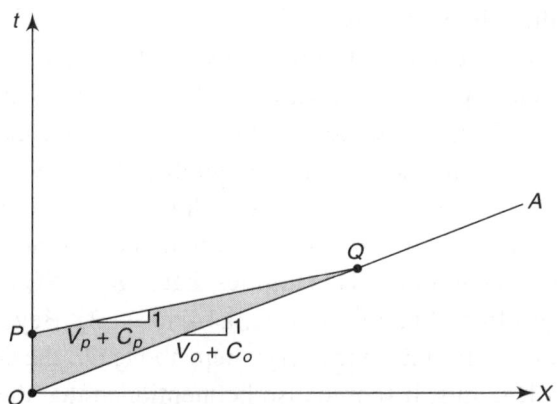

Fig. 7.15 The X-t plane for increasing depth at the upstream end

Let the initial velocity be V_0 and depth y_0 (i.e., celerity $C_0 = \sqrt{gy_0}$) and let OA be the first C^+ characteristic, which has $dX/dt = V_0 + C_0$ and along which $V_0 + 2C_0$ is constant. Since both $V + C$ and $V + 2C$ are constant along OA, it follows that both V and C are constant along OA (and are equal to V_0 and C_0 respectively). The point P which is at the upstream boundary but has a higher flow depth (since the depth is increasing with time) will have a higher celerity and the C^+ line from P intersects OA at Q. Knowing the celerity at P, C_P, we can compute the velocity (since P would lie on a C^- line originating in the zone of quiet) as $V_P = V_0 + 2C_0 - 2C_P$. As argued for OA, along PQ also the velocity and celerity should be constant (equal to V_P and C_P respectively). Since Q lies on both OA and PQ, there is a contradiction unless V and y (and, therefore, C) are discontinuous at this point. The method of characteristics, with its assumption of smooth and differentiable velocity and depth, thus breaks down. We can, of course, apply the method on either side of this shock and use an additional relationship between the depths (or velocities) at either side of the shock. However, we consider these to be too advanced for this book and do not discuss them. It is interesting to note, however, that an approximate solution may be obtained by assuming the method of characteristics to be valid. For example, in Example 7.4 we considered the sudden and complete closure of a sluice gate, which resulted in a positive surge. If we ignore the discontinuity, using the condition $V_P = V_0 + 2C_0 - 2C_P$, with $V_P = 0$, $V_0 = 0.5$ m/s and $C_0 = 3.13$ m/s, we get $C_P = 3.38$ m/s and a depth after the surge as 1.165 m, almost identical to the one obtained by the equivalent steady-state solution (and involving the solution of a nonlinear equation). We should be aware, however, that it is only an approximate solution.

If a positive wave is generated without the formation of a surge, we can still use the method of characteristics to solve the problem. One such example is the *dam-break problem*, in which water impounded behind a wall is suddenly released into a downstream dry bed. It causes a negative wave moving upstream on the upstream side of the wall and a positive wave moving downstream on the downstream side. If the downstream bed is dry, no surge would be formed and the method of characteristics can be used as shown below. This problem was first analysed by a German engineer, A. Ritter, in 1890s after failure of several large dams in the previous decades.

Figure 7.16 shows a wall (representing a dam) holding still water ($V_0 = 0$) at a depth of y_0 ($C_0 = \sqrt{gy_0}$) with no water on the downstream side. At time

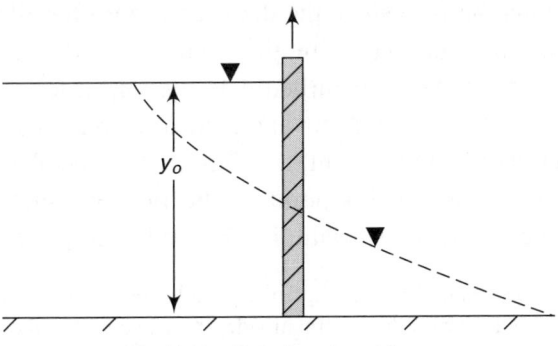

Fig. 7.16   Dam-break problem

$t = 0$, the wall is suddenly removed to simulate the failure of a dam. At the removal of the wall, a negative wave will start travelling upstream and a positive wave will move downstream. We again neglect the bed slope and the friction slope and assume that the channel is rectangular.

Figure 7.17 shows the X-t plane for this problem. Since the initial depth and velocity are uniform, the characteristics would be straight lines in the undisturbed zone (i.e., below OA), with the slope of the C^+ lines being C_0 and that of C^- lines being $-C_0$. As shown before, the fact that the C^+ lines originating in the undisturbed zone would intersect the C^- lines in the region beyond OA, implies that all C^- lines would be straight lines. Moreover, the invariance of $V + 2C$ along C^+ lines implies that $V + 2C = 2C_0$ in the entire X-t plane (recall that $V_0 = 0$). The slope of C^- lines would, therefore, be given by $dX/dt = V - C = 2C_0 - 3C$. For the line OA, of course, $C = C_0$, and its slope is $-C_0$. This characteristic represents the movement of the top of the negative wave (water depth of y_0) and shows that it travels upstream with a velocity of C_0.

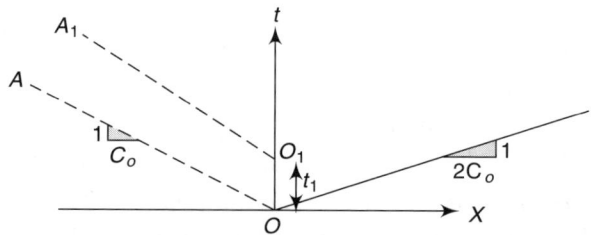

Fig. 7.17   Characteristic lines for the dam-break problem

To find out the shape of this wave and speed of movement of different particles on it, let us assume that there is a C^- line, O_1A_1, which originates from the point $(0, t_1)$, i.e., at the dam location and at time t_1. Its slope would be $2C_0 - 3C$ and, at any time t, it would be at a distance of $(t - t_1)(2C_0 - 3C)$ from the dam (this distance would come out to be negative as the wave is moving upstream). Considering the wave profile at the dam itself ($X = 0$), it is obvious that $2C_0 - 3C$ must be zero since the distance is zero for all times. This implies that the celerity (and, therefore, the depth) at the dam is constant and is equal to two-thirds of the undisturbed celerity. Or, in terms of flow depth, when the dam breaks, the flow depth at that location remains constant at four-ninths of the depth of impoundment, y_0. Thus the flow depth remains constant and the wave rotates about this point.[24] The flow velocity at this point is obtained from the Riemann invariant, $V + 2C$, as $V = 2C_0 - 2C = 2C_0/3$ and the

---

[24]We assume that the channel extends infinitely both upstream and downstream of the dam. If there is a boundary on the upstream side, the wave will strike that boundary and will be reflected towards the dam. When this reflected wave reaches the dam location, the flow depth would change accordingly.

slope of the C^- line is equal to $V - C = 0$. This shows that the vertical line $X = 0$ (i.e., the t-axis) is a C^- line and implies that all C^- lines (including O_1A_1) would start from O. Therefore, t_1 is zero and the profile of the wave is given by $X_w = t\left(2C_0 - 3\sqrt{gy}\right)$ using which the shape of the wave may be obtained at any time. Note that y would vary from y_0 to $4y_0/9$ and corresponding variation in X_w would be from $-C_0t$ to 0. The discharge at the dam would be constant (since the depth and velocity are constant) and is given by

$$q = Vy = \frac{2}{3}\sqrt{gy_0}\,\frac{4}{9}y_0 = \frac{8}{27}\sqrt{gy_0^3}$$

The downstream positive wave velocity is obtained by setting the flow depth (and celerity) equal to zero and using the invariance of $V + 2C$ as $2C_0$, indicating that it moves twice as fast as the top of the negative wave.

Effect of friction, presence of water on the downstream side, and steep slope of channel bed, are some of the factors which have not been accounted for in this analysis. It has been found experimentally that the positive wavefront is not tapered but is rounded and has a finite depth at the end. Its speed is less than that predicted by theory $(2C_0)$. The negative wave velocity is close to the theoretical value and the flow depth at the dam remains constant. We will not discuss these cases as the method of characteristics is not easily applicable to them.

We have seen that in the application of the method of characteristics, some simplifying assumptions have to be made in order to enable us to solve them analytically. Most practical problems would involve flow and channel properties that may not allow these assumptions to be made. If we still want to use this method and do not want to solve the set of partial differential equations, we would have to use numerical methods applied to Eqs (7.30) as described in the next subsection.

7.4.2   Numerical Solution

The numerical solution of Eqs (7.30) involves writing these as[25] (see Fig. 7.18)

$$V_P + 2C_P = V_Q + 2C_Q + g\int_{t_Q}^{t_P}(S_0 - S_f)\,dt\,, \\[2em] X_P = X_Q + \int_{t_Q}^{t_P}(V + C)\,dt \tag{7.31a}$$

---

[25] Again a rectangular channel is considered but the bed slope and friction terms are included.

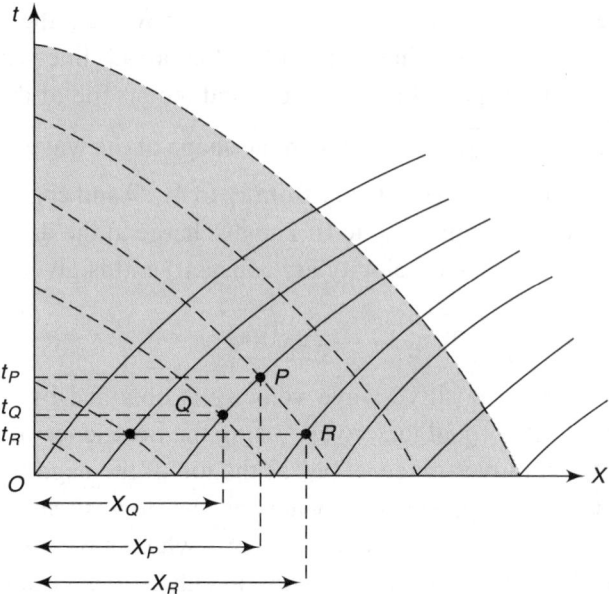

Fig. 7.18 Characteristic grid solution using the method of characteristics

$$V_P - 2C_P = V_R - 2C_R + g \int_{t_R}^{t_P} (S_0 - S_f)\,dt,$$

$$X_P = X_R + \int_{t_R}^{t_P} (V - C)\,dt$$

(7.31b)

in which P is the intersection of C^+ line from Q and the C^- line from R, and V, C, and S_f without a subscript indicate that these quantities are varying over the time period.

If the conditions at Q and R are known (e.g., they are specified as the initial conditions or have been already computed by the numerical method), Eqs (7.31) can be used to obtain the four unknowns X, t, V, and C at P. Complications arise because of the integrals in these equations, which have to be approximated since their temporal variation is not known. The easiest option is to assume that the value of the integrand is constant and is equal to the value at the *known* points, i.e., S_f, V, and C are taken equal to their values at Q for the first equation and at R for the second equation. We would then have four *linear* equations in four unknowns, which can be readily solved. For higher accuracy, we may replace the integrands by their average values over the interval (t_Q, t_P) or (t_R, t_P). For example, S_f in the first equation can be taken as $0.5(S_{fQ} + S_{fP})$. This will result in a system of four *nonlinear* equations for which an iterative solution process has to be used. Once we obtain the values at P, and similar other points, we can march forward in time using P as one of the *known* points. The process can be repeated till the desired time level is reached and the X-t plane is filled with a *characteristic grid* with velocity and celerity computed at each node.

While the solution is available at all characteristic grid nodes, it is not very convenient since most of the times the information is needed at either a specified location or at a specified time or both. As is clear from Fig. 7.18, the nodes of the characteristic grid are rather irregularly placed and we would need some form of interpolation to determine the flow conditions at a specified location and/or a specified time. The alternative method of specified time intervals (or the rectangular grid method) uses a rectangular grid in the X-t plane as described below. Figure 7.19 shows a rectangular grid in the X-t plane with conditions known at the line O_1A_1 (either through initial conditions or obtained numerically). Our aim is to compute the flow parameters at all nodes on the line O_2A_2.

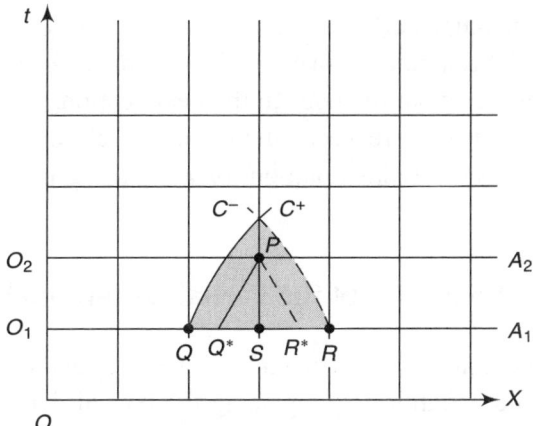

Fig. 7.19 Rectangular grid solution using the method of characteristics

We again take two points, Q and R, where conditions are known. However, the C^+ characteristic from Q and C^- characteristic from R will not, in general, intersect at P. Therefore, we need to find out the points Q^* and R^*, which are located in such a way that the characteristic lines through them intersect at P. Equations (7.31) are then written in terms of Q^* and R^* in place of Q and R and contain the two unknowns V_P and C_P. Note that X_P and t_P, which were unknown in the characteristic grid method are now known. However, the locations of Q^* and R^* are now unknown (once the location is known, the velocity and celerity may be obtained by interpolation between Q and S and S and

R respectively). The four equations of Eqs (7.31) along with the four interpolation equations (two each at Q^* and R^*, one for velocity and the other for celerity) enable us to obtain the eight unknowns: V and C at P, Q^*, and R^*; and X at Q^* and R^*. Iterative techniques have to be used to obtain the solution. Once the solution is obtained at all the grid points, it is easy to ascertain the variation of flow parameters at a particular location or at a particular time.

Another method of solution is based on the partial differential equations along the characteristics, i.e., we still use the characteristic lines but use Eqs (7.27) and (7.28) by replacing the partial derivatives with their finite difference equivalents. We do not describe this method here since the finite difference equations applied to the Saint-Venant's equation would be easier to apply.

Although the numerical solution of the ordinary differential equations obtained through the method of characteristics is convenient and accurate, the need to interpolate either after the solution (in the method of characteristic grid) or during the solution process (in the rectangular grid method) makes it a little tedious. With the advent of powerful computers and readily available software for solving partial differential equations, it has now become common to solve the partial differential form of the Saint-Venant's equations using finite difference or finite element method. In the next section, we describe the application of the finite difference method (the finite element method is not described but results in similar equations on a regular grid as the finite difference method).

7.5 Numerical Solution of the Partial Differential Equations

The partial differential equations given by Eqs (7.22) and (7.25) may be converted to algebraic equations by replacing the partial derivatives with their finite difference approximations. There are a number of ways in which the partial derivatives can be approximated (see Appendix C) and also a number of different ways in which the variables at any point (X, t) can be expressed. For example, a forward, backward, or central difference can be used for the spatial and temporal derivatives, and the parameter values at a point can be taken as some weighted average of parameter values at the surrounding points.[26] Depending on the methodology used for this purpose, a number of different schemes are possible. These schemes can be broadly classified into two categories—*explicit*, which result in an explicit expression for each unknown in

---

[26] The surrounding points in the *X-t* plane may involve points at the same time at nearby locations, or at the same location at nearby times, or a combination of both.

terms of known quantities, and *implicit,* which express the unknowns in terms of implicit equations involving other unknowns also (in addition to the known quantities). Clearly, the explicit schemes are much easier to solve in comparison to the implicit schemes, which, in general, require the solution of a set of nonlinear equations. The advantage of an implicit scheme is that a larger time step may be used compared to the explicit schemes. We briefly describe some of the commonly used explicit and implicit schemes here. In all numerical schemes, a rectangular grid has been used and the subscript (i, j) refer to the ith point in X and j^{th} point in time (Fig. 7.20). The aim is to obtain the flow parameters (velocity and depth) at the time level $j + 1$ for all i (i.e., spanning the whole length of the channel under consideration) given the values of these parameters at all i and for all time levels up to j. Of course, we assume that the appropriate boundary conditions are specified (for subcritical flow, one at the upstream boundary and the other at the downstream boundary; and for supercritical flow, both at the upstream boundary).

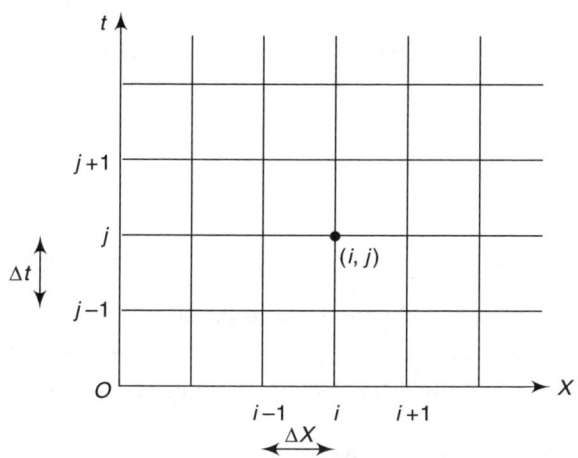

Fig. 7.20 Computational grid for the finite difference methods

For convenience, we rewrite the Saint-Venant's equations for rectangular channels as

$$\frac{\partial y}{\partial t} + y\frac{\partial V}{\partial X} + V\frac{\partial y}{\partial X} = 0 \tag{7.32a}$$

$$\frac{\partial V}{\partial t} + V\frac{\partial V}{\partial X} + g\frac{\partial y}{\partial X} = g\,(S_0 - S_f) \tag{7.32b}$$

7.5.1 Explicit Schemes

In Eqs (7.32), if we express only the time derivatives in terms of parameter value at the *unknown* time level $j + 1$, and express all other terms at the known time levels (j or lower), it is easily seen that Eq. (7.32a) would directly give us the flow depth at the time level $j + 1$ and Eq. (7.32b) would provide the velocity. The three most common schemes for doing this are mentioned below.

Diffusive scheme

In this method, the temporal derivatives are evaluated using a modified backward difference [using points $(i, j + 1)$, $(i - 1, j)$, and $(i + 1, j)$],[27] the spatial derivatives are expressed in terms of central difference [i.e., using points $(i - 1, j)$ and $(i + 1, j)$] and the parameter values are evaluated at the point (i, j). For example, considering the flow depth, we have

$$\frac{\partial y}{\partial t} = \frac{y_{i, j+1} - \dfrac{y_{i-1, j} + y_{i+1, j}}{2}}{\Delta t}, \quad \frac{\partial y}{\partial X} = \frac{y_{i+1, j} - y_{i-1, j}}{2 \Delta X}, \quad y = y_{i,j} \quad (7.33)$$

Similar expressions are used for the flow velocity. The bed slope and friction slope are also used at the point (i, j).[28] The flow depth at $(i, j + 1)$ is then obtained from Eq. (7.32a) as

$$y_{i, j+1} = \frac{y_{i-1, j} + y_{i+1, j}}{2} - \frac{\Delta t}{2 \Delta X}$$
$$\times [\, y_{i, j} (V_{i+1, j} - V_{i-1, j}) + V_{i, j} (y_{i+1, j} - y_{i-1, j})] \quad (7.34)$$

An alternative form can also be written based on the fact that the last two terms on the left hand side of Eq. (7.32a) can be combined as $\partial q / \partial X$. Similarly, the velocity is obtained from the following equations:

$$V_{i, j+1} = \frac{V_{i-1, j} + V_{i+1, j}}{2} - \frac{\Delta t}{2 \Delta X}$$
$$\times [V_{i, j} (V_{i+1, j} - V_{i-1, j}) + g (y_{i+1, j} - y_{i-1, j})]$$
$$+ g \Delta t (S_{0(i, j)} - S_{f(i, j)}) \quad (7.35)$$

Since the partial derivatives have been replaced by a finite difference, there would be a truncation error associated with the scheme (any good book on

---

[27]Liggett and Cunge in 1975 also used the point (i, j) and assigned it a weight of α, such that it reduced to the diffusive scheme for $\alpha = 0$. We do not discuss it since the methodology is identical.

[28]Some studies have suggested using the parameter values as mean of the values at $(i - 1, j)$ and $(i + 1, j)$. We prefer using the point (i, j).

Numerical Methods can be consulted to get more details of the truncation error analysis). Using the Taylor series, it is seen that the error in the spatial derivative is $O(\Delta X^2)$ and the error in the temporal derivative is $O[\Delta t + (\Delta X^2 / \Delta t)]$. If the numerical grid is refined in such a way that the ratio $\Delta X^2 / \Delta t$ is constant, the truncation error does not vanish and introduces an error term in the temporal derivative $\partial V / \partial t$, which is proportional to $\partial^2 V / \partial X^2$. Due to the similarity with the diffusion equation, $\partial \zeta / \partial t = D(\partial^2 \zeta / \partial X^2)$, it is called the diffusive scheme indicating that the scheme adds an artificial diffusion term to the governing equation.

Since the diffusive scheme is second-order accurate in space and only first-order accurate in time, a method can be devised that would be second-order accurate in both space and time. One such method, known as the *leap-frog method* due to its similarity with the children's game, is described next.

Leap-frog method

In this method, both the temporal and the spatial derivatives are evaluated using a central difference scheme and the parameter values are written at the point (i, j). The truncation error in the spatial derivative is $O(\Delta X^2)$ and that in the temporal derivative is $O(\Delta t^2)$. Thus the only difference from Eq. (7.33) is in the time derivative, which is now given as

$$\frac{\partial y}{\partial t} = \frac{y_{i,j+1} - y_{i,j-1}}{2\Delta t} \tag{7.36}$$

which may be thought of as leaping over the point (i, j). Equations (7.34) and (7.35), with the first term on the right hand side replaced by $y_{i,j-1}$ and $V_{i,j-1}$ respectively and Δt replaced by $2\Delta t$, are used to obtain the depth and velocity at $(i, j + 1)$.

While both the schemes described above work well in general, the presence of shocks is not very well handled by them. The explicit scheme proposed by Lax and Wendroff in 1960 was specifically designed to handle discontinuities and is described next.

Lax–Wendroff method

The governing differential equations are reformulated in what is known as the *conservative form*, using the discharge and the flow area as the dependent variables, as follows:

$$\frac{\partial A}{\partial t} + \frac{\partial Q}{\partial X} = 0 \tag{7.37a}$$

$$\frac{\partial Q}{\partial t} + \frac{\partial\left(\frac{Q^2}{A} + gA\overline{y}\right)}{\partial X} = gA(S_0 - S_f) \tag{7.37b}$$

A parameter (A or Q) at time level $j + 1$ is written in terms of a Taylor series including the temporal derivatives up to the second order. For example,

$$A_{i,j+1} \simeq A_{i,j} + \Delta t \frac{\partial A}{\partial t}\bigg|_{i,j} + \frac{\Delta t^2}{2} \frac{\partial^2 A}{\partial t^2}\bigg|_{i,j} \tag{7.38}$$

The time derivatives may be replaced by spatial derivatives using Eq. (7.37a) as

$$\frac{\partial A}{\partial t} = -\frac{\partial Q}{\partial X} = -\frac{\partial Q}{\partial A}\frac{\partial A}{\partial X} \tag{7.39a}$$

$$\frac{\partial^2 A}{\partial t^2} = \frac{\partial}{\partial t}\left(-\frac{\partial Q}{\partial X}\right) = -\frac{\partial}{\partial X}\left(\frac{\partial Q}{\partial A}\frac{\partial A}{\partial t}\right) = \frac{\partial}{\partial X}\left(\frac{\partial Q}{\partial A}\frac{\partial Q}{\partial X}\right) \tag{7.39b}$$

Similar (but more complicated) expressions can be written for the discharge and its time derivatives. The spatial derivatives are then replaced by central difference approximations and the discharge and area at $(i, j + 1)$ obtained explicitly using Eqs (7.38) and (7.39) and analogous expressions for the discharge. Sometimes an artificial viscosity term is added to remove oscillations exhibited by this scheme near a sharp front. The details are not described here. Interested readers may consult Chaudhary (1979, 1993) for a detailed discussion of various numerical methods.

All the explicit schemes compute the value of parameters at a point that is Δt ahead in the future based on the conditions known at the present time. It can be shown that stable results are obtained only when this time difference, Δt, is smaller than $\Delta X/|V + C|$, which is known as the Courant condition (or the CFL condition after Courant, Friedrichs, and Lewy who proposed it in 1928 as a necessary condition for convergence of numerical solution of hyperbolic equations, *http://www.stanford.edu/class/cme302/classics/courant-friedrichs-lewy.pdf*). In terms of the characteristics, it is clearly seen that it implies that the point in the future $(i, j + 1)$ must not be outside the region of influence of the points $(i - 1, j)$ and $(i + 1, j)$, see Fig. 7.13. Since V and C change from point to point, the maximum value of $V + C$ should be chosen to limit the time step size. An adaptive time-stepping scheme may also be used in which the permissible Δt is changed at every (or every few) time steps. The implicit schemes described next are generally free from this restriction on the length of the time step.

7.5.2 Implicit Schemes

In Eqs (7.32), if we express the spatial derivatives also in terms of parameter value at the *unknown* time level $j + 1$, we would not be able to write an explicit expression for computing the velocity and depth at the point $(i, j + 1)$. We would get a system of nonlinear equations, which is solved (with the additional information from the boundary conditions) to obtain the flow depth and velocity at all points at time level $j + 1$. The additional effort spent in the formulation and solution is compensated by the better stability and larger time steps[29] and may result in an overall economy. Here we describe one such scheme proposed by the French engineer A. Preissmann in 1961 and based on a four-point implicit formulation.

Preissmann scheme

The governing differential equations are reformulated using the discharge and flow depth as the dependent variables and are written for a rectangular channel[30] as follows:

$$\frac{\partial y}{\partial t} + \frac{\partial q}{\partial X} = 0 \tag{7.40a}$$

$$\frac{\partial q}{\partial t} + \frac{\partial}{\partial X}\left(\frac{q^2}{y}\right) + gy\frac{\partial y}{\partial X} - gy\left(S_0 - \frac{n^2 q |q|}{y^2 R^{4/3}}\right) = 0 \tag{7.40b}$$

with the assumption that the friction slope is same as that under uniform flow at the same depth. The derivatives and parameter values are specified as

$$\frac{\partial y}{\partial t} = \frac{\dfrac{y_{i,j+1} + y_{i+1,j+1}}{2} - \dfrac{y_{i,j} + y_{i+1,j}}{2}}{\Delta t} \tag{7.41a}$$

$$\frac{\partial y}{\partial X} = \alpha\frac{y_{i+1,j+1} - y_{i,j+1}}{\Delta X} + (1-\alpha)\frac{y_{i+1,j} - y_{i,j}}{\Delta X} \tag{7.41b}$$

$$y = \alpha\frac{y_{i+1,j+1} + y_{i,j+1}}{2} + (1-\alpha)\frac{y_{i+1,j} + y_{i,j}}{2} \tag{7.41c}$$

---

[29]The accuracy would still depend on the time step size and we cannot use very large Δt. However, reasonably accurate results can be obtained using time steps much larger than those dictated by the CFL condition.

[30]For other shapes, the discharge, Q, the top width, T, and the flow area, A, would appear in the equations.

Similar expressions are used for the specific discharge q also. Figure 7.21 shows a grid and a point P, which is spatially centred but temporally skewed (if $\alpha = 0.5$, P would be centred temporally also). Some studies use a spatially skewed location by assigning a weight, say β, in the spatial direction also, but we will consider the original Preissmann scheme only.

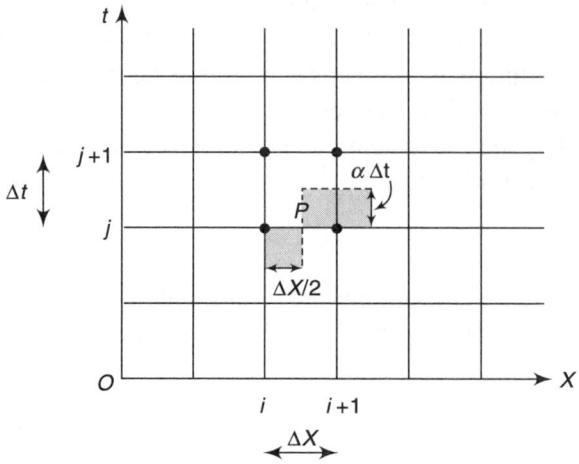

Fig. 7.21 Computational grid for the Preissmann method

It is seen that the Preissmann method writes the finite difference form of the governing equations at the point P, using a central difference for the time derivative, a *weighted central difference* for the spatial derivative, and a four-point linear interpolation for the parameter values. It has been found to be unconditionally stable (i.e., stable for all values of Δt) if α is between 0.5 and 1, with the value of 0.6 used most commonly. While values close to 0.5 would cause the scheme to become unstable, values close to 1 are also not desirable since they may cause spurious oscillations in the solution.

The discretized form of Eqs (7.33) involves the unknown depth and discharge at both nodes $(i, j + 1)$ and $(i + 1, j + 1)$ in both equations and is nonlinear. If there are $N + 1$ nodes (N segments), we would get $2N$ equations, one for each segment, involving $2(N + 1)$ unknowns (the values of depth and discharge at each node at the time level $j + 1$). The two additional equations needed to solve this system of equations are supplied by the boundary conditions. More details can be seen in Chaudhary (1993). Venutelli (2002) provides a detailed analysis of the stability and accuracy of the four-point implicit schemes with weights assigned to both spatial and temporal directions.

7.6 Approximate Solution of the Partial Differential Equations

In the previous section, we have seen that the numerical methods are very versatile, though a little complicated, in solving the unsteady flow problems in an open channel. However, sometimes, an engineer may be interested in getting an approximate idea about the solution rather than a very accurate solution. In some cases, the input data itself may not be known accurately and it would not be prudent to use a highly efficient and accurate but complicated technique. In this section, we describe these approximate methods and discuss their application to two practical problems—development of a stage-discharge (rating) curve for a channel and the routing of a flood through a given reach of a channel.

7.6.1 Kinematic Wave

The waves obtained by solution of the continuity and momentum equations (i.e., Saint-Venant's equations) are called *dynamic waves*. As stated earlier, generally the friction slope and bed slope are of the same order of magnitude but the other terms are much smaller. If we ignore the other terms from the equation of motion, the resulting model is called the *kinematic wave* model and the governing equations are written as

$$\frac{\partial A}{\partial t} + \frac{\partial Q}{\partial X} = 0 \tag{7.42a}$$

$$S_0 - \frac{n^2 Q^2}{A^2 R^{4/3}} = 0 \tag{7.42b}$$

From Eq. (7.42b), it is inferred that there is a unique value of discharge for a given flow depth (which is not true for the dynamic wave as it involves the water surface slope and acceleration terms also). Therefore, the continuity equation, Eq. (7.42a), can be written as

$$\frac{\partial A}{\partial t} + \frac{\partial Q}{\partial A} \frac{\partial A}{\partial X} = 0 \tag{7.43}$$

which shows that in a frame of reference moving with a velocity $\partial Q/\partial A$, the area of flow (and, therefore, the discharge) would appear to be constant. This velocity can be obtained from Eq. (7.42b) as (using the Manning's equation for discharge)

$$V_w = \frac{\partial Q}{\partial A} = \frac{5}{3} V \left(1 - \frac{2R}{5T} \frac{dP}{dy} \right) \tag{7.44}$$

For a very wide channel ($R \ll T$), the kinematic wave velocity would be five-thirds times the average velocity,[31] while for most natural channels, this ratio has been found to be 1.3. By considering that the crest of the wave has maximum flow depth, implying that $dy/dt = 0$, it can be shown that the crest also moves with the velocity V_w. However, since other portions of the wave will have smaller depths, they would move at a slower speed. (We do not consider the channels with closing top. Therefore, dQ/dA would be an increasing function of flow depth.) This tends to make the wave steeper as it moves downstream. However, as the wave becomes steep the water surface slope, which was ignored in the equation of motion, starts to affect the flow and tends to make the wavefront flatter. These two opposite effects have been found to have a tendency to reach an equilibrium such that the wave retains its shape and moves with a constant velocity. One such wave, called the monoclinal wave, is useful in the development of stage-discharge curve and is described next.

Monoclinal wave

A monoclinal wave is a flood wave joining two different uniform flow levels by a single rising or falling limb (Fig. 7.22) and moving with a uniform velocity while retaining its shape.

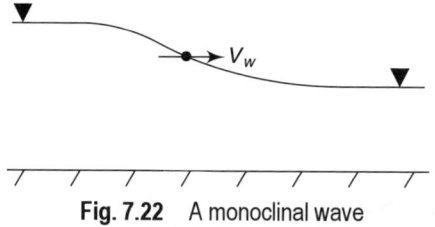

Fig. 7.22　A monoclinal wave

Since the wave is moving at a constant speed, V_w, we can use the concept of equivalent steady flow to define an overrun [see Eq. (7.1)] as $Q_r = (V_w - V) A$. Moreover, conversion into a steady state means that the total derivatives of all parameters would be zero, i.e.,

$$\frac{dy}{dt} = \frac{\partial y}{\partial t} + V_w \frac{\partial y}{\partial X} = 0 \quad \Rightarrow \quad \frac{\partial y}{\partial t} = -V_w \frac{\partial y}{\partial X} \tag{7.45}$$

and, similarly, $\dfrac{\partial V}{\partial t} = -V_w \dfrac{\partial V}{\partial X}$. From the equation of motion, Eq. (7.25), we then have

---

[31]For a triangular channel, the ratio would be 4/3.

$$\frac{\partial y}{\partial X} + \frac{V - V_w}{g} \frac{\partial V}{\partial X} = S_0 - S_f \tag{7.46}$$

Expressing the flow velocity in terms of the overrun as $V = V_w - Q_r/A$, we

get $\dfrac{\partial V}{\partial X} = \dfrac{Q_r T}{A^2} \dfrac{\partial y}{\partial X}$ and Eq. (7.46) may be written as

$$\frac{\partial y}{\partial X} = \frac{S_0 - S_f}{1 - \dfrac{Q_r^2 T}{g A^3}} \tag{7.47}$$

which is similar to the dynamic equation for the GVF and may be solved in a similar manner to obtain the wave profile. For many practical cases, the Froude number term in the denominator of the right hand side would be quite small and can be neglected. The friction slope would then be equal to $S_0 - \partial y/\partial X$, and the discharge at any depth would be given by

$$Q = \frac{1}{n} A R^{2/3} S_f^{1/2} = Q_n \sqrt{1 - \frac{1}{S_0} \frac{\partial y}{\partial X}} \tag{7.48}$$

in which Q_n is the normal discharge at the flow depth of y, i.e., the discharge if the flow occurs under uniform flow conditions with a normal depth of y. Equation (7.48) clearly indicates the possibility of multiple discharge values for the same flow depth. This discharge would be more or less than the normal discharge depending on whether $\partial y/\partial X$ is negative or positive. Typically, the stage data is available at a particular location at different times and it would be more convenient to use the temporal derivative of the depth rather than the spatial derivative. Using Eq. (7.45), we write

$$Q = Q_n \sqrt{1 + \frac{1}{V_w S_0} \frac{\partial y}{\partial t}} \tag{7.49}$$

From this equation, it is obvious that during the rising stage ($\partial y/\partial t > 0$), the discharge would be more than the normal discharge and during the falling stage, it would be less. If V_w is not known, we can take it as 1.3 times the average flow velocity under uniform flow condition for natural channels (as shown earlier, it would be 1.67 times the average velocity for rectangular channels). A typical stage-discharge curve is shown in Fig. 7.23.

Another application of approximate analysis of unsteady flow is the routing of a flood wave through a given reach of a channel, i.e., given the discharge (or depth) variation with time at an upstream location, to determine the same at a downstream location. Traditionally, flood routing is discussed

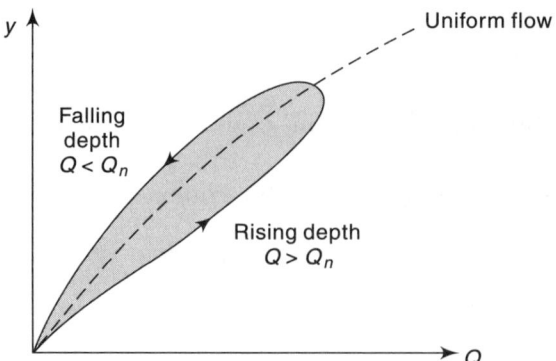

Fig. 7.23 A typical stage-discharge curve for unsteady flow

in books on hydrology but we provide a very brief description here. More details can be seen, for example, in Chow, et al. (1988).

7.6.2 Flood Routing

The continuity equation is utilized in the flood routing by writing it as

$$\frac{dS}{dt} = I - O \quad \Rightarrow \quad \frac{\Delta S}{\Delta t} = \bar{I} - \bar{O} \tag{7.50}$$

where S denotes the storage (volume of water) within the reach, and $\bar{I}$ and $\bar{O}$ are the average inflow and outflow discharges during the time period t to $t + \Delta t$. There are two commonly used methods of solution.

In the first method, introduced by L.G. Puls in 1920s, the average inflow and outflow are taken as the means of their values at times t and $t + \Delta t$, and the change in storage is written as $S_{t+\Delta t} - S_t$. The finite difference form of Eq. (7.50) then involves two unknowns—the values of S and O at time $t + \Delta t$ (note that the inflow is known at all times). An additional relationship is, therefore, needed to enable us to solve the equation. This is generally a plot of storage versus outflow based on past flood records. Puls suggested a graphical method of solution in which Eq. (7.50) is written as

$$\left(S + \frac{O\Delta t}{2}\right)_{t+\Delta t} = \left(S - \frac{O\Delta t}{2}\right)_t + \frac{\Delta t}{2}(I_{t+\Delta t} + I_t) \tag{7.51}$$

and the observed O–S relationship is used to plot two additional curves—O vs $S + (O\Delta t/2)$ and O vs $S - (O\Delta t/2)$. Starting from known initial conditions, we compute the right hand side of Eq. (7.51), which, clearly, is equal to the value of $S + (O\Delta t/2)$ at the next time step. From the prepared plot of O vs $S + (O\Delta t/2)$, the value of O at $t + \Delta t$ is read and from the plot of O vs $S - (O\Delta t/2)$, the corresponding value of $S - (O\Delta t/2)$ at $t + \Delta t$ is obtained. Equation (7.51) is then applied to the next time step and so on. The time step Δt should be less than the time of travel of the flood wave through the given reach (which is same as the Courant condition described earlier). Also, it should not be so large as to introduce errors in the estimate of average inflow and outflow as the arithmetic mean of end values.

In the second method, the storage is assumed to follow a specified relationship (usually linear) with the inflow and outflow. For example, a method proposed by G.T. McCarthy in 1930s and popularly known as the Muskingum method (since it was developed in connection with flood-protection works on the Muskingum river in Ohio) assumes that

$$S = K\,[\theta I + (1 - \theta)O] \tag{7.52}$$

where θ is inflow-weight factor, which generally varies from 0 to 0.5 (0 indicating the storage depends only on outflow as in Puls method, and 0.5 indicating inflow and outflow to be equally influential in determining the storage) and K is a constant (having units of time) obtained from observed flood data.[32] An alternative form of Eq. (7.52) may be written as $S = KO + K\theta(I - O)$, in which the first term on the right hand side represents the prism storage (Fig. 7.24) and the second term, the wedge storage. Note that wedge storage may become negative if outflow is more than inflow, which generally implies that the water level on the upstream side is lower than that at the downstream side.

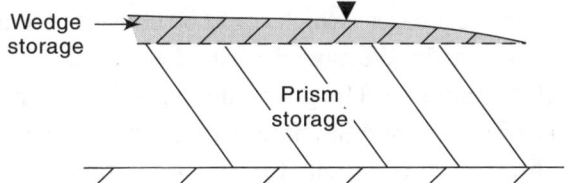

Fig. 7.24   Prism and wedge storage in a channel

Using Eq. (7.52) in the derivative form of Eq. (7.50), we get

$$\frac{d[K\theta I + K(1-\theta)O]}{dt} = I - O$$

$$\Rightarrow \quad \frac{dO}{dt} + \frac{O}{K(1-\theta)} = -\frac{\theta}{1-\theta}\frac{dI}{dt} + \frac{I}{K(1-\theta)} \tag{7.53}$$

Although this equation may be solved analytically, a finite difference solution is usually preferred, which results in

$$Q_{t+\Delta t} = \frac{2(1-\theta)-\tau}{2(1-\theta)+\tau}O_t + \frac{\tau+2\theta}{2(1-\theta)+\tau}I_t + \frac{\tau-2\theta}{2(1-\theta)+\tau}I_{t+\Delta t} \tag{7.54}$$

---

[32]Cunge (1969) proposed a method of determining K from the channel geometry and the grid spacing.

in which τ is a dimensionless time, $\Delta t/K$. Equation (7.54) applied repeatedly will give us the ordinates of the outflow graph at intervals of Δt till the desired time. A typical pair of inflow and outflow hydrograph is shown in Fig. 7.25, which illustrates the lag in peak and its attenuation as the flood moves downstream. More details of flood-routing schemes are available in Chow, et al. (1988).

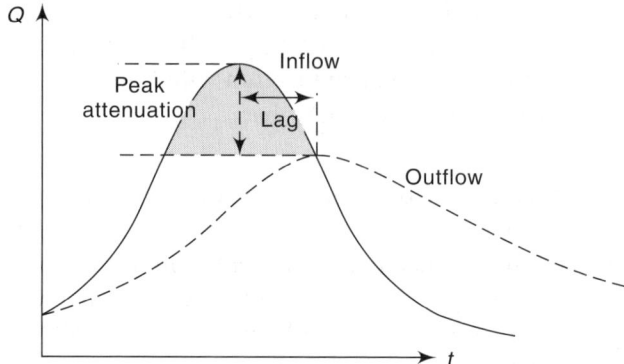

Fig. 7.25 Inflow and outflow hydrographs obtained by flood routing

SUMMARY

In this chapter, we have discussed various types of unsteady flow situations, beginning with the simplest cases, which could be converted to an equivalent steady state and solved easily. We have described waves and surges and analysed them using this technique. The governing partial differential equations have been derived and the method of characteristics has been used to convert them to ordinary differential equations for easier solution. Various methods of solution of the ordinary differential equations and the full set of partial differential equations have been discussed, including analytical (or graphical) as well as numerical methods. Application of approximate methods with particular reference to development of rating curves and flood-routing schemes has been described.

This, in effect, completes the syllabus typically followed in an undergraduate course on open channel flow and in most postgraduate level courses on the subject. However, we feel that two additional topics, flow in mobile-boundary channels and dispersion of pollutants in open channels, are becoming increasingly important in the study of open channel flow. Although books devoted fully to these topics are available, we believe that it would be desirable to provide a very brief introduction to these topics in this book since some of the undergraduate syllabi do not include these topics at all. With this in mind, the next two chapters touch upon these topics.

REFERENCES

Abramowitz, M. and Stegun, I. (1964): *Handbook of Mathematical Functions with Formulas, Graphs, and Mathematical Tables*, Dover Publications, New York.

Chaudhary, M.H. (1979): *Applied Hydraulic Transients*, Van Nostrand Reinhold, New York.

Chaudhary, M.H. (1993): *Open Channel Flow*, Prentice Hall Inc., Englewood Cliffs, New Jersey.

Chow, V.T., Maidment, D.R., and Mays, L.W. (1988): *Applied Hydrology*, McGraw-Hill, New York.

Cunge, J.A. (1969): 'On the subject of a flood propagation computation method (Muskingum method)' *J. of Hydr. Res.*, **7**(2), pp. 205-30.

Hastings, C., Jr. (1955): *Approximations for Digital Computers*, Princeton University Press, Princeton, New Jersey.

USACE (2007): 'Water wave mechanics', US Army Corps of Enginners, *Engineers Manuals, http://www.usace.army.mil/publications/eng-manuals/em1110-2-1100/PartII/Part-II-Chap_1-ppi-60.pdf* (last accessed on May 10, 2007).

Venutelli, M. (2002): 'Stability and accuracy of weighted four-point implicit finite difference schemes for open channel flow', *J. of Hydr. Eng.*, **128**(3), pp. 281-88.

Wiegel, R.L. (1960): 'A presentation of cnoidal wave theory for practical application', *J. of Flu. Mech.*, **7**, pp. 273–86.

———— (1964): *Oceanographical Engineering*, Prentice Hall Inc., Englewood Cliffs, New Jersey.

EXERCISES

7.1 A small-amplitude wave is generated in a 5 m deep still pool of water by moving a paddle. If the amplitude is measured as 10 cm, what would be the celerity of the wave? If the same amplitude wave is generated in a trapezoidal channel with a bottom width of 20 m, side slopes of 1H:1V, and water depth of 5 m, what would be the celerity?

[7.11 m/s, 6.59 m/s]

7.2 A train of wave is observed in a 5 m deep pool of water and was found to have a wavelength of 10 m and a wave height of 1 m. Assuming that the cnoidal wave theory is applicable, find the celerity of the wave. What conclusions may be drawn from the results?

[Cnoidal theory not applicable]

7.3 A 2 m wide rectangular channel carries a discharge of 1 m³/s at a flow depth of 1 m. The discharge is suddenly increased to 4 m³/s. Find the height of the resulting surge (i.e., the increase in flow depth) and its velocity. [0.34 m, 4.42 m/s]

7.4 A 2 m wide rectangular channel carries a discharge of 1 m³/s at a flow depth of 1 m. A sluice gate located in the channel is suddenly raised such that the discharge gets doubled. Find the flow depth and velocity at the gate and the wave velocity at the topmost and bottommost points of the resulting negative surge. [0.75 m, 1.33 m/s, –2.63 m/s, –1.37 m/s]

7.5 A 2 m wide rectangular channel originates from a lake ($X = 0$) and carries a discharge of 1 m³/s at a flow depth of 1 m. The lake level starts decreasing such that the flow depth in the channel varies linearly from 1 m (at $t = 0$) to 0.5 m in 1000 seconds. Using the method of characteristics, find the answer to the following questions:
  (i) How long will it take for a point 2 km downstream to feel the effect of the change in depth?
  (ii) At this time, what would be the flow depth at a point 1 km from the lake?
 (iii) When would the flow depth be 0.75 m at this point?
 [*Note*: Part (ii) involves a nonlinear equation.]
 [(i) 551 s, (ii) 0.854 m, (iii) 747 s]

7.6 A river cross section may be approximated as a rectangular section with bed width of 100 m. The bed slope is 1 in 1000 and the Manning's roughness coefficient is 0.02. It is carrying water under uniform flow conditions at a flow depth of 3 m. During a flash flood of 4 hour duration, the automatic stage recorder at a gauging station recorded water depth variation, which could be reasonably approximated by (with t in hours and y_t in m): $y_t = 3 + \exp[-(t - 2)^2]$. Prepare a rating curve by computing the discharge at every half an hour and show that it is looped. From Eq. (7.58), argue that the maximum discharge would not occur when the depth is maximum.

8 | Flow in Mobile Boundary Channels

8.1 Introduction

The study of flow through channels with erodible boundaries is much more complex than that of flow through channels with rigid boundaries.[1] Some of the complexities involved with channels with erodible boundaries are given below.

(i) Changing cross section due to erosion and deposition of sediments[2]

(ii) Additional resistance to flow due to presence of undulations on the channel bed[3]

(iii) Additional expenditure of energy in carrying the sediments

(iv) Wide variation in the size of sediments

Discussion of the process of sediment transport through open channels, therefore, requires a book on its own. The intent of this chapter is not to replace such books but provide a very basic understanding of some of the important concepts. The reader is referred to Garde and Ranga Raju (2000) and van Rijn (2007a,b) for a more thorough and updated description of the subject.

A mobile-bed channel (also called an alluvial channel)[4] will behave like a rigid-bed channel at very low flows since the *force* of water would not be enough to move the particles comprising the bed. This implies that all the theories discussed for a rigid-bed channel would also be valid for a mobile-bed channel. Since this analysis would be much simpler, it would be good to

---

[1]Sometimes, channels with rigid boundaries also carry sediments from the parent channel, overland runoff, etc. This sediment may get deposited on (and later eroded from) the boundary making it closer to a mobile-boundary channel.

[2]*Sediment* is defined as a particulate matter that can be transported by fluid flow (or, sometimes, ice) and is later deposited as a layer of solid particles on the bed or bottom of a body of water or other liquid.

[3]We will use the terms *boundary* and *bed* interchangeably but one should keep in mind the distinction.

[4]*Alluvium* is a general term for the sediment deposited by rivers, and the channels that pass through the alluvium are called *alluvial channels*.

know under what conditions the channel bed may be treated as rigid. Once the particles start moving, the bed no longer remains plane (assuming that we started with a plane bed) and undulations would start to appear. Since the form of these undulations would significantly influence the channel resistance, it would be desirable to correlate these bed forms with the flow characteristics. Another important parameter is the amount of sediment load carried by the channel since for channels joining a reservoir, the amount of sediment will have a direct bearing on the useful life of the reservoir before it gets *silted up*. Therefore, the following sections describe (i) the critical condition which leads to starting of motion of bed particles, (ii) the various bed forms observed under different flow conditions, (iii) the effect of bed forms on channel resistance, and (iv) the amount of sediment load carried by a channel. Finally, we synthesize these discussions and describe the design of a channel with mobile boundaries.

8.2 Initiation of Motion

A number of forces act on the particles on the channel boundary. Some of these are stabilizing forces in the sense that they would oppose any tendency of the particle to move while others would be trying to dislodge the particle from its place. For example, the weight of the particle would prevent it from lifting into the flow[5] while the buoyant force would try to lift it.

Similarly, the shear force near the bed and the form drag due to particle shape would try to roll the particle along the flow while the friction force at the bed and shielding and anchoring by nearby particles would try to prevent this motion. The particle will start to move when the balance of these forces favours the destabilizing forces. The point at which this happens is called the *incipient motion condition, threshold condition,* or *critical condition.*

Various theories have been proposed to correlate the critical condition with the flow and particle characteristics. Most of these are based on theoretical consideration of the forces acting on a particle but some are based on empirical observations. Due to the many complexities involved in the process of sediment motion (e.g., variation in the shape and size of particles, uncertainty in estimation of forces acting on the particles), even the theoretical approaches need to use empirical coefficients. We first look at the sediment and flow properties, which significantly affect the process of sediment movement and then consider the different approaches towards determining a critical condition.

---

[5]However, if the particle is on a steeply sloping bed or side, the weight would contribute to the destabilizing force by trying to cause it to roll down the slope.

8.2.1   Sediment Properties

The most important sediment property, from the point of view of its motion when subjected to a flow of water, is its size. If all the particles were nearly spherical, we could use the diameter to represent the size. However, in nature it would never happen, and only in a laboratory under carefully controlled conditions, we may achieve the condition of all spherical particles with the same diameter. For nonspherical particles, choosing a representative *size* becomes an important consideration. The determination of the size is also complicated by the fact that there is an assortment of particles with different sizes present on the channel bed and being carried by the water. The usual methodology for such mixtures is to use sieve analysis and obtain a distribution of the size of the particles. For example, we can use the diameter, d_{50}, which indicates that 50% of the material is finer than this size. However, the range of variation of sediment sizes is also important and some parameter representing this range (e.g., the standard deviation of particle sizes) should be incorporated in the analysis of sediment motion. The shape of a particle also considerably affects its motion. For spherical particles, of course, it does not matter how they are oriented with respect to the flow direction. For nonspherical particles, however, the orientation will have a tremendous bearing on the drag force applied by the flow on the particle. Moreover, in the determination of size by sieve analysis, one should realize that generally the sieve opening represents the size of the *smallest cross section* of a particle, i.e., a particle is likely to pass the sieve of a particular size if its smallest cross section has dimensions smaller than the sieve opening even though the size in another direction is larger (for example, a cylindrical particle with small diameter but large length could pass a sieve having an opening just larger than this diameter). The relative density of sediments is also an important parameter in analysing its motion since it directly affects the stabilizing force due to the weight and the bulk density would be important in studies of silt deposit in reservoirs since it determines the volume occupied by a given mass of sediment.

In this section, we look at some of the important properties of the sediments, considering both the cases—an individual particle and a collection of particles.

Individual particle

The particles on the boundary of a channel and those carried by it are of numerous different shapes ranging from almost spherical to nearly flat. It, therefore, becomes very difficult to define a representative size. Moreover, specifying only one size may not be sufficient for most particles (except, of course, for nearly spherical particles) and some *shape factor* has to be used to better characterize the particle. This shape factor may be based on the volume of the particle (e.g., by comparing it with the volume of a *representative* sphere), the cross-sectional area of the particle[6] or some other measure (e.g., when studying the settling of the particle in water, it is found that its size along three mutually perpendicular axes can be used to define a shape factor). The size of a particle can also be defined in various different ways. For example, it can be based on the volume, sieve analysis, settling velocity, or measurements along orthogonal axes. Here we describe only a few of these possibilities. For details, one may refer to Garde and Ranga Raju (2000) or textbooks on soil mechanics (the webpage *http://environment.uwe.ac.uk/geocal/SoilMech/classification/ soilclas.htm* offers a good description).

Triaxial size   If three orthogonal axes are taken in such a way that one is along the largest size, l_{max} (major axis), one is along the shortest particle dimension, l_{min} (minor axis), and the third is, naturally, in between, l_{mid} (intermediate axis), we may use l_{mid} to represent the sediment size. Also, a shape factor can be defined as $l_{min}/\sqrt{l_{max}\,l_{mid}}$. Note that for a spherical particles all three sizes would be same and the shape factor would be equal to 1. Typical natural sediments have a shape factor close to 0.7.

Nominal diameter   By measuring the volume of the particle, we can find the diameter of the spherical particle, which will have the same volume. This diameter is called the *nominal diameter* and is denoted by d_n. Since this is the most commonly used indicator of the size, we would drop the subscript and use d to denote the nominal diameter. A shape factor, known as *sphericity*, can be defined as the cube root of the ratio of the particle volume to the volume of the circumscribing sphere (clearly, this sphere will have a diameter equal to the largest particle dimension, l_{max}, and the sphericity would be equal to d_n/l_{max}.)

Sieve diameter   As discussed earlier, a particle may be able to pass through a sieve opening even if it has a dimension larger than the sieve opening in some direction. The sieve diameter is, therefore, based on the least cross-sectional area of the particle. For a sphere, it would be equal to the nominal diameter but for other shapes, it is generally smaller. A shape factor, based on the cross-sectional area, may be defined as the ratio of the projected area (when the particle is in the most stable position with the minor axis vertical) to the area of a circle of diameter l_{max}. This is known as the *coefficient of circularity*.

---

[6]Generally the particles would settle on the bed in its most stable configuration, i.e., with its shortest dimension being vertical. The drag force would depend on the cross-sectional area projected normal to the flow direction.

Sedimentation diameter   From the point of view of sediment transport in a channel, the velocity with which a sediment falls in water is very significant and is a good indicator of the overall size of the particle. The *sedimentation diameter* is defined as the diameter of a sphere that falls with the same *terminal* (or *fall*) velocity as that of the particle under consideration (it goes without saying that all other conditions, e.g., the relative density, the fluid, the temperature, etc. should be same in both the cases). Since the fall velocity is affected by the drag coefficient, which we will discuss in a latter section, we defer the discussion on the fall velocity till then.

Relative density   The relative density of a sediment particle would depend on its mineral composition. For example, quartz has a relative density of 2.7 while magnetite has 5.2 and organic matter has only about 1. Thus a particle containing more quartz would be lighter than one containing more magnetite. However, the sediments carried by rivers have been found to have a predominance of quartz and their relative density typically varies from 2.6 to 2.7. Hence, a constant relative density of 2.65 is assumed in most cases.

Group of particles

After discussing the properties of an individual particle, we now look at the behaviour of a group of particles, in terms of their size distribution, bulk density, and strength.

Size distribution   A typical sediment sample from a channel would contain a wide distribution of particle sizes and shapes. The analysis of size distribution of coarser particles (larger than one-sixteenth of a mm, e.g., sand, gravel, boulders) is usually performed by sieves while the smaller particles (silt and clay) are analysed based on their settling velocities. The results are plotted in terms of a *size-distribution curve*, which is a cumulative-frequency curve plotting the sediment size versus the percentage of material (by weight) finer than that size. The median size, d_{50}, represents the size that has equal weight of the sample finer than it and coarser than it, and is the most commonly used indicator of the size of the sample. Other indicators, e.g., d_{10} or d_{90} (the diameters that have 10% and 90% of the material finer than these) are also used in specific situations.[7] If we assume that the size distribution follows a *normal distribution*, the standard deviation for the sample is obtained as $\sigma = d_{84} - d_{50} = d_{50} - d_{16}$. Some other alternatives for representing the spread in

---

[7]For example, if the channel bed is such that the coarser material is on the top, d_{90} may be a better representative of the sediment motion.

the size range are the uniformity coefficient (d_{60}/d_{10}), sorting coefficient ($\sqrt{d_{75}/d_{25}}$), and Kramer's uniformity coefficient, M, defined as

$$M = \left(\sum_{i=0}^{50} p_i d_i \right) \Bigg/ \left(\sum_{i=50}^{100} p_i d_i \right) \tag{8.1}$$

with p_i representing the percentage of material corresponding to (not finer than) the diameter d_i, which has $i\%$ material finer than it.[8]

Bulk density   The *bulk density* is defined as the mass of dry sediment per unit total volume (of sediment and water/air in the pores) and indicates how tightly the sediments are packed. It is needed in the study of reservoir silting since generally the weight of the sediment carried by the river is estimated and the volume of deposition in a reservoir would be obtained using the bulk density. In a reservoir, the bulk density of deposited sediments varies with time as more consolidation takes place. For coarser particles, the consolidation effect would be small. Even for finer particles such as silt and clay, the consolidation would be negligible if the reservoir is empty most of the times (e.g., in a reservoir used only for flood control). However, if the reservoir is filled most of the times (e.g., a hydro-power project), there can be a high degree of consolidation for fine sediments. The pre-consolidation bulk density depends on the grain size and varies from about 800 kg/m³ for particle size of 1/1000 mm to about 2000 kg/m³ for a size of 1 mm.

Strength   The strength of a sediment deposit as depicted by its resistance to applied forces would mainly depend on its cohesion and the angle of internal friction, ϕ. Most studies on sediment transport involve coarser sediments and the cohesion is assumed to be negligible. The angle of internal friction is then equal to the angle of repose, which is important in the design of the side slopes of an erodible channel. The value of this angle generally increases with the grain size and also increases with decrease in the *roundedness* of the particles (due to better interlocking of angular particles). A range of ϕ of about 30° to 40° for particle size of 1 to 100 mm has been observed.

The sediment properties discussed in this subsection provide us an idea of the resistance offered by the particles. To determine whether the particles will move or not, we need to look at the various forces acting on it and whether the particle (or the group of particles) will be able to withstand these without moving.

---

[8]Strictly speaking, there is no such thing as a weight corresponding to a given diameter. One can, however, take a narrow width about a particular diameter to compute this weight. This width may or may not correspond to the sieve sizes.

8.2.2 Forces Acting on a Sediment Particle

The forces acting on a single particle on the bed of a channel include its weight, buoyancy, drag and lift due to the flowing water, and the frictional force between the channel bed and the particle. When a group of particles is present, the cohesion and the reactions of surrounding grains should also be considered. However, due to complex nature of this interaction, the general methodology used in analysis of sediment transport is to consider a single grain and then use empirical coefficients to account for the inter-grain interactions.

Weight and buoyancy

The gravitational and buoyancy forces acting on a particle are probably the easiest to determine and are often combined into a single term, called the *submerged weight*. Knowing the mass density of the particle[9] and its volume,[10] the submerged weight is written as $(\pi d^3/6) g(\rho_s - \rho_f)$ in which the subscripts s and f denote the sediment and the fluid (water) respectively and d is the nominal diameter of the particle. This force acts vertically down at the centre of gravity of the particle.

Drag force

The drag force on an object, F_D, is written as

$$F_D = C_D \frac{\rho_f V_f^2}{2} A_p \tag{8.2}$$

where C_D is the drag coefficient, A_p is the area of the particle projected perpendicular to the flow direction (also see footnote 13), and V_f is the flow velocity. Since the flow velocity is different at different locations and the sediment particle lies on the bed, a velocity at a point close to the channel bed should be used rather than the average velocity of flow over the cross section. However, near the channel bed, there is a significant change in the velocity over a very short distance and the choice of the elevation at which this velocity is taken becomes very critical. Generally the velocity at the top of the grain or at the centre is used. The projected area may be taken as proportional to the

---

[9]The *mass density* of sediments is generally taken as 2650 kg/m³. Since the fluid is invariably water and there is not much variation in its temperature, we would take the mass density of the fluid as being constant at 1000 kg/m³.

[10]We will use the nominal diameter, d, as representing the sediment size, unless mentioned otherwise. The volume of the particle will then be $\pi d^3/6$.

square of the nominal diameter, with the constant of proportionality obtained empirically (for a spherical particle, it would be $\pi/4$). The drag coefficient is a function of the Reynolds number, Re, defined as $Re = \rho_f V_f d/\mu$, in which μ is the dynamic viscosity of water. For spherical particles kept in a flow with a uniform velocity field, the drag coefficient has been analysed by numerous researchers. Figure 8.1 shows the variation of the drag coefficient with the Reynolds number of the particles.

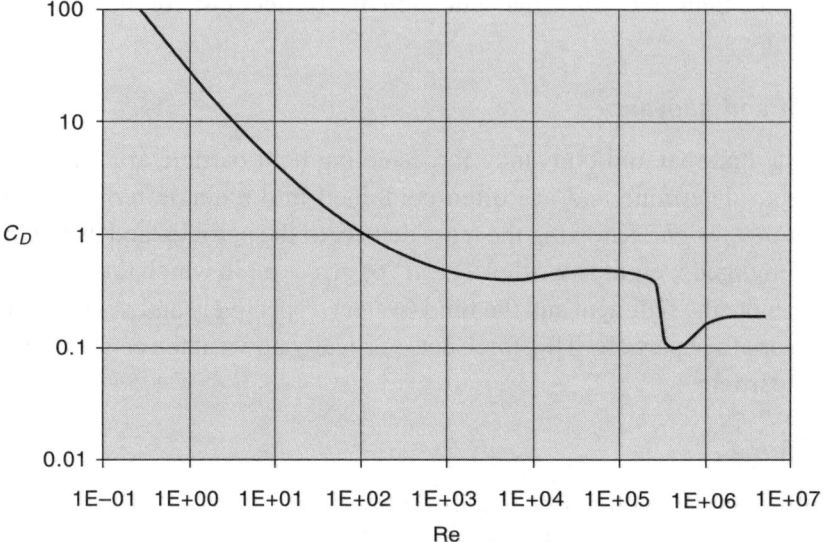

Fig. 8.1　Drag coefficient for spherical particles

At very low values of Re (less than about 0.1), the viscous forces are predominant and the drag coefficient is equal to 24/Re. The sharp dip in the coefficient at Re of about 3×10^5 occurs due to the change in the nature of the boundary layer from laminar to turbulent, which delays the separation and reduces the form drag.[11] Although the curve is useful, it is desirable to have expressions relating the drag coefficient to the Reynolds number. Approximate expressions relating C_D to Re have been suggested in several studies, e.g., Swamee and Ojha (1991). Brown and Lawler (2003) provide a recent review, summary of previous work, and propose simple approximations. Based on these and considering the errors inherent in the available data, the following approximation has been found to work well for Reynolds number up to 2×10^5:

$$C_D = \frac{24}{Re}\left(1 + 0.0888\sqrt{Re}\right)^{2.24} \tag{8.3}$$

---

[11]The reader is assumed to be familiar with fluid mechanics and detailed explanations are not given here.

Thus knowing the flow velocity at the reference point, one is able to compute the drag force on a spherical particle. Obviously the computed drag force has to be modified since it does not account for (i) velocity variation over the height of the particle, (ii) effect of nearby particles, and (iii) shape of the particle. It has also been found that the drag coefficient of a sphere rolling on a surface is higher than that of a similar spherical particle either kept in a uniform flow stream or falling in a fluid. Empirical coefficients are used to account for all these effects, as we will see in the next section.

Example 8.1   At a particular time, a spherical particle (mass density 2650 kg/m^3) of 1 mm diameter was observed to be falling in a stationary fluid (mass density 1000 kg/m^3, dynamic viscosity 0.001 Ns/m^2) with a velocity of 5 cm/s. Find the drag force exerted by the fluid on the particle. What is the net force on the particle?

Solution
Although the fluid is stationary and the sphere is moving, it is equivalent to the case of the fluid flowing past a stationary sphere. The Reynolds number is computed as

$$\mathrm{Re} = \frac{1000 \times 0.05 \times 0.001}{0.001} = 50$$

and the drag coefficient is obtained from Fig. 8.1 as about 2. To get a more accurate value, we use Eq. (8.3) to obtain C_D as 1.43. The drag force, acting upward, is obtained from Eq. (8.2) as

$$F_D = 1.43 \frac{1000 \times 0.05^2}{2} \frac{\pi 0.001^2}{4} = 2.0 \times 10^{-6} \text{ N}$$

The submerged weight acting downward is

$$F_s = \frac{\pi 0.001^3}{6} 9.81(2650 - 1000) \text{ N} = 8.5 \times 10^{-6} \text{ N}$$

which indicates a net downward force on the particle of **6.5 × 10^{-6} N**. Thus the particle will accelerate till the drag force balances the submerged weight. This leads us to a description of the fall velocity.

When we discussed the sediment properties, we defined a sedimentation diameter as the diameter of a spherical particle that has the same fall velocity as the given sediment particle. Moreover, when a sediment particle is carried by water, whether it would stay in suspension or settle down on the bed is largely dependent on its fall velocity. With the foregoing discussion of drag, we are now in a position to analyse the fall velocity in more details. As before, we consider a spherical particle and assume that it is falling in a liquid of large extent such that boundary effects are negligible.

As a particle starts falling in a liquid from rest, its velocity increases because its submerged weight is the only force acting on it. However, as the velocity increases, the drag force on the particle also increases, which causes a decrease in the net force acting downward and a consequent decrease in acceleration. After some time, the submerged weight and the drag force become equal in magnitude and the particle attains a constant velocity, known as the *fall velocity* (or *terminal fall velocity*, *terminal velocity*, or *settling velocity*). This velocity can be obtained by equating the drag force and the submerged weight as follows:

$$C_D \frac{\rho_f \omega^2}{2} \frac{\pi d^2}{4} = g(\rho_s - \rho_f)\frac{\pi d^3}{6} \tag{8.4}$$

in which we have denoted the fall velocity by its commonly used symbol, ω. Since the drag coefficient is a function of the Reynolds number, $\mathrm{Re} = \rho_f \omega d/\mu$, it is more convenient to write this equation in a dimensionless form as

$$\frac{3}{4}C_D \, \mathrm{Re}^2 = \frac{g\rho_f \, (\rho_s - \rho_f)d^3}{\mu^2} = d_*^3 \tag{8.5}$$

in which d_* is a dimensionless grain diameter defined as[12]

$$d_* = \left[\frac{g\rho_f \, (\rho_s - \rho_f)}{\mu^2}\right]^{1/3} d \tag{8.6}$$

Knowing the nondimensional diameter, we can obtain the Reynolds number (and, therefore, the fall velocity) from the implicit equation, Eq. (8.5), by an iterative process. A graph of the Reynolds number versus the dimensionless grain diameter is shown in Fig. 8.2 and may be used to avoid the iterative process. However, to avoid the errors associated with reading graphs and for convenience in mathematical modelling, explicit equations have been derived to evaluate the Reynolds number directly (Swamee and Ojha 1991, Brown and Lawler 2003). Based on these studies, the following approximation has

---

[12]In keeping with the current convention, we prefer to use the mass density and dynamic viscosity rather than the specific weight and the kinematic viscosity. Using the specific gravity of the sediment, s, and the kinematic viscosity of water, v, we may write $d_* =$

$$\left[\frac{g(s-1)}{v^2}\right]^{1/3} d \, .$$

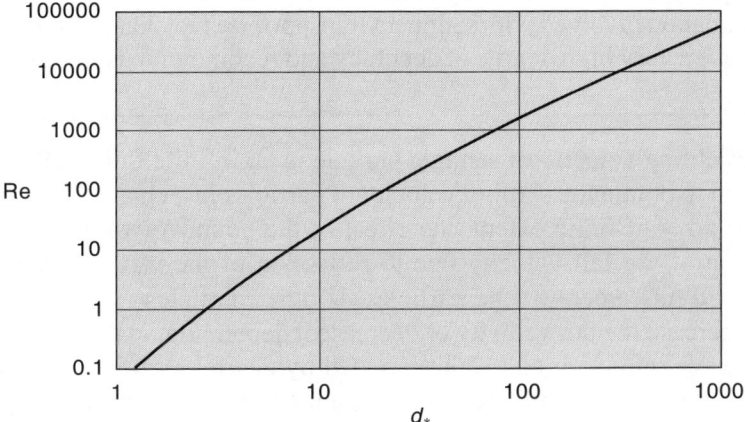

Fig. 8.2 Fall velocity for spherical particles

been found to work well for the estimation of fall velocity for $d_* < 2500$, which covers the Reynolds number range up to 2×10^5:

$$\text{Re} = \frac{d_*^3}{18 + 0.616 d_*^{3/2}} \qquad (8.7)$$

Example 8.2 Find the terminal fall velocity of a spherical particle (mass density 2650 kg/m³) of 1 mm diameter falling in a stationary fluid (mass density 1000 kg/m³, dynamic viscosity 0.001 Ns/m²). Compute the drag force exerted by the fluid on the particle. What is the net force on the particle?

Solution
We first compute the dimensionless grain diameter using Eq. (8.6) as

$$d_* = \left[\frac{9.81 \times 1000 \times (2650 - 1000)}{0.001^2} \right]^{1/3} 0.001 = 25.3$$

From Fig. 8.2, the Reynolds number is about 150. To get a more accurate value, we use Eq. (8.7) to obtain Re as 168, resulting in a fall velocity equal to $168 \times 0.001/1000 \times 0.001$ m/s, i.e., **16.8 cm/s**. The drag coefficient for this Reynolds number is obtained from Eq. (8.3) as 0.79, and the drag force acting upwards is obtained from Eq. (8.2) as

$$F_D = 0.79 \frac{1000 \times 0.168^2}{2} \frac{\pi 0.001^2}{4} = 8.7 \times 10^{-6} \text{ N}$$

This value is close to the submerged weight obtained earlier in Example 8.1. The slight discrepancy is due to the use of graphs/approximations. More accurate approximations for the drag coefficient and fall velocity are available in the literature, as mentioned earlier. However, considering the degree of approximation involved with sediment transport, we feel that it is not warranted to achieve a very high degree of accuracy and recommend using Eqs (8.3) and (8.7).

The settling velocity of a single particle in an infinite fluid will have to be modified to obtain the settling velocity of particles in a channel. For example, if the fluid is of finite extent, the effect of the boundary would be seen in a reduction of the fall velocity due to reflection of the waves created by the downward movement of the particles. If other particles are present, it may either increase the fall velocity or decrease it depending on the relative effects of the wave reflection and wakes created by nearby particles. Similarly, the shape of the particle will have a significant effect on the fall velocity. Again, empirical coefficients would be used to account for all these factors. Typical values of fall velocity range from about 5 mm/s to about 20 cm/s for particle sizes ranging from one-tenth of a mm to about 2 mm.

Lift force

Similar to the drag force, the lift force on an object is written as

$$F_L = C_L \frac{\rho_f V_f^2}{2} A_p \tag{8.8}$$

where C_L is the lift coefficient and A_p is the projected area of the particle in the horizontal plane.[13] The reference flow velocity, V_f, is again taken at an appropriate location (e.g., some researchers take it at a distance of d_{35} from the bed for a group of particles). For a single spherical particle in a uniform flow field, the lift coefficient has been extensively studied. However, a number of studies have shown the lift force to be of secondary importance in comparison with the drag force in field-scale sediment transport conditions and we will not discuss it further.

Frictional force

The frictional resistance offered by a sediment particle will depend on the angle of internal friction and, for cohesive materials, on the cohesion. We will concentrate on non-cohesive sediments and ignore the cohesion. Then, the friction force may be written as $F_f = (\pi d^3/6) g(\rho_s - \rho_f) \tan \phi$. Note that we have ignored the lift force. If it is substantial, it should be subtracted from the

---

[13]Sometimes, the projected area in the definition of drag and lift forces is taken as the maximum projected area of the particle. The value of C_L would be different depending on which area is used as A_p.

submerged weight to obtain the normal force for computation of the friction force.

As already discussed, presence of other particles causes the forces on a particle to be different from those obtained by assuming it to be isolated. For example, nonuniform size distribution may result in shielding of some particles behind larger particles. Thus after an initial period in which the fine material is eroded, the coarser particles form a protective layer and prevent further erosion (this process is known as *armouring*). To establish the conditions for initiation of motion in a practical situation, therefore, we will have to use various empirical coefficients for the forces acting on a particle and the resistance offered by it. Various theories concerning this critical condition have been proposed and some of them are described next.

8.2.3 Theories of Initiation of Motion

There are two different ways to define the critical condition. In the first one, we may consider an initially motionless bed, keep on increasing the flow rate, and see when the particles on the bed start to move. The second approach is to start from a large flow rate at which the particles are already moving, keep reducing the flow rate, and note the point at which the particles stop moving (or we can measure the amount of sediment being transported by the flow under different conditions and extrapolate to obtain the *zero transport* condition). Clearly, the first approach will result in a lower flow rate (or velocity or shear stress) and is the one preferred to define the critical conditions. Even in this approach, there can be different options of defining the *initiation of motion*. For example, some researchers may take the critical condition as the movement of a *single particle*, some others may look for the movement of a *significant amount* of material, and still others may look for a *general motion* of the entire bed material. Thus an amount of subjectivity creeps in the experimental observations. The theories that we describe here involve considering a single particle on the bed and obtaining the flow conditions when this particle just starts to move. Broadly, three theories are described here—based on the (i) flow velocity, (ii) vertical motion of particles, and (iii) horizontal movement of the particles.

Theory based on flow velocity

This theory is probably the earliest theory in which the critical velocity was correlated with the sediment size based on the laboratory and field observations. Later, researchers attempted to provide some theoretical justification to these empirical observations based on a balance of driving and resisting forces on a particle. For example, assuming the drag coefficient to be a constant, the drag force would be proportional to the square of the velocity[14] and also to the square of the grain diameter.[15] If it has to balance the frictional resistance, which is proportional to the cube of the diameter,[16] it follows that the critical velocity should be proportional to the square root of the grain diameter as observed in a number of studies. R.J. Garde in 1970 expressed this relationship, after analysing a large set of data, as

$$\frac{v_c}{\sqrt{g(s-1)d}} = 1.51 \tag{8.9}$$

in which v_c is the critical value of the grain-level velocity (i.e., at the top of the grain), and s is the specific gravity of the sediment. The following relationship for the critical velocity (also called the *competent velocity* due to it being competent in moving a particle) was also proposed:

$$\frac{V_c}{\sqrt{g(s-1)d}} = 0.5\log\frac{y}{d} + 1.63 \tag{8.10}$$

in which V_c is the critical value of the cross-sectional average flow velocity and y is the flow depth. The assumption of a constant drag coefficient, as we already know, is not justifiable (except, may be over a small range of the Reynolds number, see Fig. 8.1). Therefore, we do not discuss this theory further.

Example 8.3   Find the competent velocity of water for moving a spherical particle (mass density 2650 kg/m³) of 1 mm diameter placed on the channel bed. The flow depth may be taken as close to 10 cm at the critical point.

Solution
Using Eq. (8.10), we obtain the competent velocity as

$$V_c = \sqrt{9.81(2.65-1)0.001}\left(0.5\log\frac{0.1}{0.001} + 1.63\right) = \textbf{33 cm/s}$$

The critical velocity at the grain level, i.e., 1 mm above the bed is obtained from Eq. (8.9) as $v_c = 1.51\sqrt{9.81(2.65-1)0.001} = 19$ cm/s.

---

[14]This velocity is at the grain level, i.e., close to the channel bed. However, it is expected that it would be a function of the average velocity of the flow and the flow depth. Thus if we use the grain-level velocity, we would expect the relationship to be invariant for different flow depths. However, if we use the cross-sectional average velocity, an additional factor involving the flow depth would enter the analysis.

[15]The projected area is proportional to the square of the grain diameter.

[16]The frictional resistance is proportional to the submerged weight, which, in turn, is proportional to d^3.

Theory based on the vertical motion

This theory was proposed in the 1920s and considered the critical condition as the point at which the lift force acting on the particle is just equal to its submerged weight so that a slight increase in the flow rate would lift the particle up. The lift is created by the difference in velocities at the bottom level of the grain (where it may be taken as zero since the particle is not moving) and that at the top level. Analysing the case of a cylindrical particle using the ideal fluid flow theory, it was shown by H. Jeffreys in 1920s that the critical value of the free-stream velocity is given by

$$\frac{V_c}{\sqrt{g(s-1)d}} = 0.59 \tag{8.11}$$

For spherical particles, the lift force will be smaller since the fluid will also move around the bottom. Therefore, the critical velocity will have to be higher to be able to lift the particle. Also, the application of the ideal fluid results to the real case may not be justified because viscosity will have a significant influence on the drag and lift forces. Some other studies have also used the lift theory but there is still no well-accepted theory concerning the critical conditions. In fact, some of these studies have concluded that the lift force is insignificant in comparison to the drag. The theory based on the horizontal movement described next has been the most successful in defining a critical condition for the incipient motion.

Example 8.4   Find the competent velocity of water for moving a 1 cm long cylindrical particle (mass density 2650 kg/m^3) of 1 mm diameter placed on the channel bed. The flow depth may be taken as close to 10 cm at the critical point.

Solution
Using Eq. (8.11), we obtain the competent velocity as

$$V_c = 0.59\sqrt{9.81(2.65-1)0.001} = \textbf{8 cm/s}$$

which is much smaller than that required for a spherical particle of the same diameter, as shown in Example 8.3.

Theory based on the horizontal motion

This theory defines the critical condition as the point at which the driving force (also called the *tractive force*) at the bed is just equal to the resisting force acting on a particle due to friction and a slight increase in the flow rate would cause the particle to move. The tractive force can be defined in terms of the drag force acting on a particle. However, it is more convenient to define it in terms of the shear stress applied by the flowing water. Since the shear stress on a channel boundary varies from point to point (Appendix B), one should ideally take the maximum shear stress over the boundary to obtain the critical condition. However, if the channel is very wide, the maximum shear stress and average shear stress are not significantly different. The earliest studies attempted to obtain an empirical relation between the bed shear stress, τ_0, and the grain size of the sediments. For example, the critical shear stress was found to increase from about 2 N/m^2 for a grain size of half-a-mm to about 50 N/m^2 for a 50 mm size. Further empirical studies included the sediment properties and the size distribution also in the equation. For example, H. Kramer in 1930s proposed the following equation to determine the critical tractive force:

$$\frac{\tau_{0c}}{g(\rho_s - \rho_f)d} = \frac{1.66 \times 10^{-5}}{M} \tag{8.12}$$

where M is the uniformity coefficient defined earlier [Eq. (8.1)] and the subscript c denotes the critical condition. Other equations proposed also account for the shape of the particles. However, we will not consider these empirical equations and look at some theoretical studies (although these are also not completely free from the use of empirical coefficients).

Shields (1936) proposed a theory based on the tractive-force approach that attempted to address the issue of the initiation of sediment motion on more rational grounds than previously used. Various modifications and improvements have been suggested in this theory but it remains essentially the same as that originally proposed. We, therefore, start with a description of this theory and then consider a few modifications.

Shields considered the resistance offered by a particle to be proportional to its submerged weight and the coefficient of friction between the particle and the channel bed. The submerged weight was considered to be proportional to the submerged specific weight and the cube of the grain diameter, with the constant of proportionality dependent on the grain shape. Merging the constants in one, we may write the resisting frictional force as

$$F_f = \alpha_1 g (\rho_s - \rho_f)d^3 \tag{8.13}$$

in which d is the nominal diameter of the particle and α_1 is an empirical constant. The drag force applied by the flow on the particle is written as

$$F_D = \alpha_2 C_D \frac{\rho_f v^2}{2} d^2 \tag{8.14}$$

in which α_2 is another empirical constant that depends on the shape of the particle (such that $\alpha_2 d^2$ is the projected area of the particle) and v is the velocity at the grain level. The drag coefficient, C_D, will be a function of the Reynolds number, $\rho_f vd/\mu$, as well as the shape of the particle (e.g., Fig. 8.1 for spherical particles). The flow velocity at the grain level (considered at a height above the channel bed equal to the grain diameter) may be related to the shear stress at the bed by using the well-established velocity distribution laws for rough pipes (Appendix A) as[17]

$$\frac{v}{v_*} = f\left(\frac{\rho_f v_* d}{\mu}\right) \tag{8.15}$$

in which v_* is the shear velocity defined as $\sqrt{\tau_0/\rho}$. Thus the drag coefficient also becomes a function of the shear Reynolds number (also called the particle Reynolds number), Re_* $(= \rho_f v_* d/\mu)$, in addition to the particle shape, and Eq. (8.14) may be written as

$$F_D = \alpha_3 f_1\left(\frac{\rho_f v_* d}{\mu}\right)\rho_f v_*^2 d^2 = \alpha_3 f_1(\mathrm{Re}_*)\tau_0 d^2 \tag{8.16}$$

α_3 being another empirical constant and the function f_1 also depending on the particle shape. At critical conditions, the drag force would be just equal to the resistance, and we get

$$\alpha_3 f_1(\mathrm{Re}_*)\tau_0 d^2 = \alpha_1 g(\rho_s - \rho_f)d^3 \tag{8.17}$$

which provides us the classical Shields relationship

$$\tau_{*c}\left[= \frac{\tau_{0c}}{g(\rho_s - \rho_f)d}\right] = f_2(\mathrm{Re}_{*c}) \tag{8.18}$$

with the parameter τ_{*c} known as the Shields parameter. Shields plotted (the Shields diagram) a number of laboratory and field data sets in terms of the variables, τ_* and Re_* and delineated the regions of no sediment motion and appreciable motion through a line,[18] known as the Shields curve. Figure 8.3

---

[17]Note that the actual relationship is not important since we would be using various empirical coefficients later on.

[18]In the original Shields study, the demarcation is done by a *region* and not a line. Thus there is a narrow band in which the sediment particles may or may not be in motion. However, it has gradually been replaced by a single line, first suggested by Rouse in 1930s and then modified by various researchers as more data became available. Figure 8.3 shows the currently accepted line. The constant value of the Shields parameter at large Reynolds number has been given different values. Rouse suggested 0.06 and some other researchers found it to be as low as 0.03. A value of 0.045 to 0.052 is now accepted to be more representative.

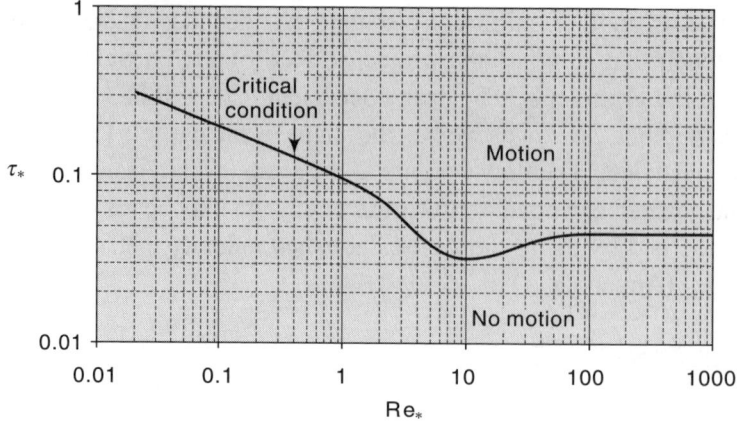

Fig. 8.3 Critical shear stress as a function of shear Reynolds number

shows a plot representing Eq. (8.18), which is based on slight modification of the original Shields curve by Yalin and Karahan (1979). Once the critical shear stress is obtained, the corresponding flow depth can be obtained using the relationship between the shear stress and the hydraulic radius, $\tau_0 = \rho_f gRS_0$.

Out of these three different approaches, we would describe the approach based on the horizontal movement, also called the tractive-force approach, which is the most commonly used approach. As seen in the Shields theory, the critical value of the nondimensional shear stress, i.e., the Shields parameter, is a function of the grain shear Reynolds number. The Shields parameter may be interpreted as being proportional to the ratio of the boundary shear force acting on the particle (note that the area on which the shear force is acting would be proportional to the square of the nominal diameter) to its submerged weight. The Reynolds number, Re_*, may be thought of as representing the ratio of the particle diameter and the thickness of the viscous sublayer (which is proportional to $\mu/\rho_f v_*$, see Appendix A). For small values of Re_* (less than 1), the particle is completely within the laminar sublayer, the boundary is considered smooth, and the viscous drag predominates.[19] The drag coefficient is inversely proportional to the Reynolds number [which implies that the drag force is directly proportional to the velocity, see Eq. (8.2)]. On the other hand, for very large Re_* (more than 100), the flow is fully rough and the drag coefficient becomes more or less constant (indicating that the drag force is proportional to the square of the velocity). For intermediate values, there would be a transition from smooth boundary to rough, as seen in Fig. 8.3.

---

[19]This is true irrespective of whether the flow itself is laminar or turbulent. However, once the particle size becomes of the same order as the sublayer thickness, the drag would depend on the nature of the flow. Since most open channel flows are turbulent, we do not consider the cases when the flow is laminar.

A number of studies have been done related to the tractive-force approach for predicting the critical condition corresponding to the initiation of motion. Buffington and Montgomery (1997) provide an excellent summary of these. These studies aimed at improving the prediction of critical conditions by considering additional factors in the ananlysis. For example, C.M. White in 1940 considered a packing coefficient for the particles, Y. Iwagaki in 1950s included the lift force (by subtracting it from the submerged weight), I.V. Egiazaroff in 1960s assumed that the velocity at a height of 0.63 d from the bed is equal to the fall velocity, and V.A. Vanoni in 1960s considered the conditions in which the flow was not fully developed. The end result of all these studies is a diagram similar to Shields diagram with slight variations in the location of the critical curve. Therefore, we would not go deeper into the subject and take Fig. 8.3 as the accepted criterion for determining the critical condition.

Example 8.5 Particles of uniform size (mass density 2650 kg/m^3, diameter 1 mm) comprise the bed of a wide rectangular channel in which the flow depth is 10 cm and the bed slope is 1 in 1000. Would the channel behave as a rigid-boundary channel or a mobile-boundary channel? What depth of flow would be just sufficient to move the particles?

Solution

The shear stress at the bed is obtained as (for wide rectangular channels, the hydraulic radius is equal to the flow depth)

$$\tau_0 = \rho_f g R S_0 = 1000 \times 9.81 \times 0.1 \times 0.001 \text{ N/m}^2 = 0.981 \text{ N/m}^2$$

Therefore, the Shields parameter is

$$\tau_* = \frac{\tau_0}{g(\rho_s - \rho_f)d} = 0.06$$

The shear velocity is

$$v_* = \sqrt{\tau_0/\rho_f} = 3.1 \text{ cm/s}$$

and the Reynolds number is Re$_*$ = $\rho_f v_* d/\mu$ = 31.3

From Fig. 8.3, it is seen that this combination of shear stress and Reynolds number lies above the Shields curve and, therefore, **the bed particles will move**.

To find the depth just sufficient to move the particles, we see that the Shields parameter should be about 0.04 for a Reynolds number of 31.3. However, since the Reynolds number is based on the shear velocity and it will change with the flow depth, the Reynolds number will not stay as 31.3. A rough estimate may, however, be based on the Shields parameter of 0.04, which will result in a flow depth of about **6.7 cm.**

We saw that the form of Fig. 8.3 is suitable for predicting whether the particles will move or not *given the particle size and the shear stress at the bed*. One can then plot the point on the Shields diagram and find, depending on whether it plots above or below the critical line, whether the particles will move or not. However, if one is interested in finding out a critical shear stress at the bed which will move particles of a given size, Fig. 8.3 is not very convenient to use since both the Reynolds number and the Shields parameter involve the unknown (shear stress and shear velocity). Shields suggested an iterative procedure and found that a single iteration was generally sufficient. However, later studies proposed alternative ways of plotting the diagram that enabled direct determination of the critical shear stress. Julien (1995) used the nondimensional diameter defined in Eq. (8.6) (and, sometimes, called Rouse's auxiliary parameter) and obtained

$$d_* = \left[\frac{g \rho_f (\rho_s - \rho_f)}{\mu^2} \right]^{1/3} d = \left[\frac{g (\rho_s - \rho_f) d^3 \rho_f^2 v_*^2}{\tau_0 \mu^2} \right]^{1/3} = \frac{Re_*^{2/3}}{\tau_{0*}^{1/3}}$$

(8.19)

Then, using the values of the Reynolds number and the nondimensional shear stress at the critical condition from Fig. 8.3, a plot of d_* versus the critical shear stress was prepared and is shown in Fig. 8.4 (computed with the modified Shields diagram).

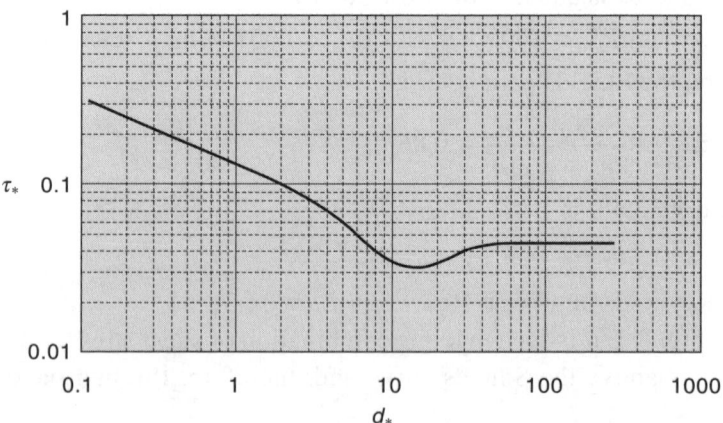

Fig. 8.4 Critical shear stress as a function of dimensionless diameter

From this figure, for a given particle size, the critical shear stress may be directly obtained. Explicit approximations for this purpose are also available in literature (Swamee and Mittal 1976, Brownlie 1981). The following approximation provides a reasonably accurate estimate of the critical shear stress:

$$\tau_{*_c} = \frac{3630 - 126 d_* + 0.045 d_*^4}{27000 d_*^{0.4} + d_*^4} \qquad (8.20)$$

Example 8.6 Particles of uniform size (mass density 2650 kg/m³, diameter 1 mm) comprise the bed of a wide rectangular channel in which the flow depth is 10 cm and the bed slope is 1 in 1000. What depth of flow would be just sufficient to move the particles?

Solution

The nondimensional particle diameter is

$$d_* = \left[\frac{g \rho_f (\rho_s - \rho_f)}{\mu^2} \right]^{1/3} d = 25.3$$

From Fig. 8.4, we get the critical shear stress parameter as about 0.035, or using Eq. (8.20), we get a value of 0.037, and the corresponding flow depth as **6.2 cm**.

In Example 8.6, once the depth increases beyond 6.2 cm, the particles will start moving. Once the motion starts, the bed will no longer remain plane. The resulting bed forms will strongly influence the resistance to flow and should be considered in the analysis of sediment transport through the channel. This topic is discussed in the next section.

8.3 Bed Forms

Once the particles on the bed start moving, small ripples are formed on the bed [Fig. 8.5(a)], which increase the resistance to flow. The ripples generally have a long and flat upstream slope and a much steeper downstream slope. There are various theories that explain the mechanism of formation of these ripples. For example, A.G. Anderson in 1950s hypothesized that water waves are the primary cause of ripple formation, H.K. Liu in 1950s suggested that these are formed due to the instability of the interface between the water and the bed, while later many researchers attributed the formation of ripples to amplification of slight perturbations in the bed profile (which are created due to nonuniform and random motion of the eroded sediment) and some attribute it to turbulence (see Coleman and Fenton 2000 for a recent study). Due to the continuous erosion and deposition of sediments over these ripples, it is observed that these ripples move downstream, although at a very slow speed. As the flow rate is increased further, the size of these ripples increases and they are

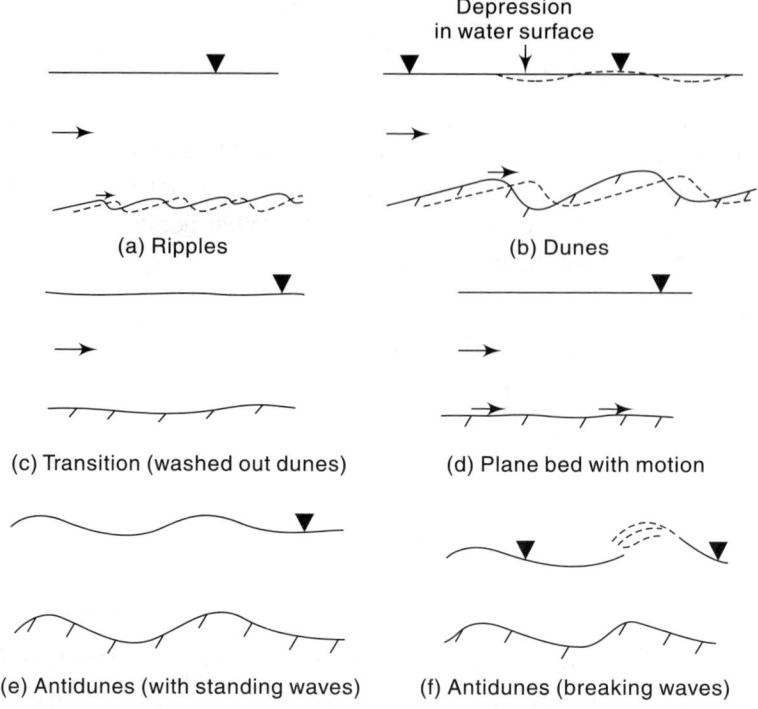

Fig. 8.5 Typical bed forms in an alluvial stream

no longer small ripples and are classified as *dunes* [Fig. 8.5(b)]. A dune may also have small ripples present on its surface.

Further increase in flow may lead to a washout of the dunes (if the bed material is fine enough) and will lead to the formation of a plane bed but with considerable sediment motion [Figs 8.5(c) and (d)]. For coarser material, the flow may not have enough *power* to wash out the dunes. The presence of dunes causes the water surface to be wavy with the crest and trough in phase with those on the bed. It has been observed that if a disturbance is created on the water surface, scouring occurs in the troughs and deposition on the crest and it appears that the dunes are moving upstream. These bed forms are called antidunes[20] and have a steeper upstream slope and flatter downstream profile (Fig. 8.5e). The growth of the size of antidunes leads to further perturbation in water velocities which, in turn, increases the size of the bed forms. However, as the antidunes reach a certain size, the water surface waves tend to break (Fig. 8.5f). Since the flow resistance is significantly influenced by these bed forms (also called *regimes of flow* with ripples and dunes included in *lower*

---

[20]The antidunes do not necessarily move in the upstream directions. Sometimes, they have been observed to move downstream or even stationary.

regime and occurring under subcritical flow conditions, and the others in *upper regime* occurring at Froude numbers near or greater than 1), it is important to predict the type of bed form that may occur in a channel, based on the sediment and flow properties. A number of studies are available for this purpose and some of these are described below.

G.K. Gilbert in early 20th century related the regimes of flow to the slope of the channel and its capacity (defined as the product of velocity and hydraulic radius). In 1950s, M.L. Albertson and coworkers used the shear Reynolds number and the ratio of shear velocity to the fall velocity as the two parameters influencing the regime. J. Bogardi in 1950s used the sediment diameter and a shear Froude number ($v_*/\sqrt{gd}$) as the relevant parameters. D.B. Simons in 1960s proposed a stream power (defined as the product of velocity and bed shear stress) versus diameter plot to obtain the regime. Most of these studies are empirical and relied on plotting the observed data with the proposed parameters used as the coordinate axes, and then demarcating different regimes. Due to the nature of the process, empiricism cannot be completely avoided. However, to choose the important parameters, we may use a little more theoretical approaches.

Since the bed forms develop when the shear stress at the boundary increases beyond the critical shear stress, it is logical to relate the bedforms to a parameter representing the *additional* shear stress beyond its critical value. The two most obvious choices are: (i) using the ratio of the actual shear stress and the critical shear stress and (ii) using the difference of the actual shear stress and the critical shear stress. We describe both these techniques without comparing them since a comparison of different methods of predicting the regimes of flow in the past has shown it to be largely data-dependent (some methods work well on some data sets and fare poorly on other data sets).

8.3.1 Garde and Ranga Raju Method

This method, proposed in the 1960s (and a similar method proposed by S. Sugio around the same time), is based on the ratio of the actual shear stress and the critical shear stress. However, the nondimensional critical shear stress is taken constant at 0.05 rather than dependent on the particle Reynolds number. A slightly modified form of their equations is presented here. The ratio of the shear stress is defined as

$$\frac{\tau_0}{\tau_{0c}} = \frac{\rho_f \, gRS}{0.05 \, g \, (\rho_s - \rho_f) d} = \frac{1}{0.05} \frac{R}{d} \frac{S}{\left(\dfrac{\rho_s}{\rho_f} - 1\right)} \tag{8.21}$$

and is plotted against the relative flow depth, R/d, with the lines of demarcation of different regimes located by the following expressions:

$\dfrac{\tau_0}{\tau_{0c}}$	Flow regime
< 1	No motion
Between 1 and $0.28(R/d)^{0.54}$	Ripples and dunes
Between $0.28(R/d)^{0.54}$ and $1.18(R/d)^{0.46}$	Transition
$> 1.18\,(R/d)^{0.46}$	Antidunes

Alternatively, the following figure can be used to quickly estimate the regime of flow:

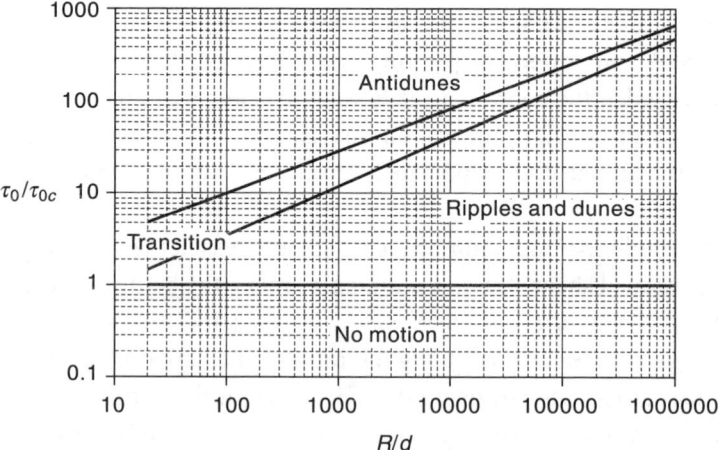

Fig. 8.6 Prediction of the regimes of flow using Garde–Ranga Raju method

8.3.2 van Rijn (1993) Method

This method, originally proposed in the 1980s, is based on the difference of the actual shear stress and the critical shear stress. The shear stress as used in this methodology, is that corresponding to the grain roughness only (i.e., effect of the bed forms is not included) and is defined as

$$\tau' = \rho_f g R S' = \frac{\rho_f g V^2}{C'^2} \tag{8.22}$$

in which the prime indicates that the values are corresponding to the grains only, R is the hydraulic radius, V is the average flow velocity, and C is the Chezy's coefficient. Based on theoretical results and experimental observations, and assuming a wide channel, the Chezy's coefficient was taken as $C' = 18$

log $4y/d_{90}$, with y being the flow depth and d_{90} the sediment size for which 90% of the material is finer.[21] A transport stage parameter, T, was defined as

$$T = \frac{\tau_0' - \tau_{0c}}{\tau_{0c}} \qquad (8.23)$$

representing the excess of shear stress nondimensionalized with the critical shear. The other parameter used by van Rijn is the nondimensional diameter, d_* [see Eq. (8.6)]. For d_* less than 10, ripples were observed when T was less than 3 and dunes were observed when T was between 3 and 15. For d_* greater than 10, dunes were observed when T was less than 15. The zone of transition was observed when T was greater than 15, irrespective of d_*, and upper regime bedforms are likely to exist when T is greater than 25 (van Rijn did not consider antidunes and the upper regime was considered to be flat bed with sediment motion). Figure 8.7 shows the classification in a graphical form.

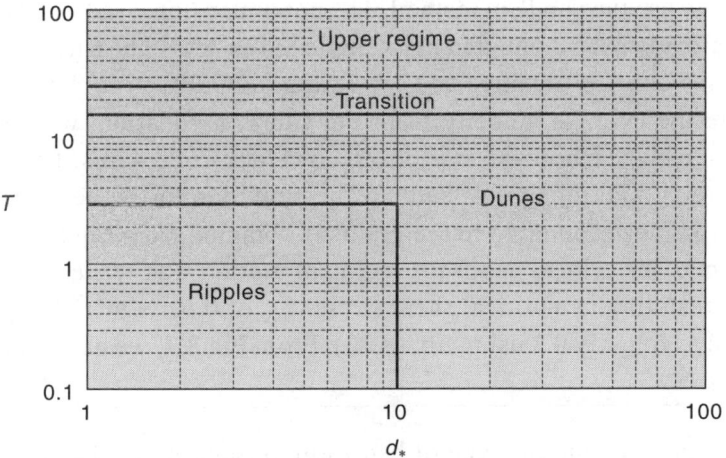

Fig. 8.7 Prediction of the regimes of flow using van Rijn method

Other techniques of the regime prediction are not described here. However, one observation (Brownlie 1981) mentioned that at bed slopes steeper than about 6 in 1000, the flow is always in the upper regime. We conclude this section by cautioning that in natural channels, more than one regimes may exist at the same section, e.g., dunes in deeper portions of a cross-section and plane bed in the shallower portions. The prediction of the regime is still more of an art rather than an exact science.

---

[21]The implicit assumption in the Chezy's coefficient is that the roughness height is equal to $3d_{90}$. Other assumptions are also possible, e.g., a roughness height equal to 2.5 or 3 times the median diameter may be used.

Example 8.7 Particles of uniform size (mass density 2650 kg/m³, diameter 1 mm) comprise the bed of a wide rectangular channel in which the flow depth is 10 cm, the bed slope is 1 in 1000, and the flow velocity was measured as 40 cm/s. What regime of flow is expected in such conditions?

Solution

We first use the Garde–Ranga Raju method. The hydraulic radius is taken as the flow depth since the channel is wide. Therefore, $R/d = 100$. The shear stress ratio is obtained from Eq. (8.21) as

$$1/0.05 \ R/d \ S/(\rho_s/\rho_f - 1) = 1.21$$

From Fig. (8.6), we see that this point lies in the ripples and dunes portion of the plot. Hence, the expected bed form would be ripples or dunes (which of these two occur in fact does not matter since the behaviours of the two are quite similar).

We now use the van Rijn method. The nondimensional sediment diameter, d_*, has already been computed in Example 8.6 as **25.3**. The Chezy's constant (for grains only) is obtained as $C' = 18 \log 0.4/0.001 = 46.8 \ \text{m}^{1/2}/\text{s}$. (If we compute the average velocity from the Chezy's equation, we obtain $V = 0.47$ m/s, which is more than the observed velocity of 0.4 m/s. This indicates that there is additional resistance to the flow due to the bed forms.) The shear stress corresponding to the grains is obtained from Eq. (8.22) as $\tau'_0 = 0.72 \ \text{N/m}^2$. The critical depth for sediment motion was earlier computed as 6.2 cm, resulting in a critical shear stress, τ_{0c}, of 0.61 N/m². Therefore, T is obtained as roughly **0.2** using Eq. (8.23). From Fig 8.7, we obtain the regime as dunes.

Although knowing the type of bed form present in the channel for given flow conditions is important, it is equally important to know the size and shape of these bed forms since it will affect the resistance to flow. Again, there are a number of studies on this subject and we describe only a few here.

For ripples, Mantz (1992) related the height (h) and length (l) of the ripples to the nondimensional shear stress and the shear Reynolds number as follows:

$$\frac{h}{d} = 195\tau_*^{0.55}\text{Re}_*^{-1.53} \tag{8.24a}$$

$$\frac{l}{d} = 2240\tau_*^{0.82}\text{Re}_*^{-1.51} \tag{8.24b}$$

Another observation made by Mantz was that the bed becomes flat for the bed shear stress larger than fifteen times the critical shear stress, similar to that observed by van Rijn (Fig. 8.7).

For dunes, van Rijn (1984) related the relative height, h/y (y being the flow depth), to the Transport parameter, T [Eq. (8.23)], as

$$\frac{h}{y} = 0.11 \left(\frac{d_{50}}{y}\right)^{0.3} [1 - \exp(-0.05T)](25 - T) \tag{8.25}$$

(Recall that for $T > 25$, a flat bed is observed.) The length of the ripples was found to be 7.3 times the flow depth. Julien and Klaassen (1995) suggested that for large rivers, a more appropriate relationship would be

$$\frac{h}{y} = 2.5 \left(\frac{d_{50}}{y}\right)^{0.3} \tag{8.26}$$

and the length of the bed forms is about 6.5 times the flow depth.

For antidunes, the geometry of the bed forms is a function of the Froude number (antidunes generally occur in a flow with Froude number range from about 0.8 to 1.8). The length of the antidunes as proposed by J.F. Kennedy in 1960s is related to the Froude number by

$$\frac{l}{y} = 2\pi F_r^2 \tag{8.27}$$

Example 8.8   Particles of uniform size (mass density 2650 kg/m^3, diameter 1 mm) comprise the bed of a wide rectangular channel in which the flow depth is 10 cm, the bed slope is 1 in 1000, and the flow velocity was measured as 40 cm/s. Estimate the height and length of the bed forms.

Solution

We have already seen in the previous example that dunes would be formed on the bed under the given conditions. Since it appears to be a small channel, we prefer to use van Rijn method over the Julien and Klaassen method (which is more appropriate for large rivers). The transport parameter, T has already been found for this case as 0.2. Using Eq. (8.25), we obtain the height of the dunes as

$$h = 0.11 \times 0.1 \times \left(\frac{0.001}{0.1}\right)^{0.3} [1 - \exp(-0.05 \times 0.2)] (25 - 0.2) \text{ m}$$

$$= \mathbf{0.7 \ mm}$$

which is negligible [using Eq. (8.26), we get the height as 6 cm]. The length of the ripples is estimated as 7.3 times the flow depth, i.e., **73 cm**.

8.4 Channel Resistance

The presence of bed forms increases the channel resistance and reduces the flow velocity in comparison to the equivalent conditions in a plane bed with no sediment motion. Thus the stage discharge curve of an alluvial channel will show a decrease in discharge as the critical conditions are reached and the sediment movement starts to occur. On the other hand, at the point of transition from the lower regime to upper regime, when the ripples and dunes are washed out and the bed becomes plane, there would be an increase in the discharge. Thus use of the rigid-bed formulae to estimate the discharge in an alluvial channel may lead to significant errors.

In order to find the velocity of flow in presence of bed forms, we can use an equation similar to the Manning's or Chezy's equation and increase the roughness coefficient (n) or decrease the value of Chezy's coefficient to account for the increased resistance to flow. For example, Karim and Kennedy (1990) proposed an equation for the computation of flow velocity, which implies that the Manning's n is increased roughly by a factor of $[1.2 + 8.92\ (h/y)]^{0.465}$ for bed forms of height h. H.A. Einstein and N. Barbarossa in 1950s separated the total resistance into a grain resistance and a form resistance such that the shear stress at the bed is written as

$$\tau_0 = \tau_0' + \tau_0'' \tag{8.28}$$

with the prime indicating surface or grain resistance and the double prime indicating the bed-form resistance. The shear stresses were obtained as $\rho g R S$, with the same slope used in all expressions but the hydraulic radius written as $R = R' + R''$. In 1960s, F. Engelund used similar subdivision but divided the slope into two components as $S = S' + S''$. Here we describe two comparatively recent methods of finding the velocity in alluvial channels.

van Rijn (1984) method

In this method, the total roughness height is taken as the sum of a grain roughness and a form roughness as $k_s = k'_s + k''_s$. The grain roughness is taken as three times the grain diameter, d_{90}, and the form roughness is taken as $1.1h[1 - \exp(-25h/l)]$, providing the following relationship for the average velocity of flow:

$$V = 5.75 \sqrt{gRS} \log \frac{12R}{3d_{90} + 1.1h[1 - \exp(-25h/l)]} \tag{8.29}$$

The geometric parameters of the bed form, h and l, are found from Eqs (8.24) given in the previous section.[22]

Karim and Keneedy method

In this method, the friction factor for flow with a moving bed is related to that for a rigid bed. The rigid-bed friction factor is computed from

$$f_0 = \frac{8}{\left(5.75\log\dfrac{12\,y}{2.5\,d_{50}}\right)^2} \tag{8.30}$$

(note that the grain roughness height is taken as 2.5 times the median diameter) and the ratio of the regime bed friction factor and f_0 is given by

$$\frac{f}{f_0} = 1.20 + 8.92\frac{h}{y} \tag{8.31}$$

The height of the bed forms, h, was related to the nondimensional shear stress as

$$\frac{h}{y} = 0.08 + 0.75\,\tau_* - 2.0\,\tau_*^2 + 2.6\,\tau_*^3 - 1.1\,\tau_*^4 \tag{8.32}$$

for τ_* up to 1.5 and zero beyond that.[23] The flow velocity was obtained by regression on available data as

$$V = 6.683\sqrt{g\,(s-1)\,d_{50}}\left(\frac{y}{d_{50}}\right)^{0.626} S^{0.503}\left(\frac{f}{f_0}\right)^{-0.465} \tag{8.33}$$

in which s is the specific gravity of the sediment and S is the bed slope. Note that we do not need to compute the friction factor, f_0, since only the friction factor ratio is required to estimate the velocity.

Example 8.9   Particles of uniform size (mass density 2650 kg/m³, diameter 1 mm) comprise the bed of a wide rectangular channel in which the flow depth is 10 cm and the bed slope is 1 in 1000. Estimate the flow velocity.

Solution

Using the van Rijn method:

We have hydraulic radius, $R = 0.1$ m, bed slope, $S = 0.001$, height of the bed form as found in the previous example, $h = 0.7$ mm, length of the bed forms,

---

[22]For antidunes, the Froude number affects the bed form geometry. Thus the process of finding the velocity would require iterations. However, antidunes are not very common and we do not discuss this issue.

[23]Another relationship for the bed form height was also suggested later, which expressed it as a function of the ratio of the shear velocity to the fall velocity. However, we mention only one of the possibilities here.

$l = 0.73$ m, and $d_{90} = 0.001$ m.

Therefore, using Eq. (8.29), we get

$$V = 5.75\sqrt{9.81 \times 0.1 \times 0.001}$$

$$\times \log \frac{12 \times 0.1}{3 \times 0.001 + 1.1 \times 0.0006[1 - \exp(-25 \times 0.0006/0.73)]}$$

i.e., **47 cm/s**.

(Recall that in the previous example, it is mentioned that the velocity was measured as 40 cm/s. So we are not very much off.)

Using the Karim and Kennedy method:

The rigid-bed friction factor is computed (although not needed) as

$$f_0 = \frac{8}{\left(5.75 \log \dfrac{12 \times 0.1}{2.5 \times 0.001}\right)^2} = 0.0337$$

The nondimensional shear stress was obtained as 0.06 (Example 8.5). Therefore, the height of the bed forms, h, is obtained as

$$h = 0.1(0.08 + 0.75\tau_* - 2.0\tau_*^2 + 2.6\tau_*^3 - 1.1\tau_*^4) = 1.18 \text{ cm}$$

The ratio of the regime-bed friction factor and f_0 is then given by

$$\frac{f}{f_0} = 1.20 + 8.92\frac{h}{y} = 2.26$$

The flow velocity is obtained from Eq. (8.33) as

$$V = 6.683\sqrt{9.81 \times 1.65 \times 0.001}\,(100)^{0.626}\,0.001^{0.503}(2.26)^{-0.465} = \textbf{32.2 cm/s}$$

which is much smaller than that obtained by van Rijn method (since the height of the bed forms was much larger than that in the van Rijn method, we do expect the velocity to be smaller).

Which of these velocities to use is a question not easy to answer. These expressions were developed by considering a wide range of data. However, there are a number of uncertainties involved in the process of bed form development and its effect on the resistance. It would seem that a velocity of 40 cm/s can be used as the *best* estimate in this case.

The procedure described here to estimate the velocity of flow can be used to develop a stage-discharge (rating) curve for an alluvial channel. As discussed earlier, starting from a low water depth, this curve will show breaks when the bed starts moving and again when the bed forms are washed out. In the first case, when ripples start to form, the resistance would increase and the water depth would show an increase while in the second case, as the dunes are washed out, the resistance decreases and the depth would decrease. However, since most alluvial channels are well past the critical stage, typical stage-discharge curves for alluvial channels show only the second break (Fig. 8.8), although the first break is also shown in the figure by a dashed line.

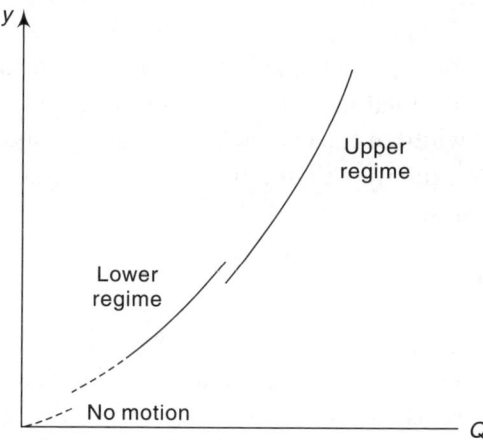

Fig. 8.8 Typical stage-discharge curve for an alluvial channel

8.5 Sediment Load

The amount of sediment carried by an alluvial channel is important in estimation of the useful life of a reservoir (through an estimate of the silting time) and in deciding the frequency of dredging of channels (through estimate of sediment-deposition rate). In this section, we describe some of the salient features of the process of sediment transport through alluvial channels.

The sediment carried by an alluvial channel may originate from the bed itself or it may be generated by the washing of the nearby catchments. The portion that has sizes close to those found in the channel bed is called the *bed-material load* while the other portion, which is typically much finer, is called the *wash load*. Out of the entire sediment load (the total *load*), some part (the finer particles) is carried by the water in suspension and is called the *suspended load*, while a major portion is carried on or near the bed and is termed the *bed load*. The bed load includes the particles that roll on the bed and maintain a continuous contact with the bed (therefore, known as the *contact load*) and those that move by bouncing along the bed, loosing contact for some time (termed the *saltation load*).[24] The mode of transport that occurs in a given flow condition is mainly dependent on the shear stress. For small shear stress (but, naturally, more than the critical shear stress) the material is

---

[24]Saltation load is more important in transport of material through air and not so much in water. Therefore, for alluvial channels, rather than distinguishing between the contact load and saltation load, we usually talk of the combined *bed load*.

transported mainly as contact load. With increase in the shear stress, some of the material may be transported as saltation load and with further increase, some material would be carried in suspension. A description of these modes of transport and the estimation of the quantities transported is given below. The quantity of sediment transported is generally expressed in terms of volume per unit time, Q_{sub} (with the subscripts being b for bed load, s for suspended load, and t for total load) or their more frequently used counterparts expressed per unit width of the channel, q_{sub}. Use of mass or weight per unit time is also common and we will use the symbols, W_{sub} and w_{sub}, and M_{sub} and m_{sub}, to represent these.

8.5.1   Bed Load

At a bed shear stress just above the critical shear stress, the sediment will move almost entirely as bed load. The amount of bed load transported by the channel will increase as the shear stress increases. Thus most researchers try to relate the transport rate, q_b, to the difference between the actual and critical shear stresses.

P. DuBoys in 1870s developed the first theory about the bed load transport. He assumed that the bed material moves in a series of layers parallel to the bed such that the velocity of the top layer (at the bed level) is maximum, that of the bottommost layer is zero, and it varies linearly in the intermediate layers. Since the lowest layer is assumed to be at rest, the resisting force and tractive force should be balanced. This led to a bed load equation of the form

$$q_b = A(\tau_0 - \tau_{oc})\tau_0 \qquad (8.34)$$

in which A is an empirical constant, which is a function of the grain size. A commonly used value of A is $6.9 \times 10^{-6} \, d_{50}^{-0.75}$, with d_{50} in mm and SI units used in Eq. (8.34), i.e., q_b in m³/s/m, τ in N/m², and A in m⁶/N²s.

Meyer-Peter and Müller in 1940s divided the slope into two parts—one corresponding to the grain resistance and other corresponding to the form resistance. A dimensionless bed load parameter, defined as

$$q_{b*} = \frac{q_b}{\sqrt{g(s-1)d^3}} \qquad (8.35)$$

was related to the nondimensional shear stress (Shields parameter) as

$$q_{b*} = 8 \, (\tau_* - 0.047)^{1.5} \qquad (8.36)$$

The sediment size, d, was an average size and was typically between d_{50} and d_{60}.

In 1940s, H.A. Einstein incorporated probabilistic analysis in the field of sediment transport and A.A. Kalinske considered the fluid force acting on a particle to be fluctuating about a mean. In 1950s, R.A. Bagnold proposed the concept of dispersion of solid particles when subjected to shear. The Einstein's theory, later modified, resulted in the following equation

$$q_{b*} = 40 \frac{\omega}{\sqrt{g(s-1)d}} \tau_*^3 \tag{8.37}$$

in which ω is the fall velocity. A more recent study by van Rijn (1984) is based on the trajectories of the saltation load and relates the bed-load transport rate to the transport parameter, T, and the dimensionless grain diameter, d_*, as

$$q_{b*} = 0.053 \frac{T^{2.1}}{d_*^{0.3}} \tag{8.38}$$

There are several other formulae derived on the basis of specific data sets. Most of the data is collected in the laboratory since measurement of bed load in canals or rivers is very difficult and quite uncertain. For practical application, one should consider the equations derived for conditions as close to the field conditions as possible. Even then, several equations should be used and a judgement has to be made about the most suitable value.

8.5.2 Suspended Load

The particles carried by a flow in suspension tend to settle down with a velocity dependent on their diameter. However, turbulent fluctuations present in the velocity field tend to cause a net upward movement of these particles because the concentration of suspended sediment is highest near the bed and decreases as we go higher. Assuming a linear variation of shear stress, τ_0 at the bed level and zero at the water level, the concentration of suspended sediments, C (generally expressed in parts per million or mg/l), at any height, Y, above the bed was obtained by H. Rouse in 1930s as

$$\frac{C}{C_a} = \left(\frac{y-Y}{Y} \frac{a}{y-a} \right)^{\omega/v_* \kappa} \tag{8.39}$$

where κ is the Karman constant for clear water flow (= 0.4), and C_a is a reference concentration at a height of a above the bed. The suspended load is negligible if the ratio of the shear velocity to fall velocity is less than about 0.5 while the concentration is nearly uniform with depth if this ratio exceeds about 80.

Assuming the velocity distribution to be given by the logarithmic law, and integrating Eq. (8.39) over the flow depth, H.A. Einstein in 1950 obtained the suspended sediment load as[25]

---

[25]We have assumed the viscous effects to be negligible. A correction factor has to be introduced, otherwise, in the logarithm.

$$q_s = 5 dv'_* C_{2d} \frac{E^{z-1}}{(1-E)^z} \left[\ln\left(\frac{30.2 \, y}{d_{65}}\right) J_1 (z, E) + J_2 (z, E) \right] \quad (8.40)$$

in which the reference concentration (as volume of sediment per unit volume of water) is taken at a height of $2d$ above the bed, v'_* is the shear velocity corresponding to the grain shear, E is the ratio of the reference height to the flow depth (i.e., $2d/y$), z is the exponent of the sediment-distribution equation ($z = \omega/\kappa v_*$), and J_1 and J_2 are called the Einstein integrals and are defined by

$$J_1(z, E) = \int_E^1 \left(\frac{1-x}{x}\right)^z dx \quad (8.41a)$$

$$J_2(z, E) = \int_E^1 \left(\frac{1-x}{x}\right)^z \ln x \, dx \quad (8.41b)$$

The integrals are obtained numerically and the plots are shown in Fig. 8.9. The following approximations can be used to obtain a very accurate estimation of these integrals:

$$J_1(z, E) = \frac{E_*^{1-z} - 1}{1-z} + 2.061 \frac{E_*^{2-z} - 1}{2-z}$$

$$-1.385 \frac{E_*^{2.6-z} - 1}{2.6-z} + \frac{0.3327}{0.6703 + z} \quad (8.42a)$$

$$J_2(z, E) = \frac{E_*^{1-z} [1 - (1-z)\ln E_*] - 1}{(1-z)^2}$$

$$-1.903 \frac{E_*^{2-z} [1 - (2-z)\ln E_*] - 1}{(2-z)^2}$$

$$+2.022 \frac{E_*^{2.6-z} [1 - (2.6-z)\ln E_*] - 1}{(2.6-z)^2} - \frac{0.2914}{1.652 + z}$$

$$(8.42b)$$

in which $E_* = E/(1-E)$. Note that singularities exist at $z = 1, 2,$ and 2.6 and the limiting values of these expressions have to be used. However, if the computed z value is *exactly* equal to one of these values, it may be changed by, say, 1%, to use Eq. (8.42), without affecting the results.

A more recent study by van Rijn (1984) has proposed the estimation of the suspended sediment load by the following equation:

$$q_s = FVyC_a \quad (8.43)$$

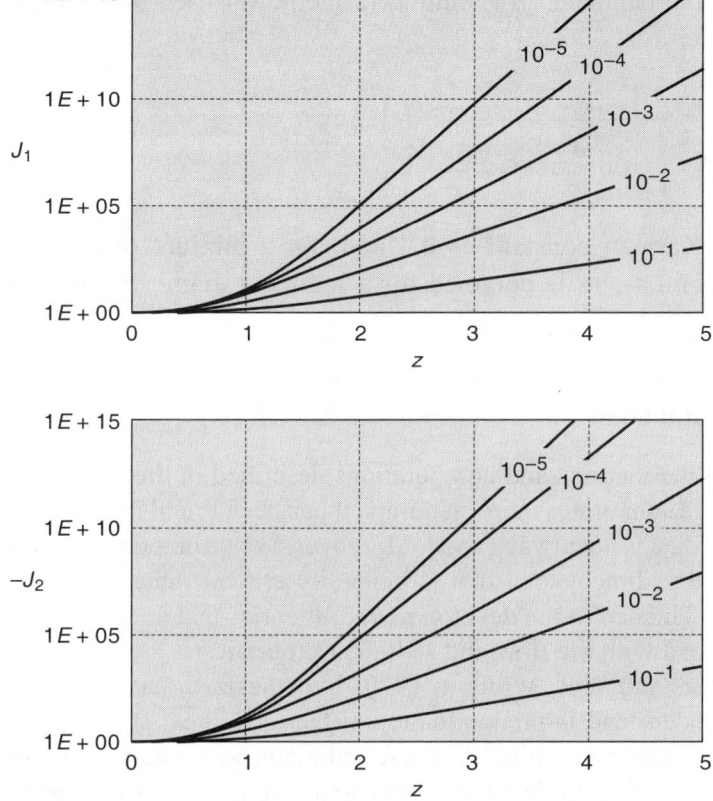

Fig. 8.9 Einstein integrals (Numbers on the curve are values of *E*.)

in which F is a correction factor and C_a is the reference concentration at a, and a is equal to the roughness height k_s or $0.01y$, whichever is greater, with k_s obtained from $V/v_* = 5.75 \log(R/k_s) + 6.25$.

The reference concentration (volume of sediment per unit volume of water) is computed from

$$C_a = 0.15 \frac{d_{50}}{a} \frac{T^{1.5}}{d_*^{0.3}}$$

using the transport parameter and nondimensional grain size. The correction factor, F, is computed from

$$F = \frac{\left(\dfrac{a}{y}\right)^p - \left(\dfrac{a}{y}\right)^{1.2}}{\left(1 - \dfrac{a}{y}\right)^p (1.2 - p)} \tag{8.44}$$

in which the parameter p is similar to the power used in Eq. (8.39) and is given by

$$p = \frac{\omega}{\left(1 + 2\frac{\omega^2}{v_*^2}\right)\kappa v_*} + 3\left(\frac{\omega}{v_*}\right)^{0.8} C_a^{0.4} \tag{8.45}$$

with the Karman constant as 0.4 and, for a mixture of sediment sizes, the fall velocity, ω, is obtained for a sediment diameter equal to $d_{50}[1 + 0.011(\sqrt{d_{84}/d_{16}} - 1)(T - 25)]$.

8.5.3 Total Load

Although the theories and computations described in the previous two sub-sections are sometimes quite elaborate, it has been found that their agreement with field data is not always good. Moreover, from a practical point of view, it is the total sediment load that is generally critical rather than its mode of transport. This led to the development of theories that attempted to correlate the total load with the flow and sediment properties.

P. Ackers and W.R. White in 1970s hypothesized that for fine sediment, the suspended load is predominant and the total shear stress is effective in moving the sediment while for coarse sediment, only a part of the shear stress is effective. C.T. Yang in 1970s postulated that the rate of sediment transport is governed by the stream power, which is the potential energy expenditure per unit weight of water. B.R. Samaga, K.G. Ranga Raju, and R.J. Garde in 1980s made the calculation of the sediment load separately for different sediment sizes and added the fractional transport to obtain the total transport. Here we list just two out of the large number of relationships proposed for estimation of total sediment load.

F. Engelund and E . Hansen in 1960s derived a total load formula based on energy considerations as

$$q_{t*} = \frac{q_t}{\sqrt{g(s-1)d_{50}^3}} = \frac{0.05\rho_f V^2 \tau_*^{2.5}}{\tau_0} \tag{8.46}$$

Karim and Kennedy (1990) proposed the following relationship (including later modifications):

$$q_{t*} = 0.00139\left[\frac{V}{\sqrt{g(s-1)d_{50}}}\right]^{2.97}\left(\frac{v_*}{\omega}\right)^{1.47} \tag{8.47}$$

Example 8.10 Particles of uniform size (mass density 2650 kg/m³, diameter 1 mm) comprise the bed of a wide rectangular channel in which the flow depth is 10 cm, the average velocity is 40 cm/s, and the bed slope is 1 in 1000. Estimate the bed load, suspended load and total load being transported by the channel.

Solution

We will use van Rijn equations for estimating the bed load and suspended load and the Karim and Kennedy equation to find the total load.

Bed load:

To use Eq. (8.38), we need the transport parameter, T, and the dimensionless sediment size, d_*. These were obtained in Examples 8.6 and 8.7 as $d_* = 25.3$ and $T = 0.2$. Therefore, the nondimensional bed load is obtained as 6.85×10^{-4}. The bed load per unit channel width is, therefore,

$$q_b = q_{b*} \sqrt{g(s-1)d^3} = \mathbf{2.75 \times 10^{-6} \, m^3/s/m}$$

Suspended load:

The shear velocity is obtained as 3.1 cm/s in Example 8.5. The height of surface roughness, k_s, is obtained from

$$\frac{V}{v_*} = 5.75 \log \frac{R}{k_s} + 6.25$$

as 7 mm, which is greater than 0.01y (i.e. 1 mm). Therefore the reference height a is taken as 7 mm. The reference concentration is obtained as

$$C_a = 0.15 \frac{d_{50}}{a} \frac{T^{1.5}}{d_*^{0.3}} = 7.27 \times 10^{-4}$$

Since the sediment is uniform, we will compute the fall velocity corresponding to this diameter which, as already seen in Example 8.2, is 16.8 cm/s. Equation (8.45) can now be used to obtain the parameter p as 0.87. The correction factor, F, is obtained from Eq. (8.44) as 0.186. Equation (8.43) then results in a suspended load of $\mathbf{5.42 \times 10^{-6} \, m^3/s/m}$.

Total load:

Using Eq. (8.47), we obtain the nondimensional total load as 3.48×10^{-3} and the total load per unit channel width as $\mathbf{4.42 \times 10^{-7} \, m^3/s/m}$. (Note that it is an order of magnitude smaller than the bed load and suspended load computed by van Rijn method.)

This example illustrates the uncertainties associated with the transport rate equations. It is strongly advised to use a number of equations and make an engineering judgement about the expected total load.

8.6 Design of Channels

The design of an alluvial channel should aim at either not allowing any erosion (by keeping the shear stress on the boundary below the critical value) or achieving a balance such that over a period of time there is no *general* erosion or deposition. The first approach, known as the *tractive-force approach*, would be useful when a channel is carrying clear water from its source and the channel would behave as a rigid-bed channel. The second approach, called the *regime approach*, is useful when the channel carries sediments (e.g., a canal taking off from a sediment-laden river without a silt control at the point of intake) and the channel will behave as an alluvial channel with bed forms and sediment load. These are briefly described below.

8.6.1 Tractive-force Approach

In this approach, it is ensured that the shear stress on any part of the boundary is not large enough to move the bed particles. Since the shear stress is not uniform on the boundary, the maximum shear stress is normally used. Also, for channels with sloping sides, the side slope and the angle of internal friction have to be considered to analyse the stability of the particle. We have previously considered the stability of a particle on the horizontal bed. If the particle is lying on a side sloping at mH:1V (Fig. 8.10), its submerged weight can be resolved into a component normal to the side, $W_{s,n}$, and the other acting downwards tangential to the side, $W_{s,t}$. The shear force acting on the particle would be $\tau_0 a$, where a is an effective area of the particle [see Eq. (8.16) for a more precise definition], and it would act in the direction of flow, i.e., perpendicular to the cross section shown in the figure. The resultant of the destabilizing force would, therefore, be $\sqrt{\tau_0^2 a^2 + W_{s,t}^2}$. The resisting or stabilizing force due to friction would be equal to $W_{s,n} \tan \phi$, where ϕ is the angle of internal friction (or angle of repose). At the critical condition corresponding to the incipient motion, these forces should balance. Therefore,

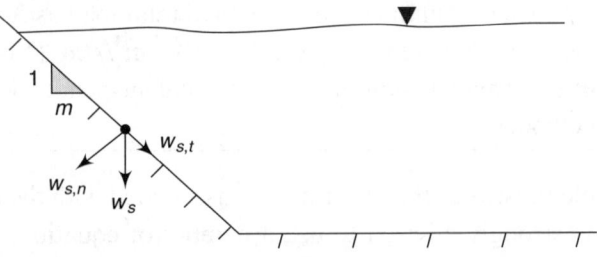

Fig. 8.10 Stability of a particle on a sloping side

$$\sqrt{\tau_0^2 a^2 + W_{s,t}^2} = W_{s,n} \tan \phi$$

$$\text{or} \quad \tau_0 a = W_s \sqrt{\frac{m^2}{1+m^2} \tan^2 \phi - \frac{1}{1+m^2}}$$

$$= W_s \tan \phi \sqrt{1 - \frac{1}{(1+m^2)\sin^2 \phi}} \tag{8.48}$$

For a particle on horizontal bed, the critical shear force, $\tau_{0h}a$, is given by W_s $\tan \phi$ [Eq. (8.48) with m tending to infinity should give the same expression]. This implies that the critical shear stress for a sloping side will be smaller than that for a horizontal side by a factor of $\sqrt{1 - \dfrac{1}{(1+m^2)\sin^2 \phi}}$. (One should note that the side slope cannot be greater than the angle of repose, otherwise the particles on the sides would be inherently unstable.) Thus a smaller shear stress on the sides would be able to move a particle as compared to that on the bed. However, for a given flow depth, the maximum shear stress on the bed is higher than that on the sides (see Appendix B). For example, a very wide rectangular channel has a maximum shear stress equal to $\rho_f gRS$ on the bed while only about $0.75\rho_f gRS$ on the sides. Therefore, the stability of particles on both the horizontal bed and the sloping sides has to be considered and the stricter of these requirements should be adopted for the design. Once the maximum permissible hydraulic radius is computed, Manning's equation is used to find the required area for a given discharge (with the Manning's roughness computed from the grain size), from which the dimensions can be obtained.[26] The critical shear stress for moving a particle is, of course, found out by using any of the techniques described in Section 8.2. Other factors that influence the channel roughness, e.g., alignment of the channel, have also to be considered. Generally this is accounted for by reducing the permissible shear stress by a factor of safety, which is a function of the curvature of the channel. Once the permissible shear stress and, therefore, the hydraulic radius, are computed, the design proceeds in a manner similar to that for the rigid-boundary channels, since the stresses are kept below the critical values.

8.6.2 Regime Approach

In this approach, it is ensured that the channel dimensions are such that it is in dynamic equilibrium. A *regime channel* is defined as an alluvial channel that

---

[26]Note that the maximum shear stress is a function of B/y for a rectangular channel. After computing the bed width and flow depth, one may have to check for the shear stress and redo the computations, if needed.

carries a constant discharge and sediment load with uniform flow depth without significant changes in its bed slope or cross section shape and size over a considerable period of time. Thus there is no significant erosion or deposition and these are also called non-silting and non-scouring conditions.[27] If a channel passing through an alluvium is designed with such characteristics that it is not in dynamic equilibrium, it would slowly move towards attaining such equilibrium by modifying its bed slope, bed width, etc. Therefore, the regime approach aims at designing the channel such that it is as close to the regime conditions as possible.

Based on observations on stable canals, R.G. Kennedy in 1890s related the average velocity of flow to the flow depth as $V = 0.55y^{0.64}$, with V in m/s and y in m. For sediment sizes different from those existing in these canals (0.32 mm), Kennedy multiplied this velocity by a critical velocity ratio, which was more than 1 for coarser particles and less for finer. G. Lacey in 1930s improved on this theory by introducing a silt factor, f_s, defined as

$$f_s = 1.76 \sqrt{d}$$

with d in mm, and expressing the velocity as[28]

$$V = 0.63 \sqrt{f_s R}$$

The silt factor, flow velocity, and discharge were found to be related by

$$A f_s^2 = 140 V^5$$

and the bed slope was observed to be related to these as

$$S = 0.0003 \frac{f_s^{5/3}}{Q^{1/6}}$$

These equations can be manipulated to directly give the elements of the design for a given discharge and bed material size as

$$A = \frac{2.28 Q^{5/6}}{f_s^{1/3}}$$

$$P = 4.75 \sqrt{Q}$$

Once the flow area, wetted perimeter, and bed slope are obtained from these equations, the channel dimensions are easily computed.

---

[27]This does not mean that there is absolutely no erosion or deposition. Since the discharge and the sediment load in a channel vary considerably with time, there would be periods of deposition when, for example, the sediment load becomes *larger than usual*. However, over a long period of time, the channel remains stable.

[28]All parameters, except the grain size, d, are in SI units. Grain diameter is in mm as mentioned.

There have been several modifications to the Lacey's theory but the essential features remain same, i.e., specification of the channel characteristics in terms of the discharge and particle size. Sometimes, the sediment load is also considered in these expressions, but we will not describe these theories here. Garde and Ranga Raju (2000) should be consulted for a more extensive treatment of the subject.

SUMMARY

In this chapter, we have described flow through erodible-boundary channels. We have discussed the flow conditions that would cause the particles on the bed to move. Various theories to determine these critical conditions for the initiation of motion have been discussed and a critical value of the bed shear stress has been related to the nondimensional sediment size. The bed forms, which are created after the sediment starts to move, have been described along with the criteria to determine the kind of bed forms likely to exist under given flow conditions. The geometry of these bed forms has also been correlated with the flow and sediment properties. Effect of the bed forms on the channel resistance has been considered and resistance equations similar to those for rigid-boundary channels have been discussed, which correlate the average flow velocity to the geometry of the bed forms. The amount of sediment carried by the channel, both in suspension and at or near the bed, has been discussed and different methods of estimation of the sediment discharge have been described. Finally, the theories of sediment motion have been synthesized to look at the design of alluvial channels.

REFERENCES

Brown, P.P. and Lawler, D.F. (2003): 'Sphere drag and settling velocity revisited', *J. of Env. Eng.*, **129**(3), pp. 222-31.

Brownlie, W.R. (1981): 'Prediction of flow depth and sediment discharge in open channels', *Report No. KH-R-43A*, W.M. Keck Laboratory of Hydraulics and Water Resources, California Institute of Technology, Pasadena, California (available at *http://caltechkhr.library.caltech.edu/7/01/KH-R-43A.pdf*, last accessed on May 23, 2007).

Buffington, J.M. and Montgomery, D.R. (1997): 'A systematic analysis of eight decades of incipient motion studies, with special reference to gravel-bedded rivers', *Water Resour. Res.*, **33**(8), pp. 1993–2029.

Coleman, S.E. and Fenton, J.D. (2000): 'Potential-flow instability theory and alluvial stream bed forms', *J. of Flu. Mech.*, **418**(1), pp. 101-17.

Garde, R.J. and Ranga Raju, K.G. (2000): *Mechanics of Sediment Transportation and Alluvial Stream Problems*, 3rd ed., New Age International, New Delhi.

Juilen, P. (1995): *Erosion and Sedimentation,* Cambridge University Press, New York.

Julien, P.Y. and Klaassen, G.J. (1995): 'Sand dune geometry of large rivers during floods', *J. of Hydr. Eng.*, **121**(9), pp. 657-63.

Karim, M.F. and Kennedy, J.F. (1990): 'Menu of coupled velocity and sediment-discharge relations for rivers', *J. of Hydr. Eng.*, **116**(8), pp. 978-96.

Mantz, P.A. (1992): 'Cohesionless, fine sediment bed forms in shallow flows', *J. of Hydr. Eng.*, **118**(5), pp. 743-64.

Shields, A. (1936): 'Application of similarity principles and turbulence research to bed-load movement' (translated from German), *Hydrodynamics Laboratory Publication no. 167*, California Institute of Technology (avaialble at *http://caltechkhr.library.caltech.edu/56/01/Sheilds.pdf*, last accessed on May 25, 2007).

Swamee, P.K. and Mittal, M.K. (1979): 'An explicit equation for the critical shear stress in alluvial streams', *J. of Irrig. Power*, CBIP, India, **33**(2), pp. 237-39.

van Rijn, L.C. (1984): 'Sediment transport: I. Bed load transport, II. Suspended load transport, III. Bed forms and alluvial roughness', *J. of Hydr. Eng.*, **110**(10), pp. 1613-41, **110**(12), pp. 1733-54.

———— (1993): *Principles of Sediment Transport in Rivers, Estuaries and Coastal*

Seas, Aqua Publications, Delft, The Netherlands.

van Rijn, L.C. (2007a): 'Unified view of sediment transport by currents and waves: I. Initiation of motion, bed roughness, and bed-load transport, *J. of Hydr. Eng.*, **133**, pp. 649-67.

———— (2007b): 'Unified view of sediment transport by currents and waves, II. Suspended transport, *J. of Hydr. Eng.*, **133**, pp. 668-89.

Yalin, M.S. and Karahan, E. (1979): 'Inception of sediment transport', *J. of Hydr. Div.*, ASCE, **105**(11), pp. 1433–43.

EXERCISES

8.1 Following sieve analysis data is available for a 200 g sediment sample taken from a channel bed:

Sieve Opening (mm)	19	12.5	9.5	4.75	2.36	1.18	0.6	0.3	0.15	0.075
Weight retained (g)	0	6.8	10.5	33.1	49.6	46.5	26.8	20.2	6.5	0

Plot the grain-size distribution curve and find the values of d_{50}, d_{90}, standard deviation, uniformity coefficient, sorting coefficient, and Kramer's uniformity coefficient (use a 5% interval).

[2.36 mm, 9.0 mm, 4.44 or 1.68 mm, 6.3, 2.1, 0.19]

8.2 For the channel of the previous exercise, assuming that the bed may be conceptualized as consisting of uniform spherical particles of diameter d_{50}, and having a specific gravity of 2.65, find the grain-level velocity competent to move the particles. [0.3 m/s]

8.3 The channel of Exercise 8.2 may be assumed to be wide and has a bed slope of 1 in 500. If the flow depth is 20 cm, would the bed particles move or not? What depth of flow would be just enough to move the bed? What regime of flow is expected in the channel when the flow depth is twice this critical value? [move, 8.84 cm, ripples and dunes]

8.4 Cross section of an alluvial channel may be approximated as a 20 m wide rectangular section. The boundary may be assumed to comprise uniform-size spherical particles with a specific gravity of 2.65. The bed slope of the channel is 1 in 2500. It was observed that the sediments start to move when the flow depth becomes 1 m. Estimate the diameter of the particles.

[4.9 mm]

8.5 Although the incipient condition is not an abrupt change, let us assume that it occurs suddenly. Compare the velocity values obtained in the channel of the previous problem at a flow depth of 0.99 m (assuming no sediment movement) and 1.01 m, when the sediment is moving. Use both van Rijn and Karim–Kennedy methods and comment on the difference.

[van Rijn: 0.98 m/s, 0.99 m/s; Karim–Kennedy: 1.36 m/s, 0.95 m/s]

8.6 For the same channel, estimate the suspended load, bed load, and total load when the flow depth is 1.5 m.

[van Rijn: $q_b = 6.68 \times 10^{-6}$ m^3/s/m, $q_s = 2.88 \times 10^{-5}$ m^3/s/m;

Karim–Kennedy: $q_t = 8.43 \times 10^{-6}$ m^3/s/m]

8.7 An alluvial channel passes through soil having a median size of 5 mm and has to be designed to carry a discharge of 20 m 3 /s. Using the Lacey's approach, what should be the bed slope and the section dimension of a trapezoidal section with side slopes of 1V:3H?

$$[S = 1.8 \times 10^{-3}, B = 15 \text{ m}, y = 1 \text{ m}]$$

9   Pollutants Transport in Open Channels

9.1   Introduction

Rivers, lakes, and other natural waters have been extensively used for disposal of municipal (and later industrial) wastes since the early days of civilization. Earlier, when the quantity of waste and its composition was such that the water bodies were able to take care of these without noticeable adverse impact, there was not much to be concerned about these practices. However, with rapid growth of population, rise in the standards of living (leading to generation of more waste per person) and swift pace of industrialization, many rivers are being fed more pollutants than they can handle causing a significant, and in most cases irreversible, damage to their environment. In absence of other reliable techniques for disposal of such a large quantity of waste, it appears that this trend will continue, at least for the next few decades. Any effort to clean up the rivers or to prevent further damage must be based on a thorough understanding of the process of pollutants transport in open channels. This is a very vast area of research and is typically covered as a part of undergraduate courses in environmental engineering. However, because of its significance to open channel flows, we provide a brief description here. For more detailed discussion, the reader should refer to one of the earliest and widely used books on this subject by Fischer, et al. (1979) or an earlier work by Fischer (1966).

The sources of pollution in streams may be classified as follows:

1. *Natural*—generally inorganic salts, available in the material on the channel boundary or in the surface soil of the catchment area
2. *Municipal*—the domestic wastewater and waste carried in the stormwater drains
3. *Industrial*—from a number of different types of industries, e.g., heavy metals such as chromium (leather industry), lead (textile and paint); organic contaminants, e.g., trichloroethylene (used in metal degreasing), pesticides; and heat (thermal power plants)

4. *Accidental*—from unintended (or sometimes intended, e.g., terrorism or war-time activities) spillage of chemical, biological, or radioactive wastes in water bodies.

The immediate effect of these pollutants would be felt by the aquatic life but the continuous absorption of these contaminants in, say, fishes, may pose a threat to the lives of people consuming these. For example, mercury contamination in Japan led to the death of hundreds of persons consuming the contaminated fish. Sometimes, the pollutants exert an oxygen demand on the river and reduce the amount of dissolved oxygen. This may seriously affect the aquatic life. In general, due to the large magnitude of flow, significant dilution of the pollutants occurs in a stream as it moves downstream from the point of discharge. Thus if we have an intake structure for a water-supply system considerably downstream of a pollutant discharge point, it may not experience noticeable effect of the discharge. However, there is a huge temporal variation in the stream flow and it may not be able to dilute the pollutants during some part of the year. In order to ensure that all the planned discharges into the stream are within safe limits and would not cause permanent damage to the ecosystem, one should be able to analyse different strategies of disposal in terms of their effect on the stream. The analysis requires an understanding of the pollutant characteristics, transport process, permissible limits (maximum, e.g., for heavy metals, or minimum, e.g., for dissolved oxygen) of different quality parameters, and treatment options. We will only discuss the transport process in this book. We start with a look at different mechanisms by which a pollutant is transported in a channel, derive the partial differential equation describing the spatial and temporal variations of the pollutant concentration, describe some typical solution methods, discuss the estimation of some significant parameters, and then mention a few advanced topics related to pollutants transport in streams.

9.2 Mechanisms of Transport

Any pollutant discharged in a stream will not remain at the point of discharge (unless it is so heavy that it cannot be moved by the flow) but will travel downstream with the flowing water. During this downstream transport, it mixes with the surrounding water, which leads to its dilution. Sometimes, it may react with the sediments suspended in the water or may decay either on its own (radioactive substance) or due to bacterial action (most organic and some inorganic pollutants). All these processes affect its concentration and some of these (some other are considered advanced and are described towards the end of the chapter) are described below.

9.2.1 Advection

The term *advection* (sometimes called convection) refers to the transport of an atmospheric property or a substance by horizontal movement of fluid. Since we are concerned with natural streams, the fluid will be taken as water and the property may be a chemical, heat, or even water with different properties (e.g., saline water). If advection is the only mechanism of transport (*pure advection*), and assuming the pollutant to be a single particle, it is carried by the water at an average velocity equal to the flow velocity.[1] Therefore, if we introduce the particle at a location and observe its position after a time t, it would move a distance of Vt from that location.[2] If a cloud of particles is introduced, it will move as a unit without changing its shape since all particles in this cloud will travel the same distance [Fig. 9.1(a)]. In reality, however, the shape of this cloud will change as the particles diffuse outwards.

9.2.2 Diffusion

Before we combine the processes of advection and diffusion, let us look at the process of diffusion separately. If we take stationary water in a tank and introduce a single particle, it would be subjected to forces applied by the surrounding water molecules due to their Brownian motion. If the particle is assumed to have the same density as that of the water, there would be no net force on the particle due to the random nature of the Brownian motion (indicating that the molecule has an equal chance of going in any direction),[3] which would cancel out *on an average*. Thus over a period of time, the particle will remain at the same location at which it was introduced. However, when a cloud of

---

[1] Sometimes, say, when the pollutant is much heavier than the water or reacts with the sediments, it may have an effective velocity smaller than the flow velocity.

[2] We are assuming the velocity to be equal to the cross-sectional average velocity, V. This assumption has served us well in most previous discussions. However, as we will see in this chapter, the velocity variation over a section becomes important in determining the spread of pollutants. Earlier, we had accounted for this velocity variation by using the momentum and energy correction factors. A similar methodology would be used later in the chapter when we use the dispersion coefficient.

[3] If we use a one-dimensional analysis, we can say that the particle has an equal chance of going either left or right.

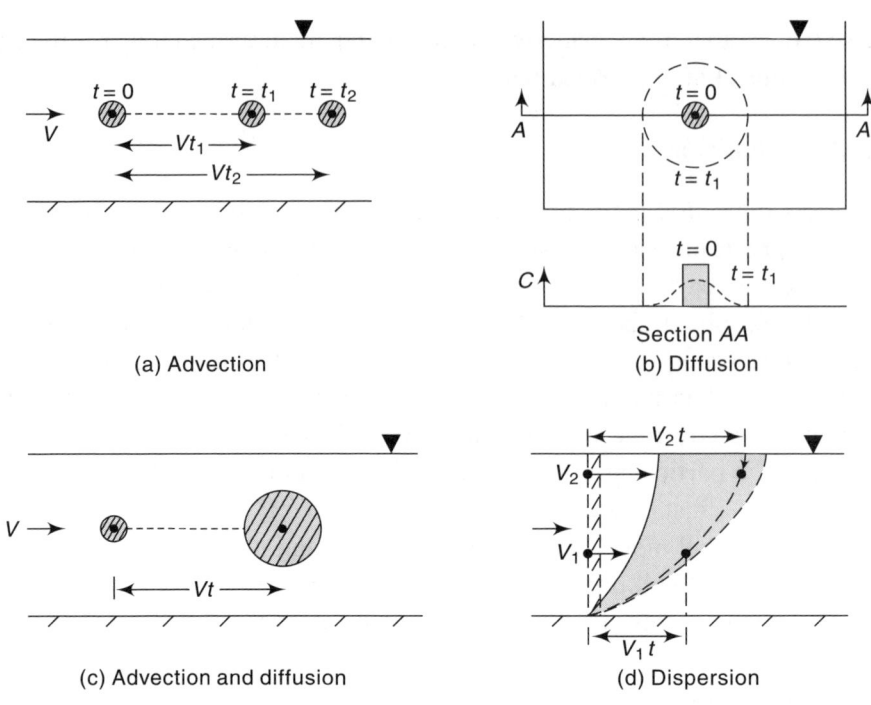

Fig. 9.1 Different mechanisms of pollutants transport

particles is introduced, the concentration[4] of the particles becomes an important characteristic in determining the particle motion. We still have an equal chance of a particle going in any direction. However, considering two nearby regions, one with a smaller concentration and the other with larger, it is obvious that on an average, more particles will move from the higher concentration region to the lower concentration region, just because the number of particles is larger. This is known as *diffusion* and may be defined as the process by which random Brownian motion causes a net movement towards regions of low concentration (with the end result being the elimination of the concentration gradient and establishment of an equilibrium with same concentration throughout the tank in this case, Fig. 9.1(b).

Instead of the tank, if we consider a flowing stream, the processes of advection and diffusion will be combined. The cloud of particles will undergo a mean displacement of Vt, but it will not retain its shape because of the diffusion of particles from within the cloud to nearby regions that do not have any particles. Thus the cloud will keep on expanding and the concentration will keep on decreasing within the cloud [Fig. 9.1(c)]. If we go one step fur-

---

[4] Here we use the term *concentration* loosely to denote the number of particles in a given volume. More rigorous definition is provided later.

ther, and consider the flow to be turbulent, the ensuing velocity fluctuations would result in a diffusion similar to that observed due to Brownian motion because the fluctuations due to turbulence may be thought of as similar to those due to Brownian motion, though with a much larger magnitude. In order to distinguish between these two mechanisms, we use the terms *molecular diffusion* to represent the diffusion due to the Brownian motion and *turbulent diffusion* to represent that due to the turbulence. We use the term *diffusion* for the combined process but mention that, for most stream flow conditions, the molecular diffusion would be negligible compared to the turbulent diffusion. However, for a very slow-moving flow, e.g., that in groundwater, the molecular diffusion would be larger than the turbulent diffusion.

Under typical field conditions, the velocity of flow varies significantly over a cross section (see Apeendix B). We have a depthwise variation of velocity from zero at the boundary to a maximum at or near the surface and a widthwise variation, which largely depends on the channel alignment and cross section shape. For example, if the cross section comprises some deep portions and some shallow ones, it is likely that the deeper portions will have a higher velocity. Similarly, at a bend there would be velocity variations across a cross section as described earlier in Chapter 5. This implies that a cloud of particles introduced uniformly[5] at a particular cross section, may show a significant lengthwise dispersion at a later time since different particles will have different velocities.

9.2.3 Hydrodynamic Dispersion

Even if we neglect diffusion and consider pure advection, there would be a spreading effect created by the velocity variation [Fig. 9.1(d)]. This is known as *hydrodynamic dispersion* and typically has a much larger magnitude than diffusion. In fact, diffusion tends to reduce the dispersion by transporting the particles from the fast-moving high-concentration regions to slow-moving low-concentration regions of a cross section. Since the overall effects of diffusion and dispersion are similar in that they both lead to spreading of a cloud, it is customary to combine these into a single dispersion component. The resulting

---

[5] We assume that the particles are introduced at once over the entire cross section. If these are introduced at a point, there would be an initial portion in which the particles would gradually expand to fill the cross section and some of our statements may not be applicable to this *zone of development*. In particular, there would be a lateral dispersion also known as the *transverse dispersion*. However, at this point, we will consider only the *longitudinal dispersion*.

process is called the *advective-dispersive transport* and is governed by what is known as the *advection-dispersion equation*, which is discussed below.

9.2.4 Governing Equation

The partial differential equation representing the spatial and temporal variations of the concentration is derived from the mass balance of the species being transported. In addition to the advection and dispersion, the transport may be affected by reactions. For example, the pollutant may get sorbed[6] on the sediments being carried by the stream as bed load or suspended load, or it may undergo a chemical, biological, or radioactive transformation. The reactions may be instantaneous (which are easier to model) or slow (in which the reaction rate becomes an additional parameter and separate differential equations need to be added, making the analysis more complicated). However, in most of this chapter, we assume that the pollutant is non-reactive.

Before deriving the mass balance equation in terms of the concentration, let us formally define the concentration. Since we are dealing with pollutants in streams, we will invariably use the terms water for the carrying fluid and pollutant for the species being carried. For a general case, the appropriate terms would be *solvent* and *solute* respectively. The concentration of a pollutant can be specified in terms of mass or volume of the pollutant per unit mass or volume of the water-pollutant mixture, i.e., the *solution* (or, sometimes, per unit mass or volume of water). In this book, we will use the concentration as mass of the pollutant per unit volume of the mixture, generally expressed in units of mg/L. Another commonly used way is the mass of the pollutant per unit mass of the mixture, generally expressed as mg/kg or parts per million (ppm). (Obviously, if the pollutant is hot water, the mass is not as important as the temperature.) Thus a 10 mg/L salt concentration indicates that there is 10 mg of salt in 1 litre of the salt-water solution.[7] We now look at the mass balance over a small element and derive the governing equation by the usual procedure of taking the limit as the element size approaches zero. For simplicity, we use one-dimensional analysis but include the effect of the cross-sectional variations in velocity through the dispersion coefficient.

---

[6] *Sorption* may be loosely defined as the chemical or physical bonding of one molecule onto the surface of another molecule.

[7] Since the mass density of water is 1 kg/L, we may use mg/kg instead of mg/L if the concentration is small and the density of the mixture is not significantly different from that of the water.

Figure 9.2 shows a reach of the channel subdivided into elements (cells) of size ΔX. Let C_i denote the concentration (mass per unit volume) of the pollutant in the element i. Taking a unit cross-sectional area, therefore, the mass of the pollutant in the ith cell would be $C_i \Delta X$ and the mass accumulation during a time period of Δt would be $[\partial(C\Delta X)/\partial t]\Delta t$. This accumulation should be equal to the net inflow of the mass over the control surface (which comprises the interfaces between the cells $i-1$ and i, and that between the cells i and $i + 1$). These terms are now analysed for the cases of advection, molecular diffusion, turbulent diffusion, and dispersion.

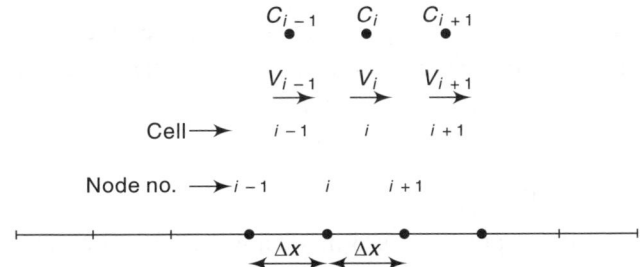

Fig. 9.2 Elemental grid for the continuity equation

Advective mass inflow

As shown in Fig. 9.2, the velocity of flow in the cell $i-1$ is V_{i-1} and the concentration is C_{i-1}. Therefore, the mass inflow across the interface between the cells $i-1$ and i during a time period of Δt is given by $V_{i-1} \Delta t\, C_{i-1}$. Similarly, the outflow across the interface between the cells i and $i+1$ is given by $V_i \Delta t\, C_i$. Therefore, the net advective mass inflow across the control surface is

$$m_a = V_{i-1}\Delta t C_{i-1} - V_i \Delta t C_i = -\frac{\partial(VC)}{\partial X}\Delta X \Delta t \qquad (9.1)$$

in which (and also in the following equations) the quantities without a subscript represent the values at the ith cell.

Mass inflow due to molecular diffusion

At the interface between the cells $i-1$ and i, there would be a number of particles going right from cell $i-1$ and some number of particles going left from the cell i. The number of particles is proportional to the concentration and we use the empirical Fick's law, proposed by A. Fick in 1850s, to write the mass flux into the ith cell across this interface as proportional to $(C_{i-1} - C_i)/\Delta X$.

Fick's law states that the diffusive mass flux rate is proportional and in the opposite direction to the concentration gradient and the constant of proportionality is called the *diffusion coefficient*. A justification of this empirical relationship can be provided by using the Brownian motion analogy. If we neglect the probability that the particle stays at its original position, the probabilities of going left and right would be equal ($=1/2$). Starting with an initial number of particles n_i in the ith cell, it is easily seen that the increase in the number of particles in cell i over a time period of Δt is given by

$$n_i^{t+\Delta t} - n_i^t = \frac{1}{2}\left(n_{i-1}^t - n_i^t\right) - \frac{1}{2}\left(n_i^t - n_{i+1}^t\right)$$

in which the first term on the right hand side represents the increase due to flow across the $i-1$, i interface and the second represents the decrease across the i, $i+1$ interface. This can be written as

$$\frac{\partial n}{\partial t}\Delta t = \frac{1}{2}\frac{\partial^2 n}{\partial X^2}\Delta X^2$$

Now, if ΔX and Δt approach zero in such a way that the value of $\Delta X^2/2\Delta t$ remains constant at D_m, we get the same result as obtained later in Eq. (9.2).

Introducing a molecular diffusion coefficient,[8] D_m, we write the mass inflow over a time period of Δt as $D_m \dfrac{C_{i-1} - C_i}{\Delta X}\Delta t$ (the dimensions of the diffusion coefficient are $L^2 T^{-1}$ and it is commonly expressed in units of cm^2/s). Similarly, the mass outflow from the other interface is written as $D_m \dfrac{C_i - C_{i+1}}{\Delta X}\Delta t$ and the net diffusive mass inflow across the control surface is written as

$$m_m = D_m \frac{C_{i-1} - C_i}{\Delta X}\Delta t - D_m \frac{C_i - C_{i+1}}{\Delta X}\Delta t$$

$$= D_m\left(\left.\frac{\partial C}{\partial X}\right|_{i-1,i} - \left.\frac{\partial C}{\partial X}\right|_{i,i+1}\right)\Delta t = D_m \frac{\partial^2 C}{\partial X^2}\Delta X \Delta t \tag{9.2}$$

Mass inflow due to turbulent diffusion

From Eq. (9.1) we see that the advective inflow is proportional to the time derivative of the product of the velocity and the concentration. In turbulent

---

[8]We assume that the molecular diffusion does not vary with time or in space.

flow, both the velocity and the concentration will have fluctuations superimposed on a mean (time-averaged) value and we may write

$$V(x, t) = \bar{V}(x) + V'(x, t)$$

in which the time-averaged mean velocity at any location is given by

$$\bar{V}(x) = \frac{1}{T} \int_0^T V(x, t) dt$$

where T is an appropriately chosen time (which is larger than the time scale of turbulent fluctuations and smaller than the time scale of any variations in the mean velocity) and the mean flow is assumed to be steady. Similar expression is written for the concentration and substituted in Eq. (9.1) to obtain the time-varying mass inflow rate (per unit length per unit time), M, due to advection and turbulent diffusion as

$$M_{a+t} = \frac{\partial \left[(\bar{V} + V')(\bar{C} + C') \right]}{\partial X} \tag{9.3}$$

and its time-averaged value as

$$\bar{M}_{a+t} = \frac{\partial (\bar{V}\bar{C})}{\partial X} + \frac{\partial (\overline{V'C'})}{\partial X} \tag{9.4}$$

since, by definition, the time average of the random fluctuations is zero.[9] The first term on the right hand side is the same as the advective inflow and the second term represents the turbulent diffusion. Analogous to the molecular diffusion, we define a turbulent diffusion coefficient, D_t, such that the net mass inflow due to the turbulent diffusion is written as

$$m_t = D_t \frac{\partial^2 C}{\partial X^2} \Delta X \Delta t \tag{9.5}$$

Mass inflow due to hydrodynamic dispersion

Similar to the analysis of turbulent diffusion, we now take Eq. (9.1) and instead of the temporal average, define an areal (cross section) average velocity. The point velocity would then be the sum of the areal average velocity and a perturbation. We then get an equation identical to Eq. (9.3) and define a hydrodynamic dispersion coefficient, D_h, in such a way that the dispersive mass inflow is given by

---

[9]However, the time average of product of two fluctuations may or may not be zero.

$$m_h = D_h \frac{\partial^2 C}{\partial X^2} \Delta X \Delta t \tag{9.6}$$

Once we have all the inflows (due to advection, molecular diffusion, turbulent diffusion, and hydrodynamic dispersion), we are in a position to write the mass balance as[10]

$$\frac{\partial (C_i \Delta X)}{\partial t} \Delta t = -\frac{\partial (VC)}{\partial X} \Delta X \Delta t + (D_m + D_t + D_h) \frac{\partial^2 C}{\partial X^2} \Delta X \Delta t$$

i.e.,

$$\frac{\partial C}{\partial t} + \frac{\partial (VC)}{\partial X} = D \frac{\partial^2 C}{\partial X^2} \tag{9.7}$$

in which D is an effective dispersion coefficient which includes the effects of dispersion and diffusion. Generally, dispersion occurs in both the longitudinal direction as well as the transverse direction, i.e., perpendicular to the flow direction, with the longitudinal dispersion coefficient, D_L, being larger than the transverse coefficient, D_T. Since water is incompressible, using the continuity equation ($\partial V/\partial X=0$), Eq. (9.7) may be written as

$$\frac{\partial C}{\partial t} + V \frac{\partial C}{\partial X} = D \frac{\partial^2 C}{\partial X^2} \tag{9.8}$$

which is called the *advection-dispersion equation* and may be solved using analytical or numerical methods to obtain the variation of concentration with space and time. Since the equation is of first order in time and second order in space, it requires one initial and two boundary conditions for its solution. To simulate the spread of a pollutant in an initially clean stream, for example, the initial condition would be a zero concentration everywhere and the boundary conditions would depend on the mode of pollutant release. If the pollutant is released as a one-time accidental spill, one of the boundary conditions would be of a *pulse input* at the point of release. However, if the pollutant is being released regularly, as is the case with a sewer outfall in a river, the boundary condition may be taken as constant concentration (though there would be variations in the concentration at different times of the day). The other boundary

---

[10] Note that, for turbulent flows, we will now use V and C to represent the time-averaged values and not the instantaneous values. In other words, we have dropped the overbar to simplify the presentation and keeping it in line with earlier chapters. One should be aware of this distinction and understand that the additional term representing turbulent diffusion is present because of the substitution of the instantaneous values of velocity and concentration by their time-averaged values [Eq. (9.4)].

condition may be taken as a zero concentration at a *large distance* from the pollutant source (this large distance may be taken as infinity in analytical solutions but as some finite large value for numerical models). Although we will mostly discuss one-dimensional flows in which the mean velocity is oriented towards the X-axis, the dispersion may occur in transverse direction also. For a general three-dimensional velocity field (with components V_X, V_Y, and V_Z in the X-, Y-, and Z-directions respectively), the advection-dispersion equation is written as

$$\frac{\partial C}{\partial t} + V_X \frac{\partial C}{\partial X} + V_Y \frac{\partial C}{\partial Y} + V_Z \frac{\partial C}{\partial Z}$$

$$= \frac{\partial}{\partial X}\left(D_X \frac{\partial C}{\partial X}\right) + \frac{\partial}{\partial Y}\left(D_Y \frac{\partial C}{\partial Y}\right) + \frac{\partial}{\partial Z}\left(D_Z \frac{\partial C}{\partial Z}\right) \tag{9.9}$$

in which the dispersion coefficient is assumed to vary spatially. It should also be noted that in its most general form, the dispersion coefficient would be a 3×3 tensor involving off-diagonal terms such as D_{XY}, D_{XZ}, D_{YZ}, which represent the dispersive flux in a direction due to concentration gradient in another direction, and each term inside the brackets on the right hand side of this equation would be the summation of three terms. However, these off-diagonal terms are generally ignored and we will follow the same convention. Also, since in this chapter we consider the flow in a stream in which V_X is the only (or predominant) velocity component, and the coefficient of dispersion in directions transverse to the flow (i.e., Y and Z) may be taken as equal, Eq. (9.9) is written as

$$\frac{\partial C}{\partial t} + V \frac{\partial C}{\partial X} = \frac{\partial}{\partial X}\left(D_L \frac{\partial C}{\partial X}\right) + \frac{\partial}{\partial Y}\left(D_T \frac{\partial C}{\partial Y}\right) + \frac{\partial}{\partial Z}\left(D_T \frac{\partial C}{\partial Z}\right) \tag{9.10}$$

with D_L and D_T being the longitudinal and transverse dispersion coefficients. Since the molecular diffusion process may be considered to be independent of the magnitude of the velocity but the turbulent diffusion and dispersion processes depend significantly on it [see Eq. (9.4)], it is customary to write the dispersion coefficient as

$$D_L = D_m + \alpha_L V \quad \text{and} \quad D_T = D_m + \alpha_T V \tag{9.11}$$

where α_L and α_T are the longitudinal and transverse dispersivities, which have dimensions of length. In most practical cases involving stream flow, the molecular diffusion will be negligible compared to the other terms. The dispersion coefficient would depend on the stream characteristics and its correct estimation is quite critical in the analysis of pollutants transport in open channels.

In the next section, we will look at some situations where analytical solutions of the advection-dispersion equation can be obtained and then discuss the more complex cases where numerical methods have to be used. This is followed by a discussion on the estimation of the dispersion coefficient in field conditions and then some advanced topics related to pollutant transport.

9.3 Simple Cases Involving Analytical Solutions of Advection-Dispersion Equation

The analytical solution of Eq. (9.10) or its one-dimensional approximation, Eq. (9.8), is possible under simplifying assumptions involving the flow, pollutant, and channel properties. For example, all of these require the velocity to be constant (and uni-directional), the pollutant to be non-reactive (or, at most, have a linear decay as in radioactive substances), the dispersion coefficients to be temporally and spatially invariable, and the pollutant discharge to be at a constant rate. Since the form of the advection-dispersion equation is identical to the convection-diffusion equation for heat transfer, most of the analytical solutions have been based on those already derived for heat conduction (Carslaw and Jaegar 1959).

9.3.1 Sudden Release of Pollutant at a Point

We use Eq. (9.10), i.e., assume the lateral dispersion to be same in the Y- and Z-directions, and consider the sudden release of a mass of pollutant, m_p, at the point $(0, 0, 0)$, which would simulate the effect of, say, an accidental discharge of a pollutant brought into the stream by a stormwater drain. If the release is over a very short period of time, we may take the initial concentration as a delta function[11] centred at the point of release

$$C(X, Y, Z, 0) = m_p \delta(0, 0, 0) \tag{9.12}$$

The boundary conditions may be taken as zero concentration at large distances ($\pm\infty$) or we may use both the conditions at $+\infty$ by stipulating that the concentration and its gradient should both be zero there. The variation of concentration is then given by

$$C(X, Y, Z, t) = \frac{m_p}{8\pi^{3/2} D_T t \sqrt{D_L t}} \exp\left[-\frac{(X - Vt)^2}{4 D_L t} - \frac{Y^2 + Z^2}{4 D_T t}\right]$$

$$\tag{9.13}$$

---

[11]The delta function $\delta(X)$ is defined in such a way that it is zero everywhere except at X, where it is infinite and its integral over its entire domain is equal to unity.

Note that the product of diffusion coefficient and time has the dimensions of L^2, and the resulting concentration has units of mass per unit volume. Also, we have taken the spatial extent of the channel to be infinite in all directions, which is not a reasonable assumption. The solution will, therefore, break down once the pollutant reaches any of the boundaries (top, bottom, or sides). Even then, this simple solution is helpful in visualizing the transport process. Since it is difficult to show the three-dimensional evolution of the plume,[12] the two most common techniques for presenting the results are: (i) breakthrough curves, which show the temporal variation of pollutant concentration at a fixed point in space and (ii) snapshots or profiles, which show the longitudinal profile (one-dimensional) or contours (two-dimensional) of concentration at a fixed point of time. The following example illustrates some of these options.

Example 9.1 A 1 km wide river has a flow depth of 5 m and the average velocity is measured as 1 m/s. The effective dispersion coefficient is estimated as 10 m²/s. A tanker carrying pesticide suddenly dumped 100 kg of pesticide through an outlet located at its base such that the point of release is in the centre of the river (i.e., 500 m from either end and at a depth of 2.5 m). Comment on the transport assuming that the pesticide is nonreactive.

Solution
Taking the point of dumping as the origin of the coordinate system (0,0,0) and the time of dumping as $t = 0$, we use Eq. (9.13)[13] to obtain the concentration at any location and at any given time. The mass of the pollutants released is m_p = 100 kg. For easier depiction, the results are presented in terms of breakthrough curves at two different locations [Fig. 9.3(a)]. Figure 9.3(b) shows the longitudinal profile along the line $Y = Z = 0$ at two different times, and Fig. 9.3(c) shows the concentration contours in the plane ($Y = 0$) 100 seconds after the pollutants are released.

From these figures, it may be noted that the spatial profiles are symmetric but the breakthrough curves are not. A cursory look at Eq. (9.13) clearly shows the spatial symmetry about the location (Vt, 0, 0). From Fig. 9.3(b), it is seen that the maximum concentration at any time occurs at the point $X = Vt$ (which is 100 m at $t = 100$ s and 200 m at $t = 200$ s). However, from Fig. 9.3(a), it is equally clear that the maximum concentration at any location does NOT occur at $t = X/V$. Both the nonsymmetric nature of the breakthrough curve and

---

[12]Plume is defined, in the context of pollutants transport, as a space in air, water, or soil containing pollutants released from a point source.

[13] Use of Eq. (9.13) implies that the river is assumed to be of infinite extent in all directions. Here, the depth is very small and this assumption will not be satisfactory.

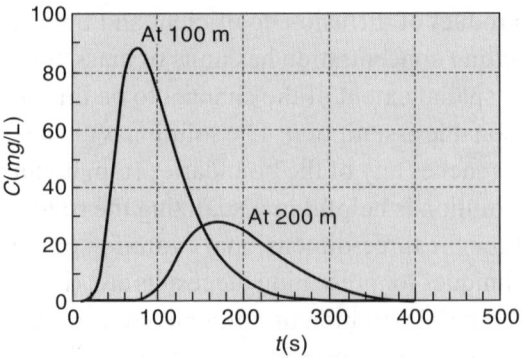

(a) Breakthrough curves at the locations (100, 0, 0) and (200, 0, 0)

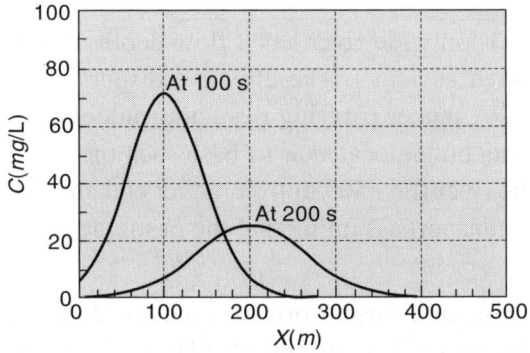

(b) Concentration profiles along the X-axis

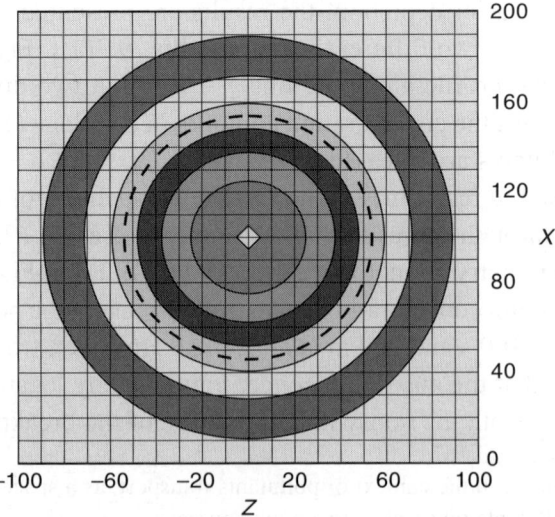

| □ 0–10 | ■ 10–20 | □ 20–30 | ▨ 30–40 | ■ 40–50 | ▨ 50–60 | ▨ 60–70 | □ 70–80 |

(c) Concentration contours in the X-Z plane at 100 s

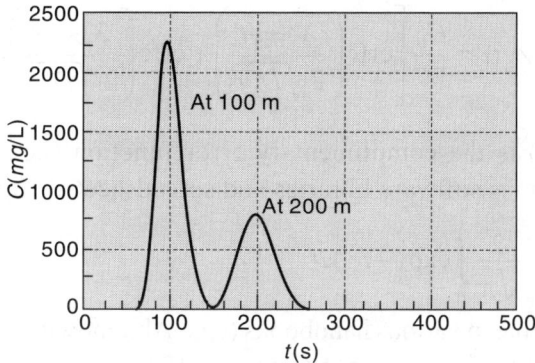

(d) Breakthrough curves at two locations for
smaller dispersion coefficient

Fig. 9.3

the early arrival of the peak, are due to the dispersion process (had it been pure advection, the curve would have been symmetric and the peak will occur at a time equal to X/V). To emphasize this fact, Fig. 9.3(d) shows the breakthrough curves at the same two locations with a ten-fold reduction in the dispersion coefficient (i.e., 1 m^2/s). The nearly symmetrical nature of the curve and the arrival of the peaks at nearly 100 s and 200 s respectively are evident. Also note the large increase in the concentration values compared to Fig. 9.3(a), which is expected since the mass remains same but the extent of the plume is much smaller due to the smaller dispersion coefficient.

9.3.2 Continuous Release of Pollutant at a Point

We now consider a continuous release of pollutant at the point $(0, 0, 0)$, which would simulate the effect of, say, a continuous stream of wastewater discharge brought into the stream through a sewer outfall. Though there would in practice be temporal variations, it would be convenient from the point of view of obtaining an analytical solution to assume a *continuous and constant* discharge. There are mainly two ways this release of pollutant can be incorporated in the boundary condition—the first by specifying it as a release of certain mass per unit time at the outfall and the second by specifying it as having a fixed concentration of pollutant at the point of release. While the first method is more realistic, the second one leads to easier solution and is employed here. The boundary condition is, therefore, written as

$$C(0, 0, 0, t) = C_0 \tag{9.14}$$

The other boundary condition is taken as zero concentration at a large distance ($+\infty$). The variation of concentration at a large distance from the source ($X \gg Y, Z$) is then given by

$$C(X, Y, Z, t) = \frac{C_0}{2}\left[\text{erfc}\left(\frac{X - Vt}{2\sqrt{D_L t}} \right) + \text{erfc}\left(\frac{X + Vt}{2\sqrt{D_L t}} \right) \exp\frac{VX}{D_L} \right] \quad (9.15)$$

in which erfc(.) is the complimentary error function (i.e., $1 - \text{erf}$) and is available in various software libraries and spreadsheets.[14] It is defined as

$$\text{erf}(x) = \frac{2}{\sqrt{\pi}} \int_0^x \exp(-t^2)\, dt$$

Tables are available in various handbooks (e.g., Abramowitz and Stegun 1964), and very accurate approximations (Hastings 1955) are also listed there.

Example 9.2 A 2 m wide channel has a flow depth of 1 m and the average velocity is measured as 1 m/s. The effective dispersion coefficient is estimated as 10 m^2/s. A pipe carrying salt water opens in the middle of a cross section and the salt concentration at this point is kept constant at 1000 mg/L. Plot typical breakthrough curves and concentration profiles. What would be the concentration at a point 100 m downstream of the outfall 100 s after the start of the release?

Solution

Taking the point of dumping as the origin of the coordinate system (0,0,0) and the time of dumping as $t = 0$, we use Eq. (9.15) to obtain the concentration at any location and at any given time (the assumption being that $X \gg Y, Z$, which is quite reasonable in this case). The breakthrough curves at two different locations are shown in Fig. 9.4(a). Figure 9.4(b) shows the longitudinal

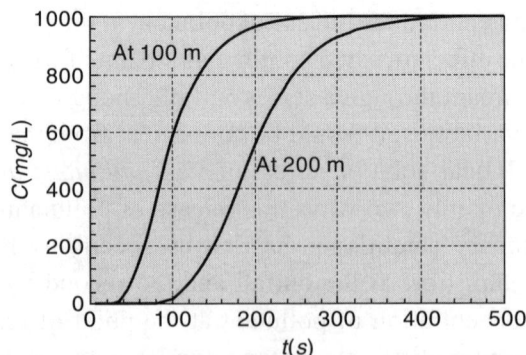

(a) Breakthrough curves at $X = 100$ m and 200 m for continuous release

---

[14]However, one should be aware that Microsoft Excel would not be able to compute the complimentary error function for negative values of the argument or for large positive values of the argument. The identity $\text{erf}(-x) = -\text{erf}(x)$ should be used in the first case and, to avoid the second problem, arguments greater than 27 should be put equal to 27.

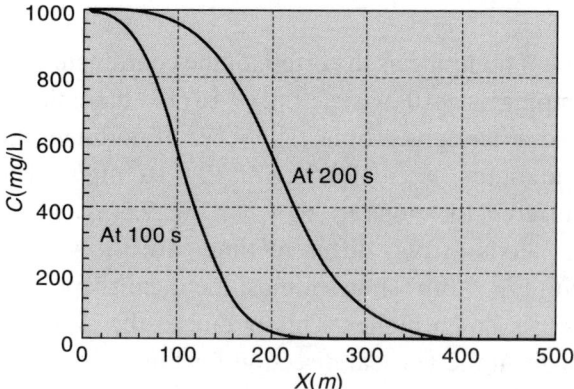

(b) Concentration profiles at 100 s and 200 s after
start of release

Fig. 9.4

profile along the line $Y = Z = 0$ at two different times. From these figures, it
may be noted that the concentration at any location ultimately approaches its
input value of 1000 mg/L since there is a continuous release of the pollutant.
The concentration at 100 m distance from the outfall 100 s after the start of
the release is about **600 mg/L**.

9.3.3 Sudden Release of Pollutant along a Line

We assume that the line is along the Y-axis and covers the entire depth of the
stream. Since the governing equation is linear, the solution for this case can
be obtained by considering the solution for a point source and integrating
over the flow depth. If the mass of the pollutant released from this line source
is m_{pl} per unit length (i.e., the total mass is $m_{pl} y$), the variation of concentration is given by

$$C(X, Y, Z, t) = \frac{m_{pl}}{4\pi \sqrt{D_L t} \sqrt{D_T t}} \exp\left[-\frac{(X - Vt)^2}{4 D_L t} - \frac{Z^2}{4 D_T t} \right] \quad (9.16)$$

Example 9.3   A 1 km wide river has a flow depth of 5 m and the average
velocity is measured as 1 m/s. The effective dispersion coefficient is estimated as 10 m²/s. A tanker carrying pesticide suddenly dumped 100 kg of
pesticide through an outlet located along the entire river depth at the centre of
the river (i.e., 500 m from either end). Compare this scenario with that in
Example 9.1, where the dumping was at a single point.

Solution

Taking the point of dumping as the origin of the coordinate system $(0,0,0)$ and the time of dumping as $t = 0$, we use Eq. (9.16) to obtain the concentration at any location and at any given time. The mass of pollutant released per unit length of the line source, $m_{pl} = 100/5$ kg/m $= 2 \times 10^7$ mg/m. The breakthrough curves at two different locations are shown in Fig. 9.5(a) and the concentration profiles along X-axis at two different times are shown in Fig. 9.5(b). A comparison with Fig. 9.3(a) shows almost identical shape but the magnitude of concentration is much higher. This is due to the assumption of infinite extent of the river in the vertical direction for Fig. 9.3(a), which results in a

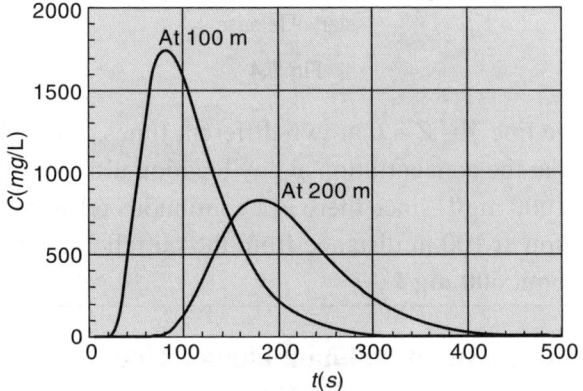

(a) Breakthrough curves at $X = 100$ m and 200 m
for line source

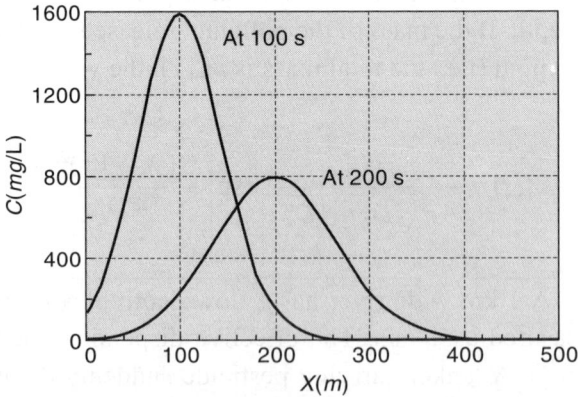

(b) Concentration profiles at 100 s and 200 s for
the line source

Fig. 9.5

significant expansion of the plume in the vertical direction, thereby reducing the concentration. Similar behaviour is observed in the profiles shown in Fig. 9.5(b).

9.3.4 Sudden Release of Pollutant along a Plane

We assume that the plane is the *Y-Z* plane and covers the entire cross section of the stream. Since the governing equation is linear, the solution for this case can be obtained by considering the solution for a line source and integrating over the channel width. If the mass of pollutant released from this plane source is m_{pp} per unit area (i.e., the total mass is $m_{pp} By$), the variation of concentration is given by

$$C(X, Y, Z, t) = \frac{m_{pp}}{\sqrt{4\pi D_L t}} \exp\left[-\frac{(X - Vt)^2}{4 D_L t}\right] \tag{9.17}$$

Example 9.4 A 1 km wide river has a flow depth of 5 m and the average velocity is measured as 1 m/s. The effective dispersion coefficient is estimated as 10 m^2/s. A pipeline carrying pesticide across the river suddenly dumped 100 kg of pesticide across the entire cross section of the river. What would be the concentration at a point 100 m from this location 100 s after the pesticide release?

Solution
Taking the point of dumping as the origin of the coordinate system (0,0,0) and the time of dumping as $t = 0$, we use Eq. (9.17) to obtain the concentration at any location and at any given time. The mass of pollutant released per unit area of the plane source is $m_{pp} = 100/(5 \times 1000)$ kg/m^2 = 20,000 mg/m^2. The breakthrough curves at two different locations are shown in Fig. 9.6(a) and the concentration profiles along *X*-axis at two different times are shown in Fig. 9.6(b). The concentration at 100 m distance from the outfall 100 s after the start of the release is about **180 mg/L**.

We see that the qualitative behaviour of an instantaneous release is similar whether it is released at a point, a line, or in a plane. The magnitude, of course, will be different. Similar to what was done for a point source, we may consider a constant release of pollutant over a line or a plane, but the solutions are more complicated and are not discussed here. Although the assumptions made in the analysis in this section are quite restrictive, a rough estimate of the concentration variation is obtained, which helps in the preliminary investigations. A more detailed analysis must be performed for cases where

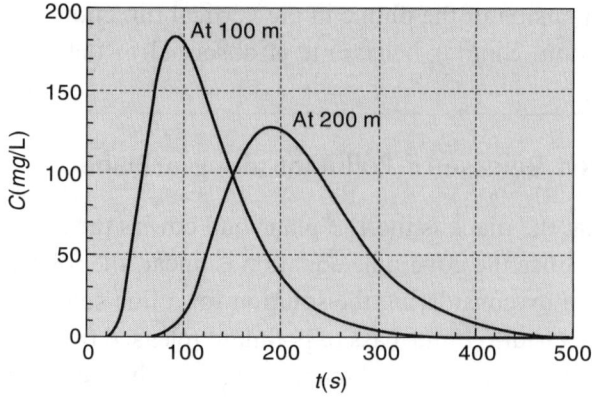

(a) Breakthrough curves at X = 100 m and 200 m for plane source

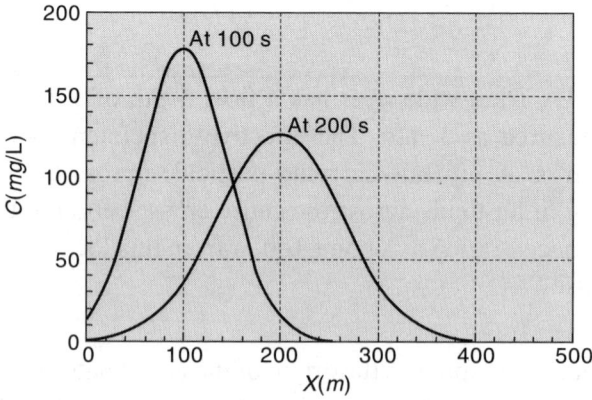

(b) Concentration profiles at 100 s and 200 s for the plane source

Fig. 9.6

these assumptions do not hold good and the results are likely to be significantly different from this simple analysis. Numerical methods are invariably used for such purpose since the analytical solutions are either not possible or are rather cumbersome. Moreover, readymade programs are freely available, which can be used with little or no modifications in a practical situation. Some of these numerical methods are described in the next section.

9.4 Complex Cases and their Numerical Solutions

The analytical solution of Eq. (9.10) or its one-dimensional approximation, Eq. (9.8), is not possible under general conditions involving the flow, pollutant, and channel properties. For example, most of the cases would involve temporal and spatial variations in the flow velocity, pollutant release pattern, and channel characteristics. A numerical method becomes the only choice available for such cases because of its versatility in incorporating these variations in the formulation. Although the numerical methods have their own limitations (e.g., artificial dispersion or smearing of results for advection-dominated transport), easy availability of fast computers have made them useful, especially if one is aware of the limitations. In this section, we will look at some of these methods and their limitations in terms of their accuracy and stability. Since the natural channels have a large width compared to the depth, generally we analyse the pollutant transport as a two-dimensional problem by averaging over the depth. Equation (9.10) is then written as

$$\frac{\partial C}{\partial t} + V \frac{\partial C}{\partial X} = \frac{\partial}{\partial X}\left(D_L \frac{\partial C}{\partial X}\right) + \frac{\partial}{\partial Z}\left(D_T \frac{\partial C}{\partial Z}\right) \tag{9.18}$$

in which C is the depth-averaged concentration.

Finite difference method

If we approximate the process as one-dimensional, the last term in Eq. (9.18) may be neglected and the resulting equation is easily solved using any of the techniques discussed in Section 7.5 for numerical solution of unsteady flow equations. The only difference is the presence of the dispersion term, which may be discretized using the central difference scheme. However, once the problem becomes two dimensional, the numerical solution becomes more complicated since the discretization involves two different directions (X and Z) in addition to the temporal discretization. A finite difference solution of Eq. (9.18) may be written by adopting a grid as shown in Fig. 9.7.

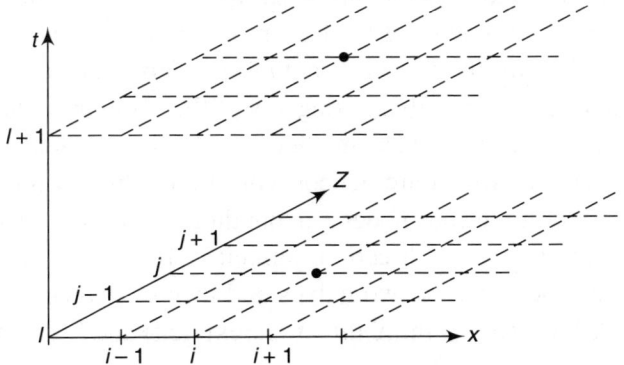

Fig. 9.7 Finite difference grid for the numerical solution

Concentration at a point (X, Z) at time t is denoted by $C_{i,j}^l$, in which (taking the origin as $X = 0$, $Z = 0$) $X = i\Delta X$ and $Z = j\Delta Z$ and l denotes the time level such that $t = l\Delta t$. The partial derivatives in Eq. (9.18) may be approximated as follows[15]:

$$\frac{\partial C}{\partial t} = \frac{C_{i,j}^{l+1} - C_{i,j}^l}{\Delta t} \tag{9.19a}$$

$$V\frac{\partial C}{\partial X} = V\frac{C_{i+1,j}^l - C_{i-1,j}^l}{2\,\Delta X} \tag{9.19b}$$

$$\frac{\partial}{\partial X}\left(D_L\frac{\partial C}{\partial X}\right) = D_L\frac{C_{i+1,j}^l - 2C_{i,j}^l + C_{i-1,j}^l}{(\Delta X)^2} \tag{9.19c}$$

$$\frac{\partial}{\partial Z}\left(D_T\frac{\partial C}{\partial Z}\right) = D_T\frac{C_{i,j+1}^l - 2C_{i,j}^l + C_{i,j-1}^l}{(\Delta Z)^2} \tag{9.19d}$$

Note that the spatial derivatives are evaluated at the known time level l, giving rise to an *explicit* scheme. Putting the discretized expressions in Eq. (9.18), knowing the concentration values at the time level l, and applying the boundary conditions, we will be able to solve for the concentration at all grid points at the time level $l + 1$. As we have seen earlier in Section 7.5, the explicit methods directly provide the concentration at each node without the need of solving a large set of linear equations. However, the maximum time step that can be taken to keep the numerical method stable is restricted by a Courant-like criterion, which may be written as $\Delta t \le \Delta X/V$ and a diffusion-based criterion of the form $D_L\Delta t/(\Delta X)^2 + D_T\Delta t/(\Delta Z)^2 \le 1/2$. Therefore, the implicit schemes are more commonly used, in which the spatial derivatives are evaluated either at time level $l + 1$ or, more commonly, at the mid-time-level of $l + 1/2$ (i.e., mean of levels l and $l + 1$). This leads to a set of linear equations, which have to be solved using any of the common techniques, e.g., Gauss elimination, Gauss–Seidel, method etc. Each equation generally involves five unknowns, which are the concentration values at time level $l + 1$ at the node i, j and at the four nodes surrounding it. In order to simplify the computations, an alternating direction implicit (ADI) scheme is often used, which divides a time step into two halves and then evaluates the spatial derivatives at time level l along the X-direction (explicit) and $l + 1/2$ along the Z-

---

[15]We make the assumption that the velocity and dispersion coefficient are constant. If these vary with time and/or space, appropriate subscripts have to be used.

direction (implicit) in the first half-step. During the second half-step, the derivatives are written at time level $l + 1$ along the X-direction (implicit) and at $l + 1/2$ along the Z-direction (explicit). This results in only three unknowns at each half-step and the equation is readily solved using a very efficient tridiagonal matrix solver. Details of the numerical methods may be seen in any good book.

Other numerical methods such as finite element method, method of characteristics, finite volume method, etc. may also be used to solve the advection-dispersion equation but are not described here.

As we have so far discussed in this chapter, the velocity of flow and the dispersion coefficient are important parameters affecting the pollutants transport. While the velocity is generally easy to measure or estimate, the estimation of the dispersion coefficient is relatively difficult. One option would be to conduct controlled release test with trace quantities of a pollutant (or an inert chemical) and measure the concentration profile or breakthrough curve. Since the shape of this observed curve would depend on the dispersion coefficient, it may be possible to estimate its value from this data. The other option is to relate the dispersion coefficient theoretically to the velocity variations in the stream. In the next section, we will look at various methods used to estimate the value of the dispersion coefficient for a given stream.

9.5 Estimation of Dispersion Coefficients

A detailed theoretical study of the dispersion/diffusion process vis-à-vis the turbulence characteristics may be seen in Fischer, et al. (1979). Since the shear stress at the bed strongly depends on the turbulence intensity, and the turbulent diffusion coefficient is dependent on the turbulence intensity, it follows that it would be strongly related to the bed shear. Using the concept of a mixing length, which is proportional to the flow depth, and the analysis performed by Taylor in early to mid-1900s, it is possible to write this diffusion coefficient as being proportional to v_*y, where v_* is the shear velocity. However, in field applications, it is the dispersion coefficient that is of interest to us, and it may also be taken as proportional to this quantity. Based on the analysis of J.W. Elder in 1959, a relationship was suggested as

$$D_L = 5.93 \, v_* y \tag{9.20}$$

which has been found to work well for laboratory data of wide rectangular channels. For natural rivers, however, it underpredicts the value and the following formula appears to work better (Fischer, et al. 1979):

$$D_L = 0.011 \frac{V^2 T^2}{v_* \, y} \tag{9.21}$$

in which T is the top width. Some other relationships have also been proposed, e.g., relating the dispersion coefficient to the discharge and the bed slope. However, most of such relationships are good only if the channel under consideration has similar characteristics to those whose data were used to develop these formulae. Variations of the order of 100% are not uncommon between the predicted and observed values of the dispersion coefficient. Hence, it may be advisable to estimate the dispersion coefficient for a particular channel directly rather than use an empirical equation. This process is described in the next paragraph.

Conceptually, one of the easiest options of estimating the longitudinal and transverse dispersion coefficients would be to introduce a pulse of a solute in a stream at a point, line, or section, and then measure its concentration (either as a spatial profile or as a temporal breakthrough curve). Then, using the appropriate analytical solution, one can obtain the expected variation of concentration for some assumed values of the dispersion coefficients. By taking different values of coefficients, we should be able to find the 'correct' values, which will minimize the difference between observed and computed concentrations. The implementation of this method, however, may not be so straightforward since the assumptions made in the analytical solutions may not hold good, the computational time required for various trials with different values of the dispersion coefficient may be too large, or a unique solution may not be obtainable from only a few measurements. The method of moments has also been used to estimate the dispersion coefficients using the fact that the longitudinal dispersion coefficient is proportional to the temporal rate of change of the normalized concentration variance[16] obtained from a breakthrough curve $[D_L = (V^2/2)\,(d\sigma_t^2/dt)]$. For evaluation of the transverse dispersion coefficient, the concentration variance across the channel width is obtained by measurement at two or more points. For field conditions where the velocity may be changing from section to section and stream characteristics are also not con-

---

[16]Recall that the variance is the second moment about the mean (the normalization here is done by the area under the curve, i.e., the zeroth moment). The mean time is found by dividing the first moment of the breakthrough curve by the zeroth moment, i.e.,

$\bar{t} = \int\limits_0^\infty C(X,Z,t)t\,dt / \int\limits_0^\infty C(X,Z,t)\,dt$. The normalized variance is given by

$$\sigma_t^2 = \int\limits_0^\infty C(X,Z,t)\left[t-\bar{t}\,\right]^2 dt / \int\limits_0^\infty C(X,Z,t)\,dt \ .$$

stant, using the analytical solution may not be possible. Numerical methods, involving trials with different values and minimizing the error between the computed and observed concentrations at a few locations, can be used to estimate the coefficients. The process of introducing a pollutant and monitoring its progress through the channel is similar to the routing of a flood through a channel where we are given an inflow hydrograph and find the outflow hydrograph by routing the flood through the channel. Therefore, this methodology of estimating the dispersion coefficients is sometimes called the *routing method*.

9.6 Advanced Topics

In addition to the processes considered in this chapter, several other processes may be significant in other conditions. Some examples of such cases are a thermal power plant discharging hot water into a stream (where buoyancy effects have to be incorporated due to the density difference), and transport of a reactive pollutant (where sorption onto the solid particles has to be considered), and presence of dead zones in the cross section (which may store the pollutants by diffusion and release it later when the concentration in the adjacent areas becomes smaller). A number of times, the pollutant may react with the bacteria present in the water or in the channel bed. Similarly, during modelling of dissolved oxygen in a stream (which is quite important from the point of view of water quality), the reaeration rate and production rate become important. All of these processes will affect the transport process to some extent and their inclusion in the formulation is desirable if dictated by the field conditions. We do not describe these in details and refer the reader to Gosink (1982) and Jirka, et al. (1981) for transport of hot water, to Habel, et al. (2002) for effect of porous streambeds on the pollutant transport process, and to Eiger (1995) for modelling of dissolved oxygen transport.

SUMMARY

In this chapter, we have looked at various processes affecting the transport of pollutants through a stream. Advection, diffusion, and dispersion have been explained, the governing equation has been derived, and the analytical and numerical solution methodologies have been discussed. Estimation of the dispersion coefficients in field conditions has been briefly described and some advanced topics related to the transport have also been mentioned.

We hope that a basic understanding of the transport process has been gained by the reader through this chapter. Those interested should refer to the book by Fischer, et al. (1979) and a large number of recent articles dealing with this subject. With the increasing emphasis on the quality of water in addition to the quantity, we expect this topic to gain more importance for engineers and hope that sufficient interest is generated in the reader to pursue this topic in greater details.

REFERENCES

Abramowitz, M. and Stegun, I. (1964): *Handbook of Mathematical Functions with Formulas, Graphs, and Mathematical Tables*, Dover Publications, New York.

Carslaw, H.S. and Jaegar, J.C. (1959): *Conduction of Heat in Solids*, Oxford University Press, New York.

Eiger, S. (1995): '2D Advective-diffusive transport of dissolved oxygen in channels,' *J. of Env. Eng., 121*(9), pp. 668–70.

Gosink, J.P. (1982). 'Thermal front formation from buoyant surface jets,' *J. of Hyd. Div., 108*(2), pp. 252–57.

Fischer, H. B. (1966): 'Longitudinal dispersion in laboratory and natural streams,' *Technical Report*, California Institute of Technology (*http:// caltechkhr. library.caltech.edu/42/01/KH-R-12.pdf*).

Fischer, H.B., List, E.J., Kob, R.C.Y., and Brooks, N.H. (1979): *Mixing in inland and coastal waters*, Academic Press, New York.

Habel, F., Mendoza, C., and Bagtzoglou, A.C. (2002): 'Solute transport in open channel flows and porous streambeds,' *Adv. Water Resour., 25*(4), pp.455–69.

Hastings, C., Jr. (1955): *Approximations for Digital Computers*, Princeton University Press, Princeton, New Jersey.

Jirka, G.H., Stolzenbach, K.D., and Adams, E.E. (1981): 'Buoyant surface jets,' *J. of Hyd. Div., 107*(11), pp.1467-87.

EXERCISES

9.1 For Example 9.1, assume that the dispersion coefficient is not known but the spatial concentration profile is measured at a time of 200 s, as shown in Fig. 9.3b. Compute the zeroth and first moments numerically and obtain the mean travel distance (which is the ratio of the first and zeroth moments). Then compute the second moment about the mean and estimate the dispersion coefficient by using the fact that the ratio of the second moment and the zeroth moment is given by $\sigma_X^2 = 2D_L t$.

$$[M_0 = 3979 \text{ mg/m}^2, M_1 = 795780 \text{ mg/m}, \overline{X} = 200 \text{ m},$$
$$M_2 = 15931872 \text{ mg}, D_L = 10.01 \text{ m}^2/\text{s}]$$

9.2 Use Eq. (9.13) to analytically compute the moments (from to $X = -\infty$ to ∞),0 which were obtained numerically in Exercise 9.1. Assume that the longitudinal and transverse dispersion coefficients are same and the profile is measured along the X-axis ($Y = Z = 0$).

$$\left[M_0 = \frac{m_p}{4\pi D_L t}, M_1 = \frac{m_p V}{4\pi D_L}, M_2 = \frac{m_p}{2\pi} \right]$$

9.3 Use Eq. (9.16) to analytically compute the spatial moments (from to $X = -\infty \, \infty$) for instantaneous release of pollutants along a line. Assume that the longitudinal and transverse dispersion coefficients are same and the profile is measured in the plane $Z = 0$. Compare these moments with those obtained in Exercise 9.2 for the release at a point.

$$\left[M_0 = \frac{m_{pl}}{2\sqrt{\pi D_L t}}, M_1 = \frac{m_{pl} V \sqrt{t}}{2\sqrt{\pi D_L}}, M_2 = \frac{m_{pl} \sqrt{D_L t}}{\sqrt{\pi}} \right]$$

9.4 Using the spatial profile shown in Fig. 9.5b at 200 s, estimate the flow velocity and dispersion coefficient assuming all other data to be known and using the results obtained in Exercise 9.3.

$$[1 \text{ m/s}, 10.01 \text{ m}^2/\text{s}]$$

9.5 Use Eq. (9.17) and the spatial profile at 200 s shown in Fig. 9.6b to analytically obtain the moments and then numerically estimate the velocity and dispersion coefficient for instantaneous release of pollutant in a plane.

$$[20000 \text{ mg/m}^2, 4000028 \text{ mg/m}, 80082325 \text{ mg}, 1 \text{ m/s}, 10.01 \text{ m}^2/\text{s}]$$

9.6 Since spatial profiles are inconvenient to measure, it would be desirable to estimate the dispersion coefficient using the breakthrough curve. From Fig. 9.3a, obtain the zeroth, first, and second (about the mean travel time) moments of the breakthrough curve at 200 m for an instantaneous point source. Estimate the flow velocity (equal to the distance divided by the mean travel time) and the dispersion coefficient[17] (related with

---

[17]Assume that the longitudinal and transverse coefficients are same.

the variance by $\sigma_t^2 = 2\,D\bar{t}/V^2 = 2XD/V^3$ in which $\bar{t}$ is the mean travel time).

[3979 mg s/m^3, 795775 mg s^2/m^3, 15916482 mg s^3/m^3, 1 m/s, 10 m^2/s]

9.7 Obtain the temporal moments (zeroth, first, and second) of the breakthrough curves at a distance of 200 m for a line source (Fig. 9.5a) and a plane source (Fig. 9.6a). Are the relations listed in Exercise 9.6 valid for these cases also? Give reasons.

[Line source: 124660 mg s/m^3, 26150211 mg s^2/m^3,
546876206 mg s^3/m^3; Plane source: 20000 mg s/m^3,
4400000 mg s^2/m^3, 96000000 mg s^3/m^3; No]

Fluid Mechanics

We briefly describe some of the important concepts of fluid mechanics that have been used in this book. We follow the usual coordinate axes notation of x, y, z (which is different from what we have used in the book) and use Z to denote the height above a horizontal datum. For cylindrical coordinates used in the analysis of flow through circular pipes, we use the r, θ, x system with x-axis along the direction of flow. The velocity components at a point are written as v_x, v_y, v_z (or v_x, v_r, v_θ) and the average velocities are written as $V_x, \ldots,$ while the vector notations for these velocities are $\vec{v}$ and $\vec{V}$.

A.1   Reynolds Transport Theorem

The laws of conservation of mass, momentum, and energy, are applicable to a *system* and not to a *control volume*. For example, mass of a specified parcel of water would be conserved but mass of water within a specified space may change. The Reynolds transport theorem lets us write these laws for a fixed control volume as follows:

$$\frac{dN_{\text{SYS}}}{dt} = \frac{dN_{\text{CV}}}{dt} + \int_{\text{CS}} \rho \eta \vec{v} \cdot d\vec{A} \tag{A.1}$$

in which N is an extensive property (e.g., mass, momentum, or energy), η is the corresponding intensive property (e.g., 'mass, momentum, or energy' per unit mass). SYS represents the system, CV the control volume, and CS the control surface. $\vec{v}$ represents the velocity vector and $d\vec{A}$ is the elemental area vector on the control surface (conventionally, the direction of the area element is along the normal pointing outwards). Equation (A.1) states that the rate of change of an extensive property for a given system is equal to the sum of the rate of change of that property within the control volume and the net efflux of the property across the control surface.

A.2 Momentum Equation

For *steady state* conditions, if we use momentum as the extensive property, N, (and, obviously, velocity as the intensive property, η), Eq. (A.1) may be written as[1]

$$\vec{F} = \int_{CS} \rho \vec{v} \, (\vec{v} \cdot d\vec{A}) \tag{A.2}$$

where $\vec{F}$ is the net force acting *on* the control volume when considered as a *free body*. Equation (A.2) is a vector equation and its x-component is written as

$$F_x = \int_{CS} \rho v_x \, (\vec{v} \cdot d\vec{A}) \tag{A.3}$$

Sometimes, it is useful to write this equation in the cylindrical coordinate system (r, θ, x). The r-direction momentum equation for an elemental control volume $(dr, rd\theta, dx)$ is written as

$$F_{r*} - \frac{1}{\rho} \frac{\partial (p + \rho g Z)}{\partial r} = (\vec{v} \cdot \nabla) v_r - \frac{v_\theta^2}{r} \tag{A.4}$$

where F_{r*} includes forces (other than the pressure and body force, which are explicitly accounted for by the second term on the left hand side) acting in the r-direction on the elemental control volume per unit mass of fluid and the $\vec{v} \cdot \nabla$ term in the convective acceleration is given by

$$\vec{v} \cdot \nabla = v_r \frac{\partial}{\partial r} + \frac{v_\theta}{r} \frac{\partial}{\partial \theta} + v_x \frac{\partial}{\partial x} \tag{A.5}$$

A.3 Energy Equation

If we use energy as the extensive property, N, (and, the energy per unit mass as the intensive property, η), Eq. (A.1) may be written as

$$\frac{dE_{SYS}}{dt} \left(= \frac{dQ}{dt} - \frac{dW}{dt} \right) = \frac{dE_{CV}}{dt} + \int_{CS} \rho e \vec{v} \cdot d\vec{A} \tag{A.6}$$

in which E represents the energy (we typically consider the kinetic, potential, and internal energies and ignore others such as nuclear, electrical, magnetic,

---

[1] We have used the Newton's law that the net force on a system is equal to the rate of change of momentum of the system.

etc.), e is the energy per unit mass, Q is the heat *added to the system*, and W is the work done *by the system*. For steady state conditions with one inlet and one outlet, writing e as the sum of *molecular internal energy, u,* kinetic energy, $v^2/2$, and potential energy, gZ, and the work W as the sum of shaft work, pressure work, and viscous work, we obtain the commonly used form of the energy equation in terms of heads (which represent energy per unit weight) as

$$\frac{p_1}{\rho g} + Z_1 + \frac{V_1^2}{2g} = \frac{p_2}{\rho g} + Z_2 + \frac{V_2^2}{2g} + \frac{u_2 - u_1 - q_*}{g} \tag{A.7}$$

in which section 1 is the inlet, section 2 is the outlet, and q_* represents the heat added to the fluid per unit mass. We have neglected the viscous work, which is generally very small. Also, the velocity distribution is assumed to be uniform, otherwise a correction factor has to be applied to the velocity head. The last term in Eq. (A.7) may be thought of as a *head loss* between the two sections. For example, frictional losses between the two sections would raise the temperature of the fluid, increasing its internal energy, and the head loss would be equal to $(u_2 - u_1)/g$. Although the pressure work and the potential head would vary within a cross section, for most pipe flow problems, the sum

remains constant at all points of the cross section. This term $\left(\dfrac{p}{\rho g} + Z \right)$ is

known as the piezometric head, P.

A.4 Velocity Distribution

For flow in circular pipes, the velocity distribution is parabolic for laminar flow conditions. However, since flow in open channels is almost always turbulent, we look at the velocity profile for turbulent flow near a boundary with specific application to a circular pipe.

For flow past a flat plate with no pressure gradient, a laminar boundary layer has a velocity profile given by the Blasius equation

$$f''' + \frac{ff''}{2} = 0$$

in which f is a function such that its derivative gives the velocity profile

$$\frac{v}{v_\infty} = f'\left(\frac{5y}{\delta} \right)$$

where v_∞ is the free-stream velocity, y is the distance from the boundary, and δ is the boundary layer thickness[2] (note that the velocities are in the x-direction but the subscript x is not written in the velocity to simplify the presentation). This profile can be closely approximated by[3]

$$\frac{v}{v_\infty} = 1.5\frac{y}{\delta} - 0.5\left(\frac{y}{\delta}\right)^3 \tag{A.8}$$

For a turbulent boundary layer on a smooth surface,[4] the velocity profile has been found to be different close to the wall (inner layer) and far away from the wall (outer layer) with a transition (or overlap) layer where both profiles would match. Based on the observed velocity profiles, the inner layer is divided into three layers—viscous sublayer (also called laminar sublayer), buffer layer, and logarithmic layer (or inertial sublayer). The length scale used in the inner layer is dependent on the wall shear stress and the fluid viscosity and is expressed as $\mu/\rho v_*$, where v_* is the shear velocity, which is equal to $\sqrt{\tau_0/\rho}$. (The shear stress at the wall, τ_0, is generally written in terms of a skin friction coefficient, C_f, multiplied to the dynamic pressure as $\tau_0 = C_f \rho v_\infty^2/2$). The velocity profile in the inner layer is then given in terms of a nondimensional distance,[5] $y* = \rho v_* y/\mu$, as

$$\frac{v}{v_*} = \begin{cases} y* & \text{for } y* \leq 5 \text{ (viscous sublayer)} \\ 5\ln y* - 3 & \text{for } 5 \leq y* \leq 30 \text{ (buffer layer)} \\ 2.5\ln y* + 5.5 & \text{for } 30 \leq y* \leq 300 \text{ (inertial sublayer)} \end{cases}$$

$$\tag{A.9}$$

Sometimes, the buffer layer is omitted in such a way that there is only a viscous sublayer up to $y* = 11.6$ (the value obtained by matching the viscous sublayer and the inertial sublayer profiles) and an inertial sublayer beyond

---

[2]The Blasius solution used another variable, $\eta = \dfrac{y\sqrt{R_{ex}}}{x}$, which turned out to be $5y/\delta$.

[3]Sometimes, the coefficients in Eq. (A.8) are changed to 1.65 and 0.65 to get a better fit at the cost of violating one of the boundary conditions that the slope of the profile should be zero at the edge of the boundary layer.

[4]We will see in Section A.5 how the smoothness of a boundary is decided.

[5]For a laminar boundary layer, the similarity hypothesis suggests the use of the boundary layer thickness as the length scale. If we use the length scale related to the shear velocity, the velocity profiles would not be similar but will depend on the Reynolds number.

that.[6] Moreover, the upper limit of the inertial sublayer may extend beyond $y* = 300$ depending on the Reynolds number.

In the outer layer, the thickness of the boundary layer is used as the length scale and the velocity profile is given by what is commonly known as the *velocity defect*[7] *law*:

$$\frac{v_\infty - v}{v_*} = \begin{cases} 2.5 - 2.45\ln\frac{y}{\delta} & \text{for } 0.01 \le \frac{y}{\delta} \le 0.1 \\ 9.6\left(1 - \frac{y}{\delta}\right)^2 & \text{for } 0.1 \le \frac{y}{\delta} \le 1 \end{cases} \tag{A.10}$$

The overlap region extends from $y*$ values of about 100 to 300 and small values of y/δ (up to about 0.1).

A.5   Rough and Smooth Boundaries

The presence of a viscous sublayer leads to the definitions of smooth and rough boundaries. For laminar flow, the roughness of the boundary has negligible influence on the frictional resistance since the disturbances are damped by the viscous effects. For turbulent flow, if the boundary is such that the roughness elements are within the viscous sublayer, it would again behave as a smooth boundary. However, if the average height of roughness is much larger, the sublayer would be broken up and the boundary will be rough. A criterion for determining the nature of the boundary is, therefore, as given below:

$$\frac{\rho v_* \varepsilon}{\mu} = \begin{cases} \le 5 & \text{Smooth (no effect of roughness)} \\ 5 \text{ to } 70 & \text{Transition (both roughness and} \\ & \text{Reynolds number affect)} \\ > 70 & \text{Rough (no effect of Reynolds number)} \end{cases} \tag{A.11}$$

where ε is the average height of the surface roughness.

---

[6]Some studies have used the inertial sublayer profile with 5.0 as the additive constant. In that case, the nondimensional thickness of the viscous sublayer would be 11. Also, the multiplicative constant of 2.5 is the inverse of the Karman constant, generally taken as κ = 0.4. Some researchers use 0.41.

[7]The term *defect* here indicates deficiency as it relates to the difference between the free-stream velocity and the point velocity.

A.6 Extension of Flat Plate Equations to Circular Pipes

The relationships derived for flow past a flat plate cannot be directly applied to the flow through a circular pipe because a pressure gradient exists and also the growth of the boundary layer is restricted to the extent of the pipe radius only. The qualitative behaviour of the velocity profile is, however, similar. The velocity and length scales, as before, are taken as v_* and $\mu/\rho\, v_*$ and the velocity profile is written in the logarithmic form as

$$\frac{v}{v_*} = A \ln r* + B \tag{A.12}$$

in which $r*$ is the nondimensional distance from the pipe wall, $\dfrac{\rho v_*(R-r)}{\mu}$, r

being measured from the centre of the pipe, and the constants A and B can be obtained experimentally.[8] This expression can be integrated over the pipe radius to obtain the average cross-sectional velocity, V, as

$$\frac{V}{v_*} = A_1 \log \frac{\rho v_* D}{\mu} + B_1 \tag{A.13}$$

Note that we have switched to the pipe diameter, D, which is more commonly used than the radius, and also changed from the natural log to the log on base 10.

To develop a resistance law, the shear stress is related to the *head loss*, which is the same as the drop in the piezometric head if the pipe is of uniform cross section. This is done by a force balance between the shear force, the pressure force, the and the gravitational force, and results in

$$\pi D \Delta x \tau_0 = \frac{\pi D^2}{4} \frac{d(p + \rho g Z)}{dx} \Delta x$$

$$\Rightarrow \qquad \tau_0 = \frac{D}{4} \frac{\rho g h_f}{L} \tag{A.14}$$

in which h_f is the head loss over a pipe length of L. By writing the shear stress in terms of the skin friction coefficient and the dynamic head,[9] Eq. (A.14) becomes

---

[8]For smooth boundaries, these should be constant but for transition or rough boundaries, the dependence on the roughness height has to be incorporated.

[9]Earlier we used the dynamic head based on the free-stream velocity. For a pipe flow, it is more convenient to use the average velocity rather than the maximum velocity.

$$C_f \frac{\rho V^2}{2} = \frac{D}{4} \frac{\rho g h_f}{L}$$

$$\Rightarrow \qquad h_f = 4 C_f \frac{L}{D} \frac{V^2}{2g} \qquad\qquad (A.15)$$

The term $4C_f$ was replaced by f, which is called the friction factor (or Darcy–Weisbach friction factor) to obtain the Darcy–Weisbach equation for head loss through circular pipes as

$$h_f = f \frac{L}{D} \frac{V^2}{2g} \qquad\qquad (A.16)$$

The only thing left now in using this resistance law is to determine the variation of the friction factor with the flow and pipe characteristics. The friction factor is related to the shear velocity as

$$f = 4 C_f = 4 \frac{\tau_0}{\frac{1}{2}\rho V^2} = 8 \left(\frac{v_*}{V}\right)^2 \qquad\qquad (A.17)$$

Using Eqs (A.13) and (A.17) and defining a Reynolds number based on the pipe diameter as $\mathrm{Re} = \rho V D / \mu$, one can write

$$\frac{1}{\sqrt{f}} = A_2 \log\left(\mathrm{Re}\sqrt{f}\right) + B_2 \qquad\qquad (A.18)$$

For smooth boundaries, L. Prandtl in 1935 suggested the values of A_2 and B_2 based on the experimental data as 2.0 and –0.8 respectively. Note, however, that it is an implicit formula since the friction factor occurs on both sides of the equation. For fully rough flow, the friction factor is independent of the Reynolds number ($A_2 = 0$) and the Karman–Nikuradse formula is used to obtain the friction factor as follows:

$$\frac{1}{\sqrt{f}} = -2.0 \log\left(\frac{\varepsilon}{3.715 D}\right) \qquad\qquad (A.19)$$

For the transitional range of flow in commercial pipes, C.F. Colebrook in 1939 proposed a formula (now commonly known as the Colebrook–White formula) for the friction factor as follows:

$$\frac{1}{\sqrt{f}} = -2.0\log\left(\frac{2.51}{Re\sqrt{f}} + \frac{\varepsilon}{3.7D}\right) \tag{A.20}$$

To estimate the friction factor directly for a given Reynolds number and relative roughness, a plot of this equation was prepared by L.F. Moody in 1944 (Moody chart), which is shown in Fig. A.1.

Various approximations have been suggested for direct evaluation of the friction factor. A commonly used equation is that proposed by Churchill (1973).[10]

$$\frac{1}{\sqrt{f}} = -2\log\left[\left(\frac{7}{Re}\right)^{0.9} + \frac{\varepsilon}{3.7D}\right] \tag{A.21}$$

From Eq. (A.16), it is obvious that the head loss varies as square of the velocity if the friction factor is constant. From the Moody chart, we see that f is virtually constant for very large Reynolds numbers. Also, the value of this limit decreases as the relative roughness increases. For example, for $\varepsilon/D = 0.001$, the rough zone starts at Re of about 10^6 while for $\varepsilon/D = 0.05$, the rough zone starts at a much smaller Re of about 10000. Therefore, for large Reynolds number, the head loss varies as square of the velocity. On the other hand, for very small values of Reynolds number, the flow would be laminar and the head loss varies linearly with the velocity (the friction factor is given by 64/Re, indicating that it is inversely proportional to the velocity). For intermediate ranges, generally the head loss is proportional to a power of the velocity about 1.7 to 1.8. Since open channel flows are generally in the completely turbulent zone, the velocity should be proportional to the square root of the gradient of the energy line (h_f/L).

Sometimes, we may be asked to estimate the discharge in a pipe for a specified head loss. An iterative technique could be used to solve such problems since the Reynolds number is not known. However, using the fact that Eq. (A.16) may be written as

$$fV^2 = \frac{2gh_fD}{L} \quad \Rightarrow \quad Re\sqrt{f} = \sqrt{\frac{2gh_f\rho^2D^3}{\mu^2L}}$$

one should be able to get the friction factor from Eq. (A.20) and then the discharge from

---

[10]An almost identical expression was used by P.K. Swamee and A.K. Jain in 1976.

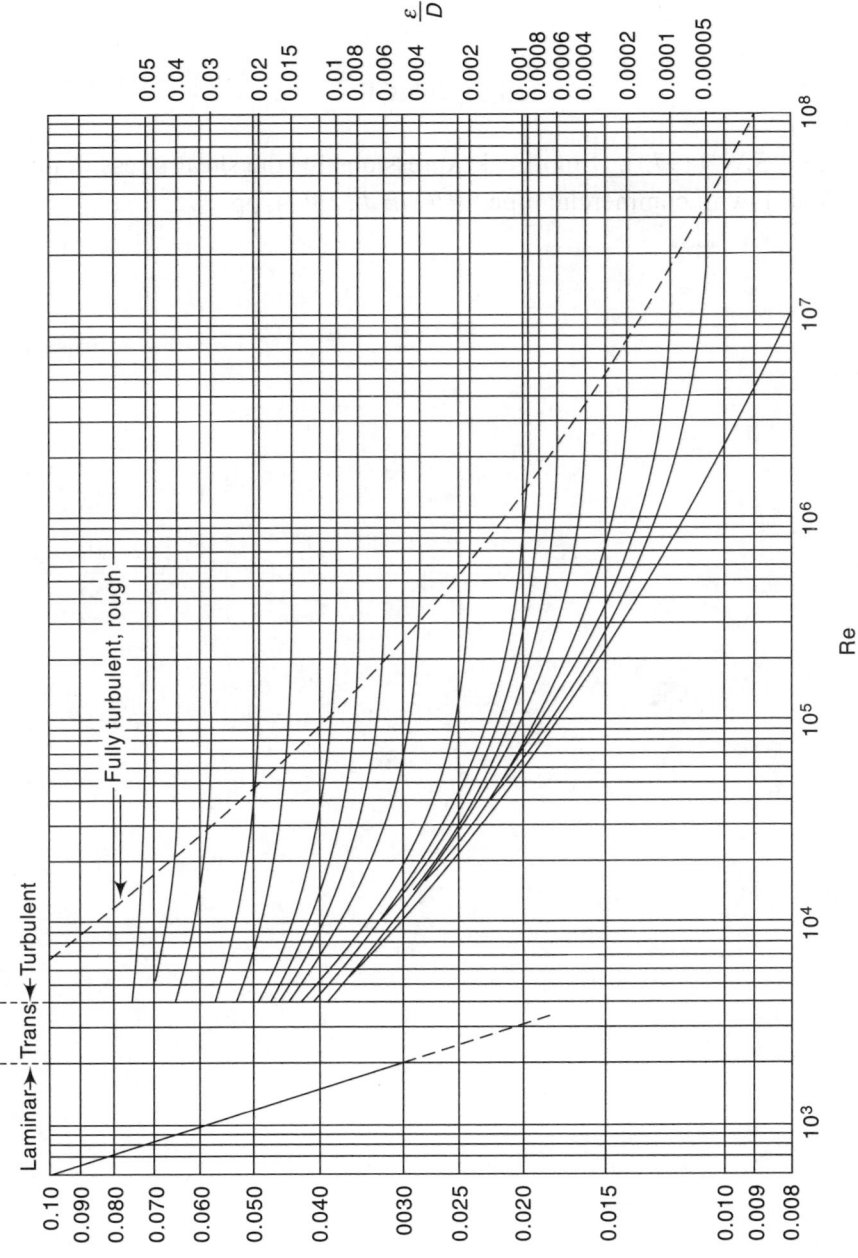

Fig. A.1 Moody chart for friction factor in circular pipes

$$Q = \sqrt{\frac{\pi^2 g h_f D^5}{8 f L}}$$ (A.22)

REFERENCE

Churchill, S.W. (1973): 'Empirical expressions for the shear stress in turbulent flow in commercial pipe,' *AIChE J.*, **19**(2), pp. 375–76.

Velocity, Pressure and Shear Stress Distributions in Channels

B.1 Velocity Distribution

The velocity in an open channel has to be zero at the boundaries because of the no-slip condition. However, as commonly happens in turbulent flows, it shows a rapid variation and the velocity gradient is generally very large near the boundaries. The velocity increases as we move away from the boundary either in the horizontal or in the vertical direction, and it would indicate that the maximum velocity at a cross section occurs at the water surface at the mid-width. However, other factors, the predominant being secondary currents, cause the point of the maximum velocity to occur a little below the water surface close to the mid-width. If we consider a rectangular channel, the point of maximum velocity on a vertical line would occur close to the water surface if the vertical line is taken near the middle. As we move closer to the banks, the point of maximum velocity shifts lower. On the other hand, if we compare a narrow rectangular channel and a wide rectangular channel, the maximum velocity will occur closer to the water surface in a wide channel. These and similar other observations may be made by looking at some typical velocity contours shown in Fig. B.1.

Some of the important observations regarding the velocity distribution in open channels are as follows:

1. The velocity at a point located at a depth of about *0.6 time the flow depth* below the water surface is close to the average velocity at that vertical line. This has important practical applications since the discharge in a channel is sometimes obtained by the area-velocity method in which the cross section is divided into several sub-areas by vertical lines and the average velocity is measured for each sub-area. Instead of measuring the velocity at a number of points in a vertical line and then computing the average, it would be more expedient to measure it at just one point, which is $0.6y$ below the water surface, where y is the flow depth at that vertical line.

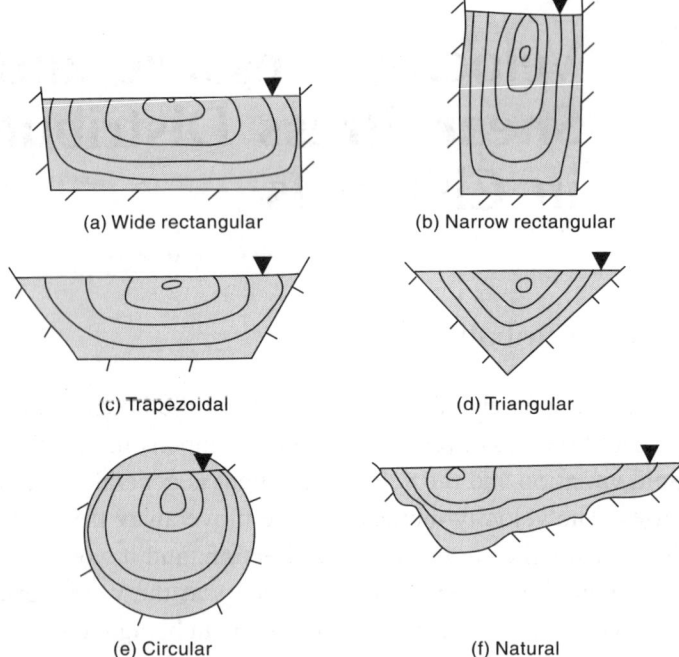

(a) Wide rectangular   (b) Narrow rectangular

(c) Trapezoidal   (d) Triangular

(e) Circular   (f) Natural

Fig. B.1 Typical velocity contours for channels with different cross-sectional shapes

2. An even more accurate estimation of the average velocity is obtained by measuring the point velocities at two points in a vertical, one at $0.2y$ and the other at $0.8y$ below the water surface.[1]

3. The magnitude of the surface velocity of a stream is generally about 5 to 25% more than that of the average velocity. The exact ratio is, of course, dependent on the channel and flow characteristics. However, for a given channel, and under comparable flow conditions, this ratio would remain more or less constant. This provides us a very easy method of estimating the discharge in a channel by observing the velocity of a floating body in the channel. This velocity would be multiplied by a pre-determined factor (which typically ranges from 0.8 to 0.95) to obtain the average velocity which, in turn, is used to obtain the discharge based on the information about the channel cross section.

---

[1]Note that these two locations are nearly identical with the locations of Gauss points, which are used to numerically integrate an arbitrary function with the best possible accuracy using two points. The exact values of the Gauss points are $0.5\left[1 \pm \left(1/\sqrt{3}\right)\right]$, i.e., 0.79 and 0.21 of the flow depth.

As discussed in the text, we use the one-dimensional analysis by using the average velocity and use the momentum and energy correction factors to account for the velocity variations within a cross section.

B.2 Pressure Distribution

In a stationary liquid kept in an open container, the pressure is zero[2] at the free surface and increases linearly with depth below the surface (known as the *hydrostatic pressure distribution*). All the points at the same depth below the surface will experience the same pressure. Even if the bed of the container is inclined, the same pressure distribution holds good as long as we have a horizontal free surface. If the fluid is moving on an inclined channel bed, the acceleration will cause the pressure distribution to deviate from the hydrostatic distribution. However, if we assume that the bed is straight, the bed slope is small, and the flow is uniform, the pressure distribution would still be hydrostatic. Deviations may occur because of a curvature in the bed or water surface (which produces normal acceleration), or a large slope of the bed (which introduces a component of weight in the direction perpendicular to the flow).

If the bed has a curvature, it introduces a normal acceleration, a_n. The value of a_n will depend on the nature of variation of velocity at different locations. For example, if we assume that the acceleration is constant at a section, it would be equal to V^2/R_a with V as the average velocity and R_a as the average radius of curvature of the flow (Fig. B.2).

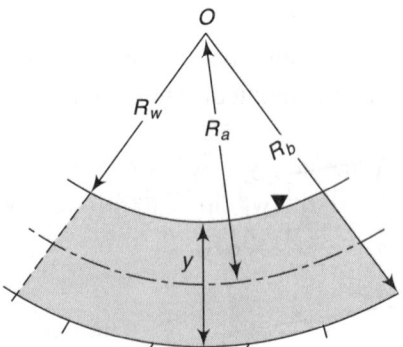

Fig. B.2 Pressure at a point in channels with large bed curvature

---

[2]The atmospheric pressure is taken as zero, implying that we are using the gage pressure and not absolute.

If we assume that the velocity is constant everywhere,[3] the normal acceleration may be found by integrating over the flow depth as

$$a_n = \frac{V^2}{y} \ln \frac{R_b}{R_w}$$

in which R_b is the radius to the bed and R_w to the water surface. The pressure distribution with a constant normal acceleration is given by

$$\frac{p}{\rho g} = d \left(1 \pm \frac{a_n}{g} \right) \tag{B.1}$$

in which d denotes the depth below the free surface. The plus sign is used for concave upward beds and the minus sign for convex upward beds (thus the pressure is more than hydrostatic pressure for concave upward beds like that shown in the figure). However, when the normal acceleration varies at different flow depths, the pressure at any point has to be obtained by considering a differential element and then performing an integration. We do not discuss these aspects here since the assumptions made in the velocity profiles may not be valid and the end result may as well be approximated by introducing a correction factor in the piezometric head.

The effect of bed slope is generally not felt in typical open channel flows since the slope is very small. However, for some cases, e.g., flow over a spillway, the channel has a large bed slope and it would affect the pressure distribution. If the bed slope is θ (Fig. B.3), a simple force balance under uniform flow conditions results in the pressure at point P as

$$p_p = \rho g d_n \cos \theta$$

in which d_n, as shown, is the depth below the water surface measured *normal to the flow direction* (i.e., perpendicular to the channel bed).

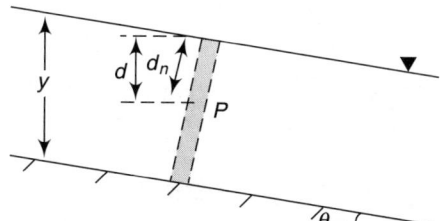

Fig. B.3   Pressure at a point in channels with large bed slope

---

[3]It is not a reasonable assumption since the velocity has to be zero at the bed.

Since the usual depth measurements are carried out vertically, we write the pressure in terms of the vertical depth below the free surface, d, as

$$p_p = \rho g d \cos^2 \theta \tag{B.2}$$

which may be significantly different from the hydrostatic value of $\rho g d$ if θ is large. The pressure force acting on this section (assuming a rectangular channel) is equal to

$$P = \frac{1}{2}\rho g y \cos^2 \theta y \cos\theta = \frac{1}{2}\rho g y^2 \cos^3 \theta \tag{B.3}$$

If the bed is steep as well as curvilinear, the effects would be combined. However, it is not described here.

B.3   Shear Stress Distribution

The average shear stress on the channel boundary has been shown (through a balance of driving and resisting forces under uniform flow conditions) to be equal to $\rho g R S$. However, the actual distribution varies considerably over the bed and the sides and the average shear stress on the bed and/or the sides may not be equal to this theoretical value. For example, a trapezoidal channel may have a shear stress distribution as shown in Fig. B.4.

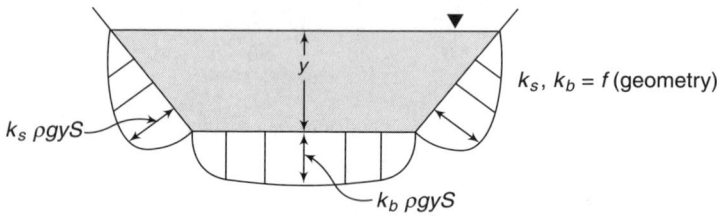

Fig. B.4   Shear stress distribution on the boundary of a trapezoidal channel

The distribution is affected by the side slope and the bed width to flow depth ratio. Similarly, the shear stress distribution on the perimeter of a rectangular channel depends on the B/y ratio. For a rectangular channel, Guo and Julien (2005) provide the graphical relationship as shown in Fig. B.5 relating the average shear to $\rho g y S$. Note that the flow depth is used and not the hydraulic radius. For wide channels, these would be almost identical.

Since we are generally concerned with the maximum shear stress at any point on the bed or sides (e.g., to analyse the stability of a sediment particle), it is desirable to look at the maximum shear stress on the bed as well as sides (since the stability analyses for particles on sides and those on the channel bed are different). For trapezoidal channels with a side slope of 2 H:1V, the

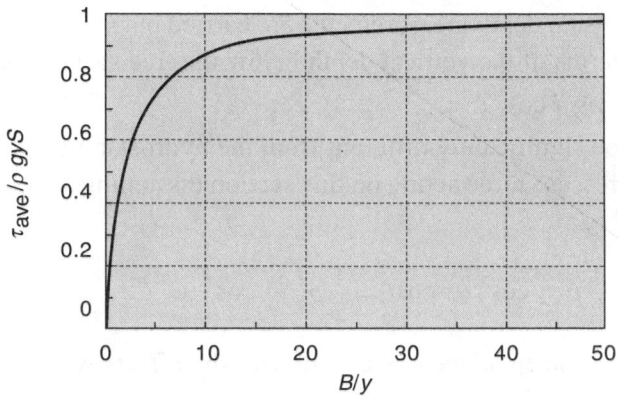

(a) Average shear stress on the bed of a rectangular channel

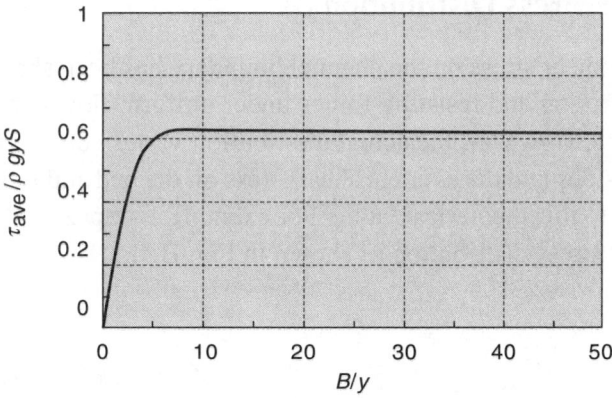

(b) Average shear stress on the sides of a rectangular channel

Fig. B.5

maximum shear stress on the sides is roughly 75% of ρgyS and that on the bed is almost equal to ρgyS.

REFERENCE

Guo, J. and Julien, P.Y. (2005): 'Shear stress in smooth rectangular open-channel flows,' *J. of Hyd. Eng.*, **131**(1), pp. 30–37.

This appendix describes various computational methods referred to in the text. The book by Chapra and Canale (2006) provides greater details of these techniques.

C.1 Solution of a Nonlinear Equation in a Single Variable

The methods used for the solution of nonlinear equations of the form $f(x) = 0$ can be classified as bracketing methods (in which a root has to bracketed by obtaining two points where the function values are of opposite signs) and open methods (where bracketing is not necessary). The open methods are preferred since they generally show faster convergence towards the root (however, they may sometimes diverge). For most of the problems we discuss in this book, we have a fairly good idea about the solution from the physics of the problem, (for example, in a problem of finding flow depth for a given specific energy, we know that the depth has to be less than the specific energy), we would invariably use the open methods with our best *starting guess* of the solution. The improvement in this guess value is done by iterative techniques, two of which are described below.

C.1.1 Fixed Point Iteration (or Successive Substitution)

We write the equation $f(x) = 0$ as $x = \phi(x)$ (for example, the nonlinear equation $\tan x - x^2 = 0$ can be written as $x = \sqrt{\tan x}$ or $x = \tan^{-1} x^2$). If a root of $f(x) = 0$ is ζ then obviously $\zeta = \phi(\zeta)$, indicating that it is a *fixed point*[1] of $\phi(x)$. Now, starting from an initial guess of x_0, we obtain $\phi(x_0)$ and compare with x_0. If their values are same, then x_0 is a root; if not, we take $x_1 = \phi(x_0)$ as our estimate of the root at the next iteration. Following this process of *successive substitution*

---

[1]The function $\phi(x)$ may be thought of as mapping the point x to a new point, say, x^*. Generally this mapping will cause the points x and x^* to be some distance apart. If there is *no movement* while mapping some x, say, x_p, [i.e., $x_p = \phi(x_p)$], x_p is called a *fixed point* of the function $\phi(x)$.

of the argument of the function $\phi(x)$, we hope to converge to the root, i.e., reach a point when two successive estimates of the root are *nearly* identical. Since the iterations require only one point to compute the next estimate, the scheme may also be called *single-point* or *one-point iteration*.

The sequence of iteration is written as $x_{i+1} = \phi(x_i)$ and the iterations are stopped when two successive iterations produce x values within a desired error tolerance. Since the root is not bracketed between two successive estimates, there is no guarantee that the method will converge. It can be shown that the error at any iteration is equal to $\phi'(x)$ times the error at the previous iteration, where $\phi'(x)$ denotes the derivative of the function $\phi(x)$. Thus, the method is linearly convergent[2] and will converge if $|\phi'(x)|$ is less than 1 in the domain of interest. If $|\phi'(x)| < 1$ for all x, then the fixed point iteration sequence is guaranteed to converge for *all starting guesses*. Also, if $\phi'(x)$ is positive, the error at any iteration would have the same sign as that at the previous iteration (monotonic sequence). For negative values of the derivative, the error will oscillate between positive and negative values.

C.1.2 Newton–Raphson Method (Tangent Method)

If we approximate the function, $f(x)$, by a straight line in the neighbourhood of our starting guess (x_0), a natural choice would be the *tangent* at that point. Thus, starting from an initial estimate of the root, we approximate the function by the tangent at this point, and obtain a new estimate as the point where this linear approximation has a root (Fig. C.1). The method was developed by *Newton* and, a little later but independently, by *Raphson*.

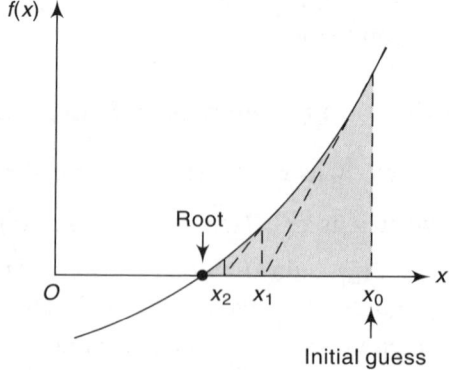

Fig. C.1 Estimate of root in the Newton–Raphson scheme

---

[2]*Linearly convergent* indicates that the error at any step is linearly related to the error at the previous iteration.

The linear approximation of the function $f(x)$ near the point x_i is given by

$$\tilde{f}(x) = f(x_i) + (x - x_i)f'(x_i) \tag{C.1}$$

and the new estimate of the root, which is a zero of $\tilde{f}(x)$, is given by

$$x_{i+1} = x_i - \frac{f(x_i)}{f'(x_i)} \tag{C.2}$$

It can be shown that the error at any iteration is proportional to the square of the error at the previous iteration and the proportionality constant is equal to $-f''(x)/2f'(x)$. Thus, the Newton–Raphson method is *quadratically convergent*, which means a faster convergence than the fixed point method. There is, however, again a chance that the method will not converge at all. Some techniques that combine the bracketing methods and Newton method to achieve fast and guaranteed convergence have been developed but will not be described here. For our purpose, the fixed point iteration and the Newton method are considered good enough.

C.2 Solution of a Set of Linear Equations

A set of linear simultaneous equations has to be solved in numerical methods involving the finite difference method, as described in Chapters 7 and 9. The methods to solve a set of linear equations can be classified as *direct methods* (which obtain the solution of the system in a finite number of computations or steps) and *indirect methods* (in which an approximate solution is assumed and iteratively improved to obtain desired accuracy). The most common technique (though not most efficient for some types of systems, e.g., those with many zero coefficients for which the indirect methods work better) of solving a set of linear equations is the direct method of Gauss elimination, which is described below.

Gauss elimination

A system of n linear equations with n unknowns[3] is written as

$$\begin{bmatrix} a_{11} & a_{12} & \cdots & a_{1n} \\ a_{21} & a_{22} & \cdots & a_{2n} \\ a_{31} & a_{32} & \cdots & a_{3n} \\ \cdot \\ \cdot \\ \cdot \\ a_{n1} & a_{n2} & \cdots & a_{nn} \end{bmatrix} \begin{bmatrix} x_1 \\ x_2 \\ x_3 \\ \cdot \\ \cdot \\ \cdot \\ x_n \end{bmatrix} = \begin{bmatrix} b_1 \\ b_2 \\ b_3 \\ \cdot \\ \cdot \\ \cdot \\ b_n \end{bmatrix} \tag{C.3}$$

[3]There may be situations where the number of equations is not equal to the number of unknowns. However, we consider that these are equal and a unique solution exists.

which can be expressed in terms of two vectors (*b* and *x*) and a matrix (*A*), as follows:

$$[A] \{x\} = \{b\} \tag{C.4}$$

where *b* is a known vector (dimension *n*), *x* is the unknown vector (dimension *n*), and *A* is the known coefficient matrix of dimension $n \times n$. In the Gauss elimination method, we first reduce the matrix to an upper triangular matrix[4] by *forward elimination* and then solve for the variables one by one, starting from the *n*th equation (which has only one unknown) and proceeding backwards (this step is called *back substitution*). A brief description of these steps is given below.

We first reduce all the elements of the first column (except the element a_{11}) to zero. This is done by keeping the first row unchanged, multiplying the first row by a_{21}/a_{11} and subtracting from the second row, multiplying the first row by a_{31}/a_{11} and subtracting from the third row and so on till the *n*th row. This effectively makes the coefficient of the variable x_1 in all subsequent equations zero, i.e., eliminates the variable. (Clearly, similar operations need to be conducted for the right hand vector *b* also.) After this step, the equations have the following form (the subscript on the matrix *A* and the vector *b* indicate that the elements have been modified once):

$$\begin{bmatrix} a_{11} & a_{12} & \cdots & a_{1n} \\ 0 & a_{22} & \cdots & a_{2n} \\ 0 & a_{32} & \cdots & a_{3n} \\ \cdot & & & \\ \cdot & & & \\ \cdot & & & \\ 0 & a_{n2} & \cdots & a_{nn} \end{bmatrix}_1 \begin{Bmatrix} x_1 \\ x_2 \\ \cdot \\ \cdot \\ \cdot \\ x_n \end{Bmatrix} = \begin{Bmatrix} b_1 \\ b_2 \\ \cdot \\ \cdot \\ \cdot \\ b_n \end{Bmatrix}_1 \tag{C.5}$$

Now, we carry out the same operation on column 2 and reduce all the elements below its main diagonal to zero. For example, we multiply the second row with (a_{32}/a_{22}) and subtract from the third row and so on. Repeating the procedure for all columns, the matrix becomes an upper triangular matrix as follows:

$$\begin{bmatrix} a_{11} & a_{12} & a_{13} & \cdots & a_{1n} \\ 0 & a_{22} & a_{23} & \cdots & a_{2n} \\ 0 & 0 & a_{33} & \cdots & a_{3n} \\ \vdots & \vdots & \vdots & \vdots & \vdots \\ 0 & 0 & 0 & \cdots & a_{nn} \end{bmatrix}_{n-1} \begin{Bmatrix} x_1 \\ x_2 \\ \vdots \\ x_n \end{Bmatrix} = \begin{Bmatrix} b_1 \\ b_2 \\ \vdots \\ b_n \end{Bmatrix}_{n-1} \tag{C.6}$$

---

[4]An upper triangular matrix has all elements below the main diagonal as zero.

and one can now easily obtain the solution of the system of equations using the *back substitution* since the last equation gives $x_n = b_n / a_{nn}$, the previous

equation gives $x_{n-1} = \dfrac{(b_{n-1} - a_{n-1,n} x_n)}{a_{n-1,n-1}}$ and x_n is already computed, and so

on till the first equation.

If the original matrix (A) is banded (i.e., all elements outside a band about the main diagonal are zero), the computational process takes advantage of that and does not perform the forward elimination outside the band. If a significant fraction of elements within the band are also zero, it may be more efficient to use an iterative method such as Gauss–Seidel method.

C.3 Solution of Ordinary Differential Equations

An ordinary differential equation of first order may be written in the form

$$\frac{dy}{dx} = f(x, y) \tag{C.7}$$

in which $f(x, y)$ is a known function and we need one boundary condition to get a unique solution. The solution of this equation aims at obtaining the value of y at $x = x_{i+1}$ given its value at x_i (and at all points prior to x_i).The methods used for the solution can be classified as explicit (which enable us to compute y_{i+1} directly) and implicit (which provide a generally nonlinear equation in y_{i+1} to be solved using any of the methods described earlier in this appendix). Another classification is based on whether we use the information at the ith point only to compute the value at the $(i + 1)$th point (single-step methods) or we use one or more points before it (multi-step methods). Since explicit methods are easier and more common (although they may need a small step length to ensure stability), we will describe only these.

C.3.1 Forward Euler's Method

In this method, we write the slope dy/dx in terms of a forward difference and get the following form of Eq. (C.7):

$$\frac{y_{i+1} - y_i}{x_{i+1} - x_i} = f(x_i, y_i) \tag{C.8}$$

from which the explicit expression for y_{i+1} is written as

$$y_{i+1} = y_i + (x_{i+1} - x_i) f(x_i, y_i) \tag{C.9}$$

The implicit counterpart of this method uses $f(x_{i+1}, y_{i+1})$ and, in general, cannot be directly solved for y_{i+1} (unless the function f is linear or quadratic[5] in y). Since the forward Euler method has a lower order of accuracy (it is first order accurate since the leading error term is proportional to the first power of the length step), we look for higher order methods while keeping the explicit nature of the scheme intact. Here we describe one such scheme—the well-known Runge–Kutta scheme.

C.3.2   Runge–Kutta Methods

Runge–Kutta methods aim at achieving a higher order of accuracy by evaluating the slope function at various intermediate points within a step and finally using a weighted average of all the slopes to compute the y-value at the next step. A general formulation of the Runge–Kutta methods can be written as

$$y_{i+1} = y_i + h \sum_{j=1}^{n} w_j s_j \qquad (C.10)$$

where h is the step size $(x_{i+1} - x_i)$, w_j are the weights and s_j are the slopes written as

$$s_1 = f(x_i, y_i) \qquad (C.11a)$$

$$s_k = f\left(x_i + \alpha_k h, y_i + \alpha_k h \sum_{l=1}^{k-1} \beta_{kl} s_l \right) \qquad \text{for } k = 2 \text{ to } n \qquad (C.11b)$$

The total number of points, n, determines the order of accuracy of the method. It is clear from the above equations that s_1 is the slope at the starting point (x_i, y_i) which is known. The other slopes (at the kth point) are obtained at points located at fractional distances of α_k from this point with y at those points predicted by using a weighted average of all previous slopes (1 to $k-$ 1). Thus it is an explicit method, which will directly give us the value of y at x_{i+1}.

As an example, the second order Runge–Kutta method can be written as

$$y_{i+1} = y_i + h(w_1 s_1 + w_2 s_2) \qquad (C.12\,a)$$
$$s_1 = f(x_i, y_i) \qquad (C.12b)$$
$$s_2 = f(x_i + \alpha_2 h, y_i + \alpha_2 h \beta_{21} s_1) \qquad (C.12c)$$

and the objective is to estimate the weights (w_1, w_2), the location of the intermediate point (α_2), and the weight assigned to the previous slope (β_{21}) in

---

[5]Even for cubic and quadratic functions, an analytical solution is possible but is quite complicated.

order to achieve maximum accuracy. This is done by comparing this approximate y_{i+1} with the Taylor's series expansion and gives rise to the following three equations in these variables:

$$w_1 + w_2 = 1, \quad w_2 a_2 = \frac{1}{2}, \quad \beta_{21} = 1 \tag{C.13}$$

A variety of second order Runge–Kutta methods can be obtained since we have one free parameter (four unknowns and three equations). Some commonly used options are given below:

$$w_1 = 0 \implies w_2 = 1, \alpha_2 = \frac{1}{2}$$

So $$y_{i+1} = y_i + hf\left[x_i + \frac{h}{2}, y_i + \frac{h}{2}f(x_i, y_i)\right] \tag{C.14a}$$

$$w_1 = \frac{1}{2} \implies w_2 = \frac{1}{2}, \alpha_2 = 1$$

So $$y_{i+1} = y_i + h\frac{f(x_i, y_i) + f[x_i + h, y_i + hf(x_i, y_i)]}{2} \tag{C.14b}$$

$$w_1 = \frac{1}{3} \implies w_2 = \frac{2}{3}, \alpha_2 = \frac{3}{4}$$

So $$y_{i+1} = y_i + h\frac{f(x_i, y_i) + 2f\left[x_i + \frac{3}{4}h, y_i + \frac{3}{4}hf(x_i, y_i)\right]}{3} \tag{C.14c}$$

The first scheme, Eq. (C.14a) is known as the mid-point scheme or modified Euler's method, the second scheme is called Heun's method or improved Euler's scheme, and the third scheme is known as Ralston's method.

By increasing the number of points, we get higher order Runge–Kutta methods. However, the gain in accuracy is not commensurate with the increase in computational effort beyond the fourth order methods. Therefore, the fourth order methods are the most commonly used Runge–Kutta method. Again, there are an infinite possible combinations of the weights and the fractions. We list below one of these, which is called the standard fourth order Runge–Kutta method:

$$y_{i+1} = y_i + h\frac{s_1 + 2s_2 + 2s_3 + s_4}{6} \tag{C.15a}$$

$$s_1 = f(x_i, y_i) \tag{C.15b}$$

$$s_2 = f\left(x_i + \frac{h}{2}, y_i + \frac{h}{2}s_1\right) \tag{C.15c}$$

$$s_3 = f\left(x_i + \frac{h}{2}, y_i + \frac{h}{2}s_2\right)$$ (C.15d)

$$s_4 = f(x_i + h, y_i + hs_3)$$ (C.15e)

Clearly, the average slope is taken as the weighted sum of four slopes, with weights of 1/6, 1/3, 1/3, and 1/6 respectively. The four slopes are obtained as follows:

The first slope is at the starting point and the second slope is at the mid-point with value of y obtained by using the slope at the starting point. The third slope is again at mid-point but is obtained using the value of y computed by using the estimated slope at the mid-point. The fourth slope is at the end point of the interval with value of y computed by using the 'improved estimate' of the slope at the mid-point.

Other variations of the method have been tried for solving the ordinary differential equations encountered in open channel flow problems. However, we believe that the standard fourth order Runge–Kutta method would be the best choice in most cases.

C.4 Solution of Partial Differential Equations

The solution of partial differential equations is frequently obtained using the finite difference method in which the partial derivatives are replaced by their finite difference approximations by dividing the continuous solution domain into a discrete computational domain using grids of finite dimensions. The equations described in the book involve a time derivative and one or more spatial derivatives. The objective of the solution is to obtain the values of the dependent variable at the *next* time step at all the spatial grid points given the values at the *current* time step. The time derivative is generally written as a forward or backward difference but some schemes may use a central difference also (which requires storing the values at more than one time steps). The process of estimating the derivatives at a point using the function values at adjacent points is known as *numerical differentiation* and is described below.

C.4.1 Numerical Differentiation

Since the computational grids are generally evenly spaced, we consider all the step lengths to be the same ($= h$). The spatial derivatives of a variable, say f, are approximated by assuming the variable to be of a particular form. For example, if we assume that the variable follows a piecewise linear variation, the first derivative at any point, x_i, may be obtained as given below (we first assume that the spatial derivatives are needed only along one coordinate direction).

(i) The slope of the line in the previous segment (backward difference):

$$f_i' = \frac{f(x_i) - f(x_{i-1})}{h} \tag{C.16}$$

(ii) The slope of the line in the next segment (forward difference):

$$f_i' = \frac{f(x_{i+1}) - f(x_i)}{h} \tag{C.17}$$

(iii) The average of these two slopes or the slope of the line joining the previous point and the next point (central difference):

$$f_i' = \frac{f(x_{i+1}) - f(x_{i-1})}{2h} \tag{C.18}$$

Clearly, the backward difference will not be valid at the first node, the forward difference will not work at the last node, and the central difference cannot be used at the first and last nodes. Now, if we want the second derivative at x_i, we may argue that the first derivative of the function at the mid-point[6] of the previous segment is given by Eq. (C.16) and that at the mid-point of the next segment is given by Eq. (C.17). Therefore, the second derivative may be approximated as

$$f_i'' = \frac{\dfrac{f(x_{i+1}) - f(x_i)}{h} - \dfrac{f(x_i) - f(x_{i-1})}{h}}{h}$$

$$= \frac{f(x_{i+1}) - 2f(x_i) + f(x_{i-1})}{h^2} \tag{C.19}$$

Another alternative may be to fit a second degree polynomial through the three points (x_{i-1}, x_i, x_{i+1}) and find its second derivative (which would be a constant over the interval), which results in the same expression. Of course, the same argument can be extended to arrive at expressions of the higher derivatives, which would generally involve an increasing number of grid points around (or on one side of) the point at which we want the derivative value. However, there is a more convenient way, based on the Taylor's series, of deriving these expressions, which also provides an estimate of the error in the approximation. Based on this analysis, Tables C.1 and C.2 list various

---

[6]Since the slope is constant throughout a segment, any other point could also be taken. However, as we will see a little later, the central difference gives a better accuracy.

approximations of the derivatives and their order of error for the forward and central difference schemes using the notation

$$f_i^n = \frac{1}{h^n} \sum_j c_{i+j} f(x_{i+j}) \tag{C.20}$$

Table C.1   Table for forward difference formulae
(The backward difference scheme is obtained with $-h$ in place of h and $i-j$ in place of $i+j$.)

Accuracy	Derivative	c_i	c_{i+1}	c_{i+2}	c_{i+3}	c_{i+4}	Error[7]
$O(h)$	f_i'	-1	1				$-\dfrac{hf''}{2}$
	f_i''	1	-2	1			$-hf'''$
	f_i'''	-1	3	-3	1		$-\dfrac{3hf^{iv}}{2}$
	f_i^{iv}	1	-4	6	-4	1	$-2hf^{v}$
$O(h^2)$	f_i'	$-3/2$	2	$-1/2$			$\dfrac{h^2 f'''}{3}$
	f_i''	2	-5	4	-1		$\dfrac{11 h^2 f^{iv}}{12}$
	f_i'''	$-5/2$	9	-12	7	$-3/2$	$\dfrac{7 h^2 f^{v}}{4}$
$O(h^3)$	f_i'	$-11/6$	3	$-3/2$	$1/3$		$-\dfrac{h^3 f^{iv}}{4}$
	f_i''	$35/12$	$-26/3$	$19/2$	$-14/3$	$11/12$	$-\dfrac{5 h^3 f^{v}}{6}$

If the problem is two dimensional in space, these expressions represent the partial derivatives with respect to that particular direction. However, there would be another subscript added to indicate the other direction. For example, using subscripts i for the x-direction and j for the y-direction, we write the first derivative approximation with central derivative as

$$\frac{\partial f}{\partial x} \simeq \frac{f_{i+1,j} - f_{i-1,j}}{2 \Delta x}, \qquad \frac{\partial f}{\partial y} \simeq \frac{f_{i,j+1} - f_{i,j-1}}{2 \Delta y} \tag{C.21}$$

---

[7]The derivatives in the error expressions are evaluated at some point in the appropriate interval.

Table C.2 Table for central difference formulae

Accuracy	Derivative	c_{i-2}	c_{i-1}	c_i	c_{i+1}	c_{i+2}	Error
$O(h^2)$	f_i'		$-1/2$	0	$1/2$		$-\dfrac{h^2 f'''}{6}$
	f_i''		1	-2	1		$-\dfrac{h^2 f^{iv}}{12}$
	f_i'''	$-1/2$	1	0	-1	$1/2$	$-\dfrac{h^2 f^v}{4}$
	f_i^{iv}	1	-4	6	-4	1	$-\dfrac{h^2 f^{vi}}{6}$
$O(h^4)$	f_i'	$1/12$	$-2/3$	0	$2/3$	$-1/12$	$\dfrac{h^4 f^v}{30}$
	f_i''	$-1/12$	$4/3$	$-5/2$	$4/3$	$-1/12$	$\dfrac{h^4 f^{vi}}{90}$

and similarly for higher derivatives. However, for unsteady problems, we should be aware of the fact that the function is time dependent and the spatial derivatives should be evaluated at *some* time level. In the explicit schemes, all these derivatives are evaluated at the *known* time level and we get the value of f_{ij} at the next (unknown) time level directly. The implicit schemes evaluate the spatial derivatives at a time level somewhere between the known and the unknown time levels (it may be at the unknown time level but NOT at the known time level). These schemes result in a set of linear equations and the Gauss elimination method discussed earlier in this appendix may be used for its solution. For two-dimensional unsteady problems, we have the option of evaluating the partial derivatives in different directions using different schemes. For example, we may use the explicit scheme along the x-direction and implicit scheme along the y-direction to reduce the number of unknowns in each equation. Details of these schemes may be seen in Chapra and Canale (2006).

REFERENCE

Chapra, S.C. and Canale, R.P. (2006): *Numerical Methods for Engineers,* 5th ed., McGraw-Hill, New York.

Index